Surveys in Number Theory

Millennial Conference on Number Theory
May 21-26, 2000

Surveys in Number Theory

Papers from the Millennial Conference on Number Theory

Edited by
M. A. Bennett, B. C. Berndt,
N. Boston, H. G. Diamond,
A. J. Hildebrand, and W. Philipp

CRC Press
Taylor & Francis Group
Boca Raton London New York

CRC Press is an imprint of the
Taylor & Francis Group, an **informa** business

AN A K PETERS BOOK

First published 2003 by A K Peters, Ltd.

Published 2018 by CRC Press
Taylor & Francis Group
6000 Broken Sound Parkway NW, Suite 300
Boca Raton, FL 33487-2742

© 2003 by Taylor & Francis Group, LLC
CRC Press is an imprint of Taylor & Francis Group, an Informa business

No claim to original U.S. Government works

ISBN 13: 978-1-56881-162-8 (pbk)

Visit the Taylor & Francis Web site at
http://www.taylorandfrancis.com

and the CRC Press Web site at
http://www.crcpress.com

Library of Congress Cataloging-in-Publication Data

Millennial Conference on Number theory (2000 : University of Illinios at Urbana-Champaign)
    Surveys in Number theory ; papers from the Millennial Conference on Number Theory /
    edited by M. A. Bennett ... [et al.].
        p. cm.
    Conference held at the University of Illinois at Urbana-Champaign, May 21-26, 2000.
    Includes bibliographical references.
    ISBN 1-56881-162-4
    1. Number theory–Congresses. I. Bennett, M. A. (Michael A.), 1966- II. Title.

QA241 .M534 2000
512'.7–dc21
                                                                        2002070231

Contents

Preface

The Millennial Conference on Number Theory was held on May 21–26, 2000 on the campus of the University of Illinois at Urbana-Champaign. The meeting was organized by M. A. Bennett, B. C. Berndt, N. Boston, H. G. Diamond, and A. J. Hildebrand of the Mathematics Department and W. Philipp of the Statistics Department at the University of Illinois. A total of 276 mathematicians from 30 countries were present at the meeting.

The proceedings of this conference, containing 72 papers based on lectures given at the conference, have been published separately in three volumes under the title *Number Theory for the Millennium*. The present volume contains fourteen of these papers which represent broad surveys of topics in number theory or related areas. We hope that these papers will help stimulate interest in these areas and will be of lasting value as references on the topics covered.

Urbana, Illinois
7 December 2001

M. A. Bennett
B. C. Berndt
N. Boston
H. G. Diamond
A. J. Hildebrand
W. Philipp

Completeness Problems and the Riemann Hypothesis: An Annotated Bibliography

Michel Balazard

1 Notation and Basic Formulas

- We consider complex functions f defined and measurable on the positive half-line $(0, +\infty)$. Functions defined on some subset E of $(0, +\infty)$ are extended to $(0, +\infty)$ by assigning the value 0 outside E.

- For $\lambda > 0$, the dilation operator K_λ is defined by

$$K_\lambda f(x) = f(\lambda x), \quad x > 0.$$

- The letter s denotes a complex variable, σ its real part and τ its imaginary part.

- The Mellin transform

$$Mf(s) = \int_0^{+\infty} f(t)t^{s-1}\, dt \qquad (1.1)$$

is defined for every s such that the integral converges absolutely; it is holomorphic in the region where it is defined.

- For f in $L^2(0, +\infty)$, $Mf(s)$ is *a priori* defined nowhere, but according to the Plancherel theory, it can be defined a.e. on the line $\sigma = 1/2$ by setting

$$Mf(s) := \underset{T \to +\infty}{\mathrm{l.i.m.}} \int_{1/T}^{T} f(t)t^{s-1}\, dt, \quad \sigma = 1/2.$$

The operator $f \mapsto (2\pi)^{-1/2} Mf$ is a unitary operator from $L^2(0, +\infty)$ to $L^2(1/2 + i\mathbb{R}) =: L^2$.

- The letter ρ denotes two different objects: the "fractional part function" $\rho(x) := x - \lfloor x \rfloor$, and a generic complex number $\rho = \beta + i\gamma$, $\beta \in \mathbb{R}$, $\gamma \in \mathbb{R}$, usually a zero of some Mellin transform. No confusion should occur from this usage.

1

- $\chi_{(0,a)}$ denotes the characteristic function in $(0, +\infty)$ of the interval $(0, a)$, and χ denotes $\chi_{(0,1)}$.

- A **subspace** of X is a *closed* vector space contained in the topological vector space X; the **span** of a subset $\mathcal{E} \subset X$ is the smallest subspace of X containing $\mathcal{E}$.

- A family $(e_\alpha)_{\alpha \in A}$ of elements of a topological vector space X is **minimal** if $e_\alpha \notin \mathrm{span}(e_\beta : \beta \in A,\ \beta \neq \alpha)$ for every $\alpha \in A$. The family is a **topological basis** of X if, moreover, $\mathrm{span}(e_\alpha : \alpha \in A) = X$. If X is locally convex, there exists a family $(e_\alpha^*)_{\alpha \in A}$ of elements of X^* that is uniquely determined by the conditions

$$e_\alpha^*(e_\beta) = \begin{cases} 1 & \text{if } \alpha = \beta, \\ 0 & \text{if } \alpha \neq \beta. \end{cases}$$

One writes symbolically $x \sim \sum_{\alpha \in A} e_\alpha^*(x) e_\alpha$.

- If H is a (closed) subspace of a Hilbert space K, the orthogonal complement of H in K is denoted by $K \ominus H$. If $H_1 \subset H_2 \subset K$, then $H_2 \ominus H_1 = H_2 \cap (K \ominus H_1)$.

- For $\Lambda \subset (0, +\infty)$, $\mathcal{B}_\Lambda$ denotes the set of functions

$$t \mapsto f(t) = \sum_{k=1}^{n} c_k \rho(\theta_k / t), \tag{1.2}$$

where $n \in \mathbb{N}$, $c_k \in \mathbb{C}$, and $\theta_k \in \Lambda$. We denote by $\mathcal{N}_\Lambda$ the subset of $\mathcal{B}_\Lambda$ obtained by adding the condition

$$\sum_{k=1}^{n} c_k \theta_k = 0.$$

We have $\mathcal{B}_{(0,+\infty)} \subset L^p(0, +\infty)$ for $1 < p \leqslant +\infty$ and $\mathcal{N}_{(0,+\infty)} \subset L^p(0, +\infty)$ for $1 \leqslant p \leqslant +\infty$. Moreover, $\mathcal{N}_{(0,1)} \subset L^\infty(0,1)$. The closure of $\mathcal{N}_{(0,1)}$ in $L^p(0,1)$ is denoted by $\mathcal{N}^p$.

- If f is given by (1.2), then

$$Mf(s) = -\frac{\zeta(s)}{s} \sum_{k=1}^{n} c_k \theta_k^s, \quad 0 < \sigma < 1$$

(cf. [21, 2.1.5]). In the case where $f \in \mathcal{N}_{(0,+\infty)}$, the Mellin transform Mf is defined and analytic in the half-plane $\Re s > 0$. Moreover, one deduces easily from this formula that the functions $t \mapsto \rho(\theta/t)$, $\theta > 0$, are linearly independent.

- The Hardy space of the half-plane $\Re s > 1/2$ (resp. $\Re s < 1/2$) is denoted by H^{2+} (resp. H^{2-}). This space can be considered (via boundary values) as a subspace of L^2, or as a set of holomorphic functions in the half-plane $\Re s > 1/2$ (resp. $\Re s < 1/2$). One has the orthogonal direct sum $L^2 = H^{2+} \oplus H^{2-}$.

- Let φ be a function defined and unimodular a.e. on $\frac{1}{2} + \mathbb{R}i$. Then the multiplication by φ is a unitary operator of L^2. In particular, if $\varphi H^{2+} \subset H^{2+}$ (i.e., if φ is inner),

$$H^{2+} \ominus \varphi H^{2+} = H^{2+} \cap (L^2 \ominus \varphi H^{2+}) = H^{2+} \cap \varphi H^{2-}.$$

- In the Appendix, Szász spaces are defined, and some of their basic properties are listed.

2 Introduction

Completeness problems arise in a natural way in the study of the zeros of a Mellin transform.[1] Assume that f belongs to some functional topological vector space X, invariant under a family of dilation operators K_λ, $\lambda \in \Lambda$, which are assumed to be continuous. Here Λ is a semigroup of positive numbers, i.e., we have $1 \in \Lambda$ and $\lambda\mu \in \Lambda$ if $\lambda \in \Lambda$ and $\mu \in \Lambda$. Let $T_\Lambda(f)$ be the (closed) span in X of the family $K_\lambda f$, $\lambda \in \Lambda$; $T_\Lambda(f)$ is an invariant subspace in the sense that

$$K_\lambda(T_\Lambda(f)) \subset T_\Lambda(f), \quad \lambda \in \Lambda.$$

In the following, when we say "the family $f(\lambda x)$, $\lambda \in \Lambda$, is complete in X" we simply mean that $T_\Lambda(f) = X$.

Now let Ω be the set of those complex numbers s such that (1.1) defines a non trivial continuous linear functional for $f \in X$. If $\rho \in \Omega$ and $Mf(\rho) = 0$, then

$$MK_\lambda f(\rho) = \int_0^{+\infty} f(\lambda t)t^{\rho-1}\,dt = \lambda^{-\rho}Mf(\rho) = 0$$

for every $\lambda > 0$, so that $T_\Lambda(f)$ lies in the null space of the linear functional defined by ρ. Therefore a zero of Mf may be viewed as an obstruction to the completeness in X of the family $f(\lambda x)$, $\lambda \in \Lambda$.[2]

Even in cases where Ω is empty, a modified form of this obstruction may hold. If, for instance, $X = L^2(0, +\infty)$, (1.1) has to be interpreted in

---

[1]We are sketching here a multiplicative theory. There is, of course, an isomorphic additive theory, where dilation is replaced by translation, Mellin by Laplace, and so forth. Note, however, that we will equip $(0, +\infty)$ with the measure dt, and not dt/t.

[2]This is true even if Λ is not a semigroup.

the Plancherel sense on the line $\Re s = \frac{1}{2}$ and the vanishing of $Mf\left(\frac{1}{2} + i\tau\right)$ a.e. on a set of positive measure, and hence that of all $MK_\lambda f\left(\frac{1}{2} + i\tau\right)$, is an obstruction to the completeness in X of the family $f(\lambda x)$, $\lambda > 0$.

If $\Lambda = [1, +\infty)$, there is another trivial obstruction, that is best understood by an example. If $X = L^p(0,1) \subset L^p(0,+\infty)$ and $f = \chi_{(0,1/2)}$, then $T_\Lambda(f) = L^p(0,1/2) \neq L^p(0,1)$, although the Mellin transform $Mf(s) = 2^{-s}/s$ has no zeros.

The inverse problem is in general a difficult one: if $f \in X$ is such that $T_\Lambda(f) \neq X$, is this deficiency due to one or both of the two types of obstructions: a zero (or a large set of zeros) of $F(s)$, or a support of f that is too small (in the case $\Lambda = [1, +\infty)$)?

The starting point of this investigation is the Hahn-Banach Theorem. Assume X is locally convex and also, for simplicity, that X^* can be identified with a function space. The fact that $T_\Lambda(f) \neq X$ means that there exists $g \in X^*$, $g \neq 0$, such that

$$\int_0^{+\infty} f(\lambda t)g(t)\, dt = 0, \quad \lambda \in \Lambda,$$

or

$$\int_0^{+\infty} f(t)g(t/\lambda)\, dt = 0, \quad \lambda \in \Lambda.$$

Thus every function in $T_{\Lambda^{-1}}(g)$ (a subspace of X^*, equipped with the weak-$*$ topology) is orthogonal to f. Observe, in the case $X = L^p(0,1)$, $\Lambda = [1, +\infty)$, say, that if f vanishes on $(a, 1)$, $0 < a < 1$, then any $g \in L^q(0,1)$ $(q = p/(p-1))$ that vanishes on $(0, a)$ is orthogonal to $T_\Lambda(f)$. Hence, assuming that g does not vanish on any neighborhood of 0, one may try to show that $T_{\Lambda^{-1}}(g)$ contains a *character*, $t \mapsto t^{\rho-1}$, in which case the first obstruction will hold. This may require a sophisticated argument.

We end the general discussion here and give criteria for the completeness of the family $f(\lambda x)$, $\lambda \in \Lambda$, in six cases.[3] It turns out that in most cases the theorems give criteria for the equality $V = X$, where V is a subspace of X that is invariant under Λ-dilation, i.e., such that $K_\lambda(V) \subset V$ for $\lambda \in \Lambda$. Note also that the equality $V = X$ is equivalent to $\varphi \in V$, where φ is any given Λ-generator of X, i.e., a function belonging to X and such that $T_\Lambda(\varphi) = X$.

Theorem 1 (Wiener's L^1 Tauberian Theorem). *Let V be a subspace of $L^1(0,+\infty)$ invariant under all dilations K_λ, $\lambda > 0$. Then $V \neq$*

---

[3]The paper [7] contains an informative introduction to these topics, as well as an important contribution to the case where X is a weighted L^p space.

$L^1(0, +\infty)$ *if and only if there is a real number* τ *such that*

$$\int_0^{+\infty} f(t) t^{i\tau} \, dt = 0$$

for every $f \in V$.

Theorem 2 (Wiener's L^2 Tauberian Theorem). *Let V be a subspace of $L^2(0, +\infty)$ invariant under all dilations K_λ, $\lambda > 0$. Then $V \neq L^2(0, +\infty)$ if and only if there is a measurable set $E \subset \mathbb{R}$ of positive measure such that for every $f \in V$ and almost every $\tau \in E$*

$$\hat{f}(\tau) := \underset{T \to +\infty}{\text{l.i.m.}} \int_{1/T}^T f(t) t^{i\tau - 1/2} \, dt$$

vanishes.

Theorem 3 (Beurling 1952 [5]). *Let $1 < p < +\infty$, and let f be a measurable complex function on $(0, +\infty)$ such that*

(i) $\displaystyle \int_0^{+\infty} x^{\alpha/p - 1} |f(x)|^\alpha \, dx < +\infty$ *for every* $\alpha \geqslant 1$;

(ii) *the set of real numbers τ such that*

$$\int_0^{+\infty} f(x) x^{-1/q + i\tau} \, dx = 0, \quad q := p/(p-1)$$

has Hausdorff dimension d, $0 \leqslant d < \min(1, 2/q)$. Then the family $f(\lambda x)$, $\lambda > 0$, is complete in $L^p(0, +\infty)$.

Theorem 4 (Nyman 1950). *Let V be a subspace of $L^1(0, 1)$ invariant under all dilations K_λ, $\lambda \geqslant 1$. Then $V = L^1(0, 1)$ if and only if*

(i) *there is no a, $0 < a < 1$, such that all $f \in V$ vanish a.e. on $(a, 1)$;*

(ii) *there is no ρ such that $\Re\rho \geqslant 1$ and $\int_0^1 f(t) t^{\rho - 1} \, dt = 0$ for all $f \in V$.*

Theorem 5. *Let V be a subspace of $L^2(0, 1)$ invariant under all dilations K_λ, $\lambda \geqslant 1$. Assume that at least one Mellin transform of a function in V is analytic in the closed half-plane $\Re s \geqslant \frac{1}{2}$. Then $V = L^2(0, 1)$ if and only if*

(i) *there is no a, $0 < a < 1$, such that all $f \in V$ vanish a.e. on $(a, 1)$;*

(ii) *there is no ρ such that $\Re\rho > \frac{1}{2}$ and $\int_0^1 f(t) t^{\rho - 1} \, dt = 0$ for all $f \in V$.*

Theorem 5 is a corollary of a general theorem due to Beurling, Karhunen, Nyman and Lax, describing the structure of a subspace of $L^2(0, 1)$ that is invariant under all dilations K_λ, $\lambda \geqslant 1$. In that theorem, a prominent role is played by the concept of inner factor of a function in the Hardy space

$H^2(\Re s > 1/2)$; we refer to 3.20 for an introduction to these matters. The general theorem implies that, in the case of analytic continuation to the closed half-plane $\Re s \geqslant 1/2$, the only obstructions to completeness are, once more, the two "trivial" ones.

Theorem 6 (Beurling 1953 [6]). *Let V be a subspace of $L^p(0,1)$, $1 < p < +\infty$, invariant under all dilations K_λ, $\lambda \geqslant 1$. Assume that V is the closure in $L^p(0,1)$ of a subset of $\bigcup_{p'>p} L^{p'}(0,1)$. Then $V = L^p(0,1)$ if and only if*

(i) for every a, $0 < a < 1$, the set of restrictions to $(a,1)$ of functions of V is dense in $L^p(a,1)$;

(ii) there is no ρ such that $\Re\rho > 1/p$ and $\int_0^1 f(t)t^{\rho-1}\,dt = 0$ for all $f \in V$.

In fact, [6] contains the dual formulation of Theorem 6: if $1 < q < +\infty$, $g \in L^q(0,1)$, and g does not vanish a.e. in any neighborhood of 0, then there is a ρ such that $\Re\rho > 1/p$ and for every r with $1 < r < q$ the character $t \mapsto t^{\rho-1}$ lies in the span in $L^r(0,1)$ of the functions

$$x \mapsto \begin{cases} g(x/\lambda) & \text{if } 0 < x \leqslant 1, \\ 0 & \text{if } x \geqslant 1, \end{cases}$$

where $\lambda \geqslant 1$.

The above theorems, dealing with the vanishing or non-vanishing of Mellin transforms on vertical lines or half-planes of $\mathbb{C}$, are relevant to the Riemann hypothesis.[4] It is therefore natural to study in detail the links between this famous unsolved question and completeness problems. The aim of the present paper is to provide an exhaustive list of papers published on this theme through the year 2001, together with some comments. The papers, as well as the corresponding subsections describing these papers are numbered from 3.1 to 3.23.

I thank Eric Saias; all remarks that follow arose in conversations with him, and Propositions 1 and 2, which may be new, are joint results. Thanks also to Luis Báez-Duarte, Gerard J. Foschini, Bernard Landreau, Nikolaï Nikolski and many others for conversations on the subject matter of this survey.

---

[4]Sometimes, we will denote the Riemann hypothesis by (RH).

3　Papers

3.1　N. Wiener, Tauberian Theorems, Ann. of Math. 33 (1932), 1–100

This classic paper is the primary source of all subsequent developments. Chapter I, "The closure of the set of translations of a given function", contains the statements and proofs of Theorems 1 and 2 above, in their additive formulation. Chapter IV, "Tauberian theorems and prime number theory", contains Wiener's proof of the prime number theorem. Let us quote Remarks (2) and (3), p. 93–94, of §24 (Some unsolved problems) of Chapter VII "Tauberian theorems and spectra".

> (2) Obviously, the power of Tauberian theorems in number theory has not been exhausted. Is there any Tauberian theorem which will reach from one complex ordinate to another, and enable us to handle the more refined forms of the prime number theorem?
>
> (3) In particular, can we make a direct study of the closure of the set of all polynomials in the functions
>
> (24.01) $$e^{-\lambda(x-\xi)}\frac{d}{dx}\frac{e^{x-\xi}}{e^{e^{x-\xi}}-1},$$
>
> and thus attack directly the problems of the zeros of the zeta-function in the critical strip?

The relevance of Remark (3) to our subject is that, if $\lambda < 1$, the statement

$$\zeta(\lambda + i\tau) \neq 0 \quad \text{for every } \tau \in \mathbb{R}$$

is equivalent to the completeness in $L^1(\mathbb{R})$ of the functions (24.01), $\xi \in \mathbb{R}$. This is an immediate application of Wiener's L^1 Tauberian theorem and the formula

$$\int_{-\infty}^{+\infty} e^{-\lambda x}\frac{d}{dx}\frac{e^x}{e^{e^x}-1}e^{-i\tau x}\,dx = (\lambda + i\tau)\zeta(1 - \lambda - i\tau)\Gamma(1 - \lambda - i\tau),$$

which can be deduced using a change of variables, partial integration, and analytic continuation, from the formula

$$\int_0^{+\infty}\left(\frac{u}{e^u - 1} - 1\right)u^{-s-1}\,du = \zeta(1 - s)\Gamma(1 - s), \quad 0 < \sigma < 1$$

(cf. [21, 2.7.1]).

If V denotes the span in $L^1(\mathbb{R})$ of the functions (24.01), $\xi \in \mathbb{R}$, the fact that $\zeta(s)$ has only finitely many zeros on the line $\sigma = \lambda$ is equivalent to

V being *complemented* in $L^1(\mathbb{R})$ (meaning that there is a subspace $W \subset L^1(\mathbb{R})$ such that the addition map $V \times W \ni (f_1, f_2) \mapsto f_1 + f_2 \in L^1(\mathbb{R})$ is a Banach space isomorphism). This follows from the characterization of complemented subspaces of $L^1(\mathbb{R})$ (cf. [1]) and density theorems for the ζ function (cf. [21, Chapter 10]).

Finally, observe that Theorem 1 implies that the set $\mathcal{N}_{(0,+\infty)}$ is dense in $L^1(0, +\infty)$ and Theorem 2 that it is dense in $L^2(0, +\infty)$.

3.2   B. Nyman, On the One-Dimensional Translation Group and Semi-Group in Certain Function Spaces, Ph.D. Thesis, Uppsala, 1950

"Tauberian theorems" by Wiener influenced to a large extent the development of harmonic analysis. Among others, Beurling pushed the theory in several directions during the thirties and the fourties, and published in 1949 what is perhaps his best known result: the description of all subspaces of $\ell^2(\mathbb{N})$ that are invariant under the shift operator (cf. [4] and [14, Chapter 17]). At the same time he proposed to his doctoral student Nyman to investigate corresponding problems in other function spaces.

In his thesis, Nyman proved Theorem 4 above, and a variant of Theorem 5, building on work by Beurling and Karhunen. As an illustration, he obtained what is the main theorem of our subject.

Theorem 7 (Nyman 1950). *The Riemann hypothesis is true if and only if $\mathcal{N}_{(0,1)}$ is dense in $L^2(0,1)$.*

This is a simple corollary of Theorem 5: the Mellin transforms of elements of $\mathcal{N}_{(0,1)}$ are entire functions, and $2\rho(1/2x) - \rho(1/x) = 1$ for $1/2 < x \leqslant 1$.

We will see a number of variants of this theorem. To begin with, since $\chi := \chi_{(0,1)}$ (the characteristic function of $(0,1)$) is a $[1, +\infty)$-generator of $L^2(0,1)$, we have the following proposition, first noted by Beurling in 3.4.

Variant 1. *The Riemann hypothesis is true if and only if χ is in the $L^2(0,1)$-closure of $\mathcal{N}_{(0,1)}$.*

At this point, it is interesting to compare Variant 5 in 3.17 with the last assertion of 3.1. We know that χ, as a function in $L^2(0, +\infty)$, is a limit of sums

$$\sum c_k \rho(\theta_k/t) \quad \text{with } \theta_k > 0.$$

Is it possible to take these approximations as starting points and, using some properties of the fractional part, construct new ones with the numbers θ_k between 0 and 1? To be able to do so would prove the Riemann hypothesis.

Completeness properties of the functions $t \mapsto \rho(\theta_k/t)$, $\theta_k > 0$, reflect properties of the zeros of ζ: the completeness of all of these functions is due to the fact that the set of zeros of ζ on $\Re s = 1/2$ has measure zero, and the Riemann hypothesis is equivalent to the completeness of the functions $\rho(\theta_k/t)$, $0 < \theta_k \leqslant 1$, in $L^2(0,1)$. Thus one may ask which completeness properties correspond to various known results or conjectures about the zeros of ζ. Here is an example.

Proposition 1. *If $0 < a < b < +\infty$, the family $\rho(\theta/x)$, $\theta > 0$, $\theta \notin [a, b]$, is complete in $L^2(0, +\infty)$.*

Proof. We will use the following result. Let $N_0(T)$ be the number of complex numbers $\beta + i\gamma$ with $\beta = 1/2$, $0 < \gamma < T$, $\zeta(\beta + i\gamma) = 0$. Then $N_0(T)/T \to +\infty$ when $T \to +\infty$. This was proved by Selberg in 1942 (see [21, Chapter 10]) in a sharper form, namely $N_0(T) \gg T \log T$ (i.e., a positive proportion of complex zeros of ζ lie on the critical line).

Let $f \in L^2(0, +\infty)$ be a function orthogonal to all functions $\rho(\theta/x)$ with $0 < \theta < a$ or $\theta > b$. By Plancherel's theorem, we get

$$\int_{-\infty}^{+\infty} Mf\left(\frac{1}{2} + i\tau\right) \frac{\zeta\left(\frac{1}{2} - i\tau\right)}{\frac{1}{2} - i\tau} \theta^{-i\tau} d\tau = 0, \quad \theta \notin [a, b].$$

In other words, the function in $L^1(\mathbb{R})$ defined by

$$g : \tau \mapsto Mf\left(\frac{1}{2} + i\tau\right) \frac{\zeta\left(\frac{1}{2} - i\tau\right)}{\frac{1}{2} - i\tau}$$

has an (additive) Fourier transform $\hat{g}$ that vanishes outside the interval $[\log a, \log b]$. By Fourier inversion,

$$g(\tau) = \frac{1}{2\pi} \int_{\log a}^{\log b} \hat{g}(x) e^{i\tau x} dx \quad \text{a.e.,}$$

which proves that g has an analytic extension to the whole complex plane and is of exponential type. Therefore Mf has a meromorphic extension to the whole complex plane and $Mf(s)\zeta(1-s)$ is an entire function of exponential type.

If $f \neq 0$, then $Mf \neq 0$ and the number of zeros $\beta + i\gamma$ of $Mf(s)\zeta(1-s)$ with $\beta = 1/2$ and $0 < \gamma \leqslant T$ is $O(T)$, by Jensen's formula. By Selberg's result, we deduce that Mf has infinitely many poles on the line $\Re s = 1/2$. This is a contradiction since $\tau \mapsto Mf(1/2 + i\tau)$ is in $L^2(\mathbb{R})$. $\square$

Our last observation on Nyman's thesis is that Theorem 4 implies that $\mathcal{N}_{(0,1)}$ is dense in $L^1(0, 1)$. One checks that the function $x \mapsto 2\rho(1/2x) -$

$\rho(1/x) = \lfloor 1/x \rfloor - 2\lfloor 1/2x \rfloor$ equals 1 on $(1/2, 1)$. Thus condition (i) of Theorem 4 is satisfied, and the Mellin transforms of functions of $\mathcal{N}_{(0,1)}$ have no common zeros in the half-plane $\Re s \geqslant 1$, since $\zeta(s)$ has no zeros there.

3.3   R. Salem, Sur une Proposition Équivalente à l'Hypothèse de Riemann, C. R. Acad. Sci. Paris 236 (1953), 1127–1128

Let

$$k_\sigma(x) = \frac{x^{\sigma-1}}{e^x + 1}, \quad x > 0.$$

Salem observes that, for $0 < \sigma < 1$, the non-existence of a number $\gamma \in \mathbb{R}$ such that $\zeta(\sigma + i\gamma) = 0$ is equivalent to the completeness in $L^1(0, +\infty)$ of the family $k_\sigma(\lambda x)$, $\lambda > 0$, or to the fact that $\varphi \equiv 0$ is the only bounded solution of the integral equation

$$\int_0^{+\infty} \frac{x^{\sigma-1}}{e^{xy} + 1} \varphi(x)\,dx = 0, \quad y > 0.$$

This follows from Wiener's L^1 Tauberian theorem and the formula

$$\Gamma(s)\left(1 - \frac{2}{2^s}\right)\zeta(s) = \int_0^{+\infty} k_\sigma(x)x^{i\tau}\,dx.$$

3.4   A. Beurling, A Closure Problem Related to the Riemann Zeta-Function, Proc. Nat. Acad. Sci. 41 (1955), 312–314

Beurling gives in this paper a generalization of Nyman's Theorem 7.

Theorem 8. *Let $p \in (1, \infty)$. The following assertions are equivalent.*
   (i)   *$\zeta(s)$ has no zeros in the half-plane $\Re s > 1/p$;*
   (ii) *$\mathcal{N}_p = L^p(0, 1)$;*
   (iii) *$\chi \in \mathcal{N}_p$.*

REMARKS.
   (a) The proof of Theorem 8 follows from Beurling's Theorem 6. We have first

$$\mathcal{N}_{(0,1)} \subset L^\infty(0,1) \subset \bigcup_{p' > p} L^{p'}(0,1).$$

To prove that the restrictions of functions in $\mathcal{N}_{(0,1)}$ to $(a, 1)$ are dense in $L^p(a, 1)$, $0 < a < 1$, Beurling uses the following dyadic argument. It is enough to take $a = 2^{-n}$, $n \geqslant 1$. Now, observe that, for $2^{-n} < b < 2^{-n+1}$,

$$2^n b\rho(2^{-n}/x) - \rho(b/x) = \begin{cases} 1 & \text{if } 2^{-n} < x < b, \\ 0 & \text{if } b < x < 1, \end{cases}$$

so that any function $g \in L^q(2^{-n}, 1)$ orthogonal to $\mathcal{N}_{(0,1)}$ on $(2^{-n}, 1)$ must vanish on $(2^{-n}, 2^{-n+1})$. The result then follows by induction.

(b) One may wonder if this property of density in $L^p(a, 1)$ for any $a > 0$ could help prove density in $L^p(0, 1)$. For the moment we can only say that this is certainly not the case for $p > 2$, in view of Beurling's Theorem 8 and the existence of zeros of $\zeta(s)$ having real part $1/2 > 1/p$. Later (in Proposition 2) we will give a precise statement showing that, for $p = 2$, there is definitely nothing to expect from this approach.

(c) The reader can find a self-contained proof of Theorem 8 in [8]. Note, however, that there is a slight error in the argument, corrected in [2]; this proof is also sketched in [10].

(d) The last sentence of the paper, i.e.,

> We finally point out that the problem of how well $k = 1$ can be approached by functions $\in C$ has a direct bearing on the distribution of the primes even in case ζ has zeros arbitrary close to the line $\sigma = 1$.

where C denotes $\mathcal{N}_{(0,1)}$, suggests a program which has been taken up in 3.10.

3.5 N. Levinson, On Closure Problems and the Zeros of the Riemann Zeta-Function, Proc. Amer. Math. Soc. 7 (1956), 838–845

This work of Levinson has two interesting features: the vector space X is no longer a Banach space, and the dilation set Λ is no longer a semigroup. For $0 < a < b < +\infty$ define

$$\|f\|_{2,a,b} := \left(\int_0^{+\infty} |f(t)|^2 \left(t^{2a-1} + t^{2b-1} \right) dt \right)^{1/2}.$$

Let $L^2_{\sigma_1,\sigma_2}(0, +\infty)$ be the Fréchet space defined by these (semi)-norms, when $\sigma_1 < a < b < \sigma_2$.

Theorem 9. *Let λ_n be a positive increasing sequence such that*

$$\sum \frac{1}{\lambda_n} = +\infty. \tag{3.1}$$

Let $1/2 \leqslant \sigma_1 < \sigma_2 \leqslant 1$ and $\varphi(x) = e^{-x}/(1 + e^{-x})$. Then the family $\varphi(\lambda_n x)$ is complete in $L^2_{\sigma_1,\sigma_2}(0, +\infty)$ if and only if $\zeta(s)$ has no zeros in the strip $\sigma_1 < \sigma < \sigma_2$.

REMARKS.

(a) In the original paper, the theorem is stated slightly differently: Levinson's necessary and sufficient condition is that $x \mapsto e^{-x}$ is in the span of the $\varphi(\lambda_n x)$.

(b) Condition (3.1) can be thought of as an analog of the Blaschke condition (3.5). It is obtained by an application of an inequality of Carleman ([20, §3.7]).

(c) Among researchers in the field, this paper is best known for the footnote on page 840, in which Levinson comments on the "zero" obstruction to completeness:

> The trivial character of all such sufficiency proofs seems to indicate that if the Riemann hypothesis is true the closure theorems do not seem to be a very promising direction to pursue.

3.6   G. Klambauer, A Note on Nyman's Function System, Proc. Amer. Math. Soc. 13 (1962), 312–314

The author proves in an elementary way that no non-zero (ordinary) polynomial can be orthogonal to $\mathcal{N}_{(0,1)}$. Of course, much more is true: we saw in Section 3.2 that $\mathcal{N}_{(0,1)}$ is dense in $L^1(0,1)$, and hence no non-zero function in $L^\infty(0,1)$ can be orthogonal to $\mathcal{N}_{(0,1)}$. We will give other properties of the orthogonal complement $L^2(0,1) \ominus \mathcal{N}^2$ in the following section.

3.7   H. Bercovici and C. Foias, A Real Variable Restatement of Riemann's Hypothesis, Israel J. Math. 48 (1984), 57–68

In this paper, the authors use the Nagy-Foias theory [18] to study the theorems of Nyman and Beurling.

> This approach provides a better understanding of the infinite dimensional geometry involved in the study of Riemann's hypothesis.[5]

The main result of the paper is the following description of $\mathcal{N}^p \cap L^2(0,1)$ for $1 \leqslant p \leqslant 2$.

Theorem 10. *Let* $1 \leqslant p \leqslant 2$. *Then*

$$\mathcal{N}^p \cap L^2(0,1) = \left\{ f \in L^2(0,1), \quad \frac{Mf(s)}{\zeta(s)} \text{ is holomorphic for } \Re s > \frac{1}{p} \right\}.$$

---

[5]Page 58.

REMARKS.

(a) In our paper 3.20 we ask whether a more general result holds, which allows values of $p > 2$ and does not require taking the intersection with $L^2(0,1)$.

Question 1. *If* $1 \leq p < +\infty$, *is it true that*

$$\mathcal{N}^p = \left\{ f \in L^p(0,1), \quad \frac{Mf(s)}{\zeta(s)} \text{ is holomorphic for } \Re s > \frac{1}{p} \right\}?$$

The paper 3.20 contains a detailed introduction to the case $p = 2$ of Theorem 10.

(b) Theorem 10 tells us that $L^2(0,1) \ominus \mathcal{N}^2 =: \mathcal{N}^{2\perp}$ is precisely the Szász space $\mathcal{S}(Z)$, where Z is the multiset of zeros of ζ in the half-plane $\Re s > 1/2$. We already saw in Section 3.6 that $\mathcal{N}^{2\perp} \cap L^\infty(0,1) = \{0\}$. Now we can add the property

$$f \in \mathcal{N}^{2\perp} \text{ and } \int_0^1 |f(t)| \frac{dt}{t} < +\infty \text{ implies } f = 0, \qquad (3.2)$$

which, according to Proposition 6 in the Appendix, is equivalent to the inclusion relation

$$Z \subset \left\{ s, \frac{1}{2} < \Re s \leq 1 \right\}.$$

There is also an elementary proof of (3.2) based on Möbius inversion. If $f \in \mathcal{N}^{2\perp}$ and $\int_0^1 |f(t)| \, dt/t < +\infty$, then

$$G(t) := \sum_{n \geq 1} F\left(\frac{t}{n}\right) = Ct,$$

where $F(t) = \int_0^t f(u)du$ and $C = \int_0^1 f(u)\lfloor 1/u \rfloor du$, and hence

$$F(t) = \lim_{y \to +\infty} \sum_{p \mid n \Rightarrow p \leq y} \mu(n) C \frac{t}{n} = 0.$$

We leave the details to the reader.

(c) The paper also introduces other function spaces, the Zorn spaces, connected with the problem; see also 3.16 below.

3.8 J. C. Carey, The Riemann Hypothesis as a Sequence of Surface to Volume Ratios, Linear Algebra Appl. 165 (1992), 131–151

The starting point of Carey's investigations is the following formulation of (RH).

Variant 2. *Let $B' \subset L^2(0,1)$ be the vector space of restrictions to $(0,1)$ of functions in $B_{(0,1)}$, and Φ the linear form on B' defined by*

$$\Phi f = \sum_{k=1}^n c_k \theta_k \text{ if } f(t) = \sum_{k=1}^n c_k \rho(\theta_k/t).$$

Then (RH) *holds if and only if Φ is unbounded.*

REMARKS.

(a) The "if" part follows from the formula

$$\int_0^1 f(t) t^{s-1} \, dt = \frac{\Phi f}{s-1} - \frac{\zeta(s)}{s} \sum_{k=1}^n c_k \theta_k^s; \tag{3.3}$$

if ρ is a zero of ζ with real part $> 1/2$, then $|\Phi f| \leqslant |\rho - 1| \|f\|/(2\beta - 1)$.

This sufficient condition was studied by Ryavec in the seventies. Apparently unaware of Nyman's criterion for (RH), he explored the possibility of constructing functions f with a large $|\Phi f|$; see [15], [16], [17] and [12]. As observed by Bombieri, the "only if" part follows from Nyman's criterion: if Φ is bounded, $\ker \Phi = \mathcal{N}_{(0,1)}$ cannot be dense in $L^2(0,1)$.

(b) Carey considers the subsets $W_1 \subset W_2 \subset \ldots$ of B' defined by

$$W_n := \left\{ f : t \mapsto \sum_{k=1}^n c_k \rho(\theta_k/t), c_k \in \mathbb{C}, 0 < \theta_n \leqslant \theta_{n-1} \leqslant \cdots \leqslant \theta_1 \leqslant 1 \right\},$$

where n is fixed. The spaces W_n are not vector spaces but one can define

$$M_n = \sup_{f \in W_n, \|f\|=1} |\Phi f|,$$

and (RH) is the assertion that $M_n \to +\infty$ when $n \to +\infty$. Carey proves the following theorem.

Theorem 11. *For every $n \geqslant 1$,*
   (i) *there exists $f_n \in W_n$, with $0 < \theta_n < \theta_{n-1} < \cdots < \theta_1 = 1$, such that $M_n = \Phi f_n$;*
   (ii) *$M_n < M_{n+1}$.*

Moreover, Carey gives a geometric argument (alluded to in the title of his paper) suggesting that, perhaps, $M_n \gg n^{1/2}$.

REMARKS.

(a) This is the first paper in our list where the functional equation of ζ is used (namely in the proof of the inequality $M_n < M_{n+1}$).

(b) Here is a problem closely related with Carey's theorem. For $n \geqslant 1$, define

$$\delta_n := \inf_{f \in \mathcal{W}_n} \|\chi - f\|,$$

where $\mathcal{W}_n$ is defined like W_n as the set of n-term linear combinations of the functions $\rho(\theta/t)$, $0 < \theta \leqslant 1$, but consists of functions on $(0, \infty)$, and not their restrictions to $(0, 1)$. Nyman's criterion tells us that (RH) is equivalent to $\lim_{n \to +\infty} \delta_n = 0$.

Question 2. *For $n \geqslant 1$, does there exist $f \in \mathcal{W}_n$ such that $\delta_n = \|\chi - f\|$?*

A positive answer would yield $\delta_n > \delta_{n+1}$. Indeed, if $\delta_n = \delta_{n+1} = \|\chi - f\|$, with $f \in \mathcal{W}_n$, all functions $t \mapsto \rho(\theta/t)$, $0 < \theta \leqslant 1$, are orthogonal to $\chi - f$. Hence, the restriction to $(0, 1)$ of $\chi - f$ is orthogonal to $\mathcal{N}^2$. Since $\chi - f \in L^\infty(0, 1)$, this implies $f = 1$ on $(0, 1)$, which is impossible by (3.3).

3.9   J. Alcántara-Bode, An Integral Equation Formulation of the Riemann Hypothesis, Integral Equations Operator Theory 17 (1993), 151–168

The author interprets Nyman's criterion within the framework of operator theory. For $f \in L^1(0, 1)$, define

$$Af: \theta \mapsto \int_0^1 \rho(\theta/t) f(t) \, dt, \quad 0 < \theta \leqslant 1.$$

Af is a continuous function on $[0, 1]$. We will consider A only as a linear operator in $L^2(0, 1)$, which is clearly Hilbert-Schmidt, and hence compact. The paper gives first two reformulations of Nyman's criterion.

Variant 3. (RH) *holds if and only if A is injective.*

Variant 4. (RH) *holds if and only if the function $\theta \mapsto \theta$ does not belong to the range of A.*

The author then proceeds to a detailed study of A and introduces the entire function

$$T(z) := 1 - z + \sum_{r=1}^{+\infty} (-1)^{r+1} \frac{\prod_{l=1}^r \zeta(l+1)}{(r+1)!(r+1)} z^{r+1}.$$

Theorem 12. *The non-zero eigenvalues of A are the reciprocals of the zeros of T. They all have algebraic multiplicity one: if $\lambda \neq 0$ is an eigenvalue, $\ker(A - \lambda I)^k$ has dimension 1 for every $k \geqslant 1$. The eigenspace corresponding to λ is generated by $x \mapsto xT'(x/\lambda)$.*

The paper contains a number of other properties of A, in particular the computation of its Fredholm determinant, which turns out to be the entire function $D^*(z) := e^z T(z)$. However, there is a gap in the proof of this last result, which has been filled in 3.22.

3.10   L. Báez-Duarte, On Beurling's Real Variable Reformulation of the Riemann Hypothesis, Adv. Math. 101 (1993), 10–30

The author observes that Möbius inversion and the prime number theorem yield the identity

$$\sum_{n\geqslant 1} \mu(n)\rho\left(\frac{1}{nx}\right) = -\chi(x), \quad x > 0,$$

so that, roughly speaking, proving (RH) amounts to getting from pointwise convergence to convergence in $L^2(0,\infty)$.[6] He then defines a *natural approximation* to $-\chi$ by

$$f_n(x) := \sum_{k=1}^{n} \mu(k)\rho\left(\frac{1}{kx}\right) - n\left(\sum_{k=1}^{n}\frac{\mu(k)}{k}\right)\rho\left(\frac{1}{nx}\right),$$

so that $f_n \in \mathcal{N}_{(0,1)}$. He proceeds to study it in detail, proving in particular that f_n converges to $-\chi$ in $L^1(0,1)$, thus providing a constructive version of the equality $\mathcal{N}^1 = L^1(0,1)$. A by-product of this study is a simple derivation of zero-free regions for ζ from order estimates of the summatory function of the Möbius function.

He also asks the two following questions, which led to further research.

Question 3. *Let $1 < p \leqslant 2$ and suppose $\zeta(s) \neq 0$ for $\sigma > 1/p$. Is it true that $\|\chi + f_n\|_p \to 0$ when $n \to +\infty$?*

Question 4. *Let $1 < p \leqslant 2$ and suppose $\zeta(s) \neq 0$ for $\sigma > 1/p$. Is it true that $\|\chi + f_n\|_r \to 0$ when $n \to +\infty$, for every r, $1 \leqslant r < p$?*

Question 3 was answered positively by Lee in 3.14. The answer to Question 4 is negative for $p = 2$ (Báez-Duarte, unpublished; see also 3.17).

The paper ends with other constructions of approximations of χ by means of functions $x \mapsto \rho(\theta/x)$, $0 < \theta \leqslant 1$.

---

[6]However, we do not know that (RH) implies that $\chi \in \mathrm{span}_{L^2(0,\infty)}\{x \mapsto \rho(1/nx), n \geqslant 1\}$; see Question 7. Assuming (RH), in order to approximate χ in $L^2(0,\infty)$ we need *a priori* all functions $x \mapsto \rho(\theta/x)$, $0 < \theta \leqslant 1$ (but see the "Note added in proof" at the end of this paper).

3.11 C. P. Chen, The Riemann Hypothesis and Gamma Conditions, J. Math. Anal. Appl. 173 (1993), 258–275

This paper, the third one involved with L^1-completeness, contains the following equivalent form of (RH).

Let $\mathcal{L}$ be the set of functions

$$x \mapsto \int_a^b (1 + e^{xy})^{-1} \varphi(y) dy,$$

where $0 < a < b$, and $\varphi \in L^1(\mathbb{R}^+, dx)$. Then (RH) is equivalent to

$$e^{-x} x^{-\lambda} \in \mathcal{L}_\sigma, \qquad (3.4)$$

the closure of $\mathcal{L}$ in $L^1(\mathbb{R}^+, x^{\sigma-1}\, dx)$, for $0 < \lambda < 1/2 < \sigma < 1$.

In fact it is enough to get (3.4) for all σ with $1/2 < \sigma < 1$ and $0 < \lambda < \sigma$ because then $x \mapsto e^{-x} x^{-\lambda}$ is a $(0, +\infty)$-generator of $L^1(\mathbb{R}^+, x^{\sigma-1}\, dx)$. The result is a consequence of Wiener's L^1-Tauberian theorem.

3.12 N. Nikolski, Distance Formulae and Invariant Subspaces, with an Application to Localization of Zeroes of the Riemann ζ-Function, Ann. Inst. Fourier (Grenoble) 45 (1995), 143–159

The author defines a general notion of "distance function" associated with a given Banach space of analytic functions on, say, the unit disk. As an application he suggests a possible approach to zero-free regions or density theorems for ζ (in the sense of [21, Chapter 9]) by proving the following result.

Theorem 13. *Let $\gamma > 0$ and $\omega \in \mathbb{C}$, $\Re \omega > -\gamma$. Let d be the distance in $L^2\left((0,1), x^{2\gamma-1}\, dx\right)$ between $x \mapsto x^\omega$ and $\mathcal{N}_{(0,1)}$. Then $\zeta(s)$ has fewer than n zeros in the disk*

$$\gamma + \left\{ z : \left| \frac{z - (\omega + \gamma)}{z + (\bar{\omega} + \gamma)} \right| < \left(1 - 2(\gamma + \Re\omega) d^2\right)^{1/2n} \right\}.$$

Note that the case $d = 0$ implies that the half-plane $\Re s > \gamma$ is free of zeros of ζ.

3.13   V.-I. Vasyunin, On a Biorthogonal System Associated
         with the Riemann Hypothesis, Algebra i Analiz 7 (1995),
         118–135 (in Russian); English translation: St. Petersburg
         Math. J. 7 (1996), 405–419.

The main purpose of this paper is an investigation of the geometry of the
vectors e_n in the Hilbert space $L^2(0, \infty)$, where $e_n(x) = \rho(1/nx)$, $x > 0$,
and n is a positive integer.[7] The following theorem summarizes Vasyunin's
results.

Theorem 14. *The family* $(e_n)_{n \geqslant 1}$ *is minimal in* $L^2(0, \infty)$. *Moreover, for*
$m \geqslant 1, n \geqslant 1,$

$$(e_m, e_n) = \frac{\log(2\pi) - \gamma}{2} \left(\frac{1}{m} + \frac{1}{n} \right) + \frac{n - m}{2mn} \log \left(\frac{m}{n} \right)$$
$$- \frac{\pi d}{2mn} \sum_{k=1}^{m_0 - 1} \rho \left(\frac{kn_0}{m_0} \right) \cot \frac{\pi k}{m_0} - \frac{\pi d}{2mn} \sum_{k=1}^{n_0 - 1} \rho \left(\frac{km_0}{n_0} \right) \cot \frac{\pi k}{n_0},$$

where γ *is Euler's constant,* $d = \gcd(m, n)$, $m_0 = m/d$ *and* $n_0 = n/d$.

REMARKS.
   (a) Vasyunin's formulas have been the starting point of various numerical experiments of Landreau and Richard; see [11].
   (b) Until now, no theoretical consequence of Vasyunin's formulas has
been published. It would be interesting to know how much of the following
strategy for proving (RH) can be achieved: compute the distance d_n from
χ to span$(e_1, \dots, e_n)$ by the formula

$$d_n^2 = \frac{\mathrm{Gram}(e_1, \dots, e_n, \chi)}{\mathrm{Gram}(e_1, \dots, e_n)},$$

and prove that $d_n = o(1)$. The computations in [11] show that all three
quantities in this equality behave fairly smoothly for $n \leqslant 10000$, say.
   For $\gcd(m, n) = 1$, define the **Vasyunin sum**

$$V(m, n) := \sum_{k=1}^{n-1} \rho \left(\frac{km}{n} \right) \cot \frac{\pi k}{n}.$$

In view of their relevance in this approach to (RH), these sums deserve a
detailed study.

---

[7]In fact Vasyunin considers the vectors $e_n - e_1/n \in L^2(0, 1)$, $n \geqslant 2$, but it seems
now more natural to study directly the vectors e_n, $n \geqslant 1$.

3.14   J. Lee, Convergence and the Riemann Hypothesis, Comm. Korean Math. Soc. 11 (1996), 57–62

The author gives a positive answer to Question 3, thus providing "an arithmetic version of the Beurling's theorem". This *almost* reproves Beurling's Theorem 8 and tells us, assuming (RH), say, how to approximate $-\chi$ in $L^{2-\varepsilon}(0,1)$ by means of the Möbius function. *Almost*, because ε is not allowed to take the value 0.

3.15   M. Balazard and E. Saias, Notes sur la Fonction ζ de Riemann, 1, Adv. Math. 139 (1998), 310–321

Lee's theorem in 3.14 is reproved in the same way (independently). Let us correct two misprints on page 314: in Remark (l), one should read

> ... chacune des assertions de (v) à (x) découle de (i) ... les preuves connues de (i) $\Rightarrow$ (ii)...

Let $e'_n := -e_n + e_1/n$. The last question in the paper was:

Question 5. *Let $g_n = \sum_{k=2}^n a_{k,n} e'_k$ be the orthogonal projection of χ on the* span$(e'_2, \ldots, e'_n) \subset L^2(0,1)$. *Is it true that*

$$\lim_{n \to +\infty} a_{k,n} = \mu(k)?$$

We did not notice then that a positive answer implies (RH). This can be proved using the system biorthogonal to $(e'_n)_{n \geqslant 2}$ constructed by Vasyunin in 3.13.

3.16   H. Bercovici and C. Foias, On the Zorn Spaces in Beurling's Approach to the Riemann Hypothesis, in: Analysis and Topology (1998), World Sci. Publishing, River Edge, NJ, 143–149

For $0 < \theta < 1$, define $\rho_\theta(x) = \rho(\theta/x) - \theta\rho(1/x)$. The Zorn spaces are defined by $Z_\theta = \text{Vect}(K_\lambda \rho_\theta, \lambda \geqslant 1) \subset L^\infty(0,1)$. One of the results of 3.7 was the following extension of Beurling's Theorem 8. Here $Z_\theta^p := \text{clos}_{L^p(0,1)} Z_\theta$.

Theorem 15. *For fixed $\theta \in (0,1)$ and $p \in [1,2)$, the following assertions are equivalent:*
  (1) *There are no zeros s of ζ with $\Re s > 1/p$;*
  (2) $\mathcal{N}_{(0,1)}$ *is dense in $L^p(0,1)$;*
  (3) $Z_\theta^p \cap L^2(0,1) = Z_\theta^1 \cap L^2(0,1)$;
  (4) $Z_\theta^1 \cap L^p(0,1) = Z_\theta^p$.

In 3.7 the proof of (3) $\Rightarrow$ (4) was incomplete. It is corrected in this paper. The authors point out that $Z^p_{1/2} = \operatorname{clos}_{L^p(0,1)} W$, where W is the space of all functions f that can be written as

$$f(x) = \sum_{n=1}^{+\infty} (-1)^{n-1} F(nx)$$

for some continuous F with bounded variation on $(0, +\infty)$ such that $F(x) = 0$ for $x > 1$.

3.17   L. Báez-Duarte, A Class of Invariant Unitary Operators, Adv. Math. 144 (1999), 1–12

The author considers the following reformulation of Nyman's criterion.

Variant 5. *The Riemann hypothesis is true if and only if χ is in the $L^2(0, +\infty)$-closure of $\mathcal{B}_{(0,1)}$.*

> Looking through the prism, as it were, of unitary operators on $L^2(0, +\infty)$ one hopes to achieve a better understanding of the implications of the above theorem.[8]

Since the subspaces under consideration are dilation invariant, one is led to *invariant* unitary operators, i.e., operators commuting with the dilation operators K_λ, $\lambda > 0$, defined in Section 1. These operators are obtained by the following recipe: Take any unimodular measurable function φ on $1/2 + i\mathbb{R}$, and define

$$U_\varphi : f \mapsto M^{-1}(\varphi M f).$$

This is the general invariant unitary operator on $L^2(0, +\infty)$. Let us call φ the *symbol* of U_φ. The author considers in particular the following unimodular symbols:

$$\varphi_1(s) := \frac{s}{1-s}; \quad \varphi_2(s) := \frac{\zeta(1-s)}{\zeta(s)} = 2(2\pi)^{-s} \cos\frac{\pi s}{2} \Gamma(s).$$

U_{φ_1} is $\mathfrak{M} - I$, where $\mathfrak{M}$ is Hardy's averaging operator[9]

$$\mathfrak{M}f(x) = \frac{1}{x} \int_0^x f(t)\, dt,$$

---

[8] Page 3.

[9] The fact that $\mathfrak{M} - I$ is unitary provides a geometric interpretation of Hardy's well known inequality $\|\mathfrak{M}f\|_2 \leqslant 2\|f\|_2$.

and U_{φ_2} is $\mathfrak{C}S$, where $Sf(x) = f(1/x)/x$ and $\mathfrak{C}$ is the Fourier cosine transform

$$\mathfrak{C}f(x) := \underset{T \to +\infty}{\text{l.i.m.}} \, 2 \int_0^T f(t) \cos(2\pi x t) \, dt.$$

Many identities can be proved by means of this symbolic calculus. For instance, one gets (see Theorem 2.5 of the paper)

$$2 \int_0^{+\infty} \cos(2\pi x t) \frac{\rho(t)}{t} \, dt = \sum_{n \leqslant x} \frac{1}{n} - \log x - \gamma,$$

an equality between two elements of $L^2(0, +\infty)$. This can be written as $\mathfrak{C}S\rho_1 = S(\mathfrak{M} - I)\rho_1$, where $\rho_1(x) = \rho(1/x)$, and equality in L^2 amounts to

$$\frac{\zeta(1-s)}{\zeta(s)} \cdot \frac{-\zeta(s)}{s} = \frac{s'}{1-s'} \frac{-\zeta(s')}{s'}, \quad s' = 1 - s$$

(observe that $MSf(s) = Mf(1-s)$).

The author uses the operator $U = (\mathfrak{M} - I)\mathfrak{C}S = U_{\varphi_1 \varphi_2}$ to prove the divergence in $L^2(0, +\infty)$ of the series $\sum_{n \geqslant 1} \mu(n)\rho(1/nx)$. However, there is an easier proof of that result:[10] if the series is convergent, the sum would be necessarily the pointwise sum $-\chi$ and

$$\int_{1/N}^{+\infty} \left| -\chi(t) + \sum_{n \leqslant N} \mu(n)\rho(1/nt) \right|^2 dt = N \left(\sum_{n \leqslant N} \frac{\mu(n)}{n} \right)^2 \neq o(1).$$

A more striking application of this operator is given in 3.19.

3.18   V. I. Vasyunin, On a System of Step Functions, Zap. Nauchn. Sem. S.-Peterburg. Otdel. Mat. Inst. Steklov (POMI) 262 (1999), 49–70, 231–232 (in Russian; English and Russian summaries.)

The author addresses the problem of finding all functions

$$f : x \mapsto \sum_{k=1}^n c_k \left\lfloor \frac{x}{m_k} \right\rfloor, \quad x > 0,$$

where the m_k are relatively prime positive integers, such that $f(x) \in \{0, 1\}$ for every $x > 0$. The relevance to (RH) of such an investigation is that, perhaps, such *idempotents* could serve as building blocks for approximations of χ in $L^2(0, \infty)$.

---

[10] I thank Luis Báez-Duarte for this remark.

If $\deg f := \sum_{k=1}^{n} |c_k|$, the author shows that $\deg f$ is an odd integer $\geqslant 3$, finds all solutions with degree 3, and gives four infinite families of solutions with degree 5 and 52 "sporadic" solutions with degree 5, 7 or 9. He conjectures that the list of these solutions is complete; in particular that there is no solution with degree > 9.

3.19 L. Báez-Duarte, M. Balazard, B. Landreau, and E. Saias, Notes sur la Fonction ζ de Riemann, 3, Adv. Math. 149 (2000), 130–144

Define

$$D(\lambda) := \operatorname{dist}_{L^2(0,+\infty)} \left(\chi, \operatorname{span}(x \mapsto \rho(\theta/x), \, \lambda \leqslant \theta \leqslant 1) \right),$$

so that (RH) holds if and only if $D(\lambda) = o(1)$ when $\lambda \to 0$. In this paper the following limitation on the rate of convergence is obtained.

Theorem 16.
$$\liminf_{\lambda \to 0} D(\lambda) \sqrt{\log(1/\lambda)} \geqslant \sum_{\rho} |\rho|^{-2},$$

where ρ denotes a generic critical zero of ζ, counted only once, regardless of its multiplicity.

The proof uses in an essential way the operator U defined by Báez-Duarte in 3.17. Moreover, numerical experiments strongly suggest that the theorem gives the correct order of magnitude of $D(\lambda)$.

REMARKS.
(a) Define

$$E(\lambda) := \operatorname{dist}_{L^2(0,+\infty)} \left(\chi, \operatorname{span}(x \mapsto \rho(\theta/x), \, \theta \geqslant \lambda) \right).$$

We know by Wiener's L^2 Tauberian theorem that $E(\lambda) = o(1)$ when $\lambda \to 0$.

Question 6. *Is it true that $E(\lambda) \ll 1/\sqrt{\log(1/\lambda)}$ when $\lambda \to 0$?*

Indeed, *any* effective upper bound implying $E(\lambda) = o(1)$ would be of interest: no effective version of Wiener's L^2 Tauberian theorem seems to be available in the literature.

(b) Let us recall one of the question asked in the paper.

Question 7. *Is (RH) a mollification problem? More precisely, is it true that*

$$(\text{RH}) \Rightarrow \lim_{n \to +\infty} \inf_{(a_1,\dots,a_n) \in \mathbb{C}^n} \int_{\Re s = 1/2} \left| 1 - \zeta(s) \sum_{k=1}^{n} a_k k^{-s} \right|^2 \frac{|ds|}{|s|^2} = 0?$$

By Plancherel's theorem and Nyman's criterion, we know that (RH) is a generalized mollification problem in the sense that

$$\text{(RH)} \iff \lim_{n \to +\infty} \inf_{\substack{(a_1,\dots,a_n) \in \mathbb{C}^n \\ 0 < \theta_1 \leqslant \cdots \leqslant \theta_n \leqslant 1}} \int_{\Re s = 1/2} \left| 1 - \zeta(s) \sum_{k=1}^{n} a_k \theta_k^s \right|^2 \frac{|ds|}{|s|^2} = 0.$$

3.20 M. Balazard and E. Saias, The Nyman-Beurling Equivalent Form for the Riemann Hypothesis, Expos. Math. 18 (2000), 131–138

This paper is an introduction to Nyman's criterion, centered on the proof of Theorem 10 for $p = 2$, and its analytical background.

3.21 G. J. Foschini, Problem 10821, Amer. Math. Monthly 107 (2000), 653

Let us reproduce the text of this problem.

Find a sequence of functions $f_1, f_2, \dots$ in $L_2[0, 1]$ that satisfies the following conditions.

(1) For all $\epsilon \in (0, 1)$, the space spanned by $\{f_{1,\epsilon}, f_{2,\epsilon}, \dots\}$ is $L_2[\epsilon, 1]$, where $f_{n,\epsilon}$ is the restriction of f_n to $[\epsilon, 1]$.

(2) The space spanned by $\{f_1, f_2, \dots\}$ has an infinite-dimensional orthogonal complement in $L_2[0, 1]$.

The following general fact about Szász spaces provides a class of solutions.

Proposition 2. *Let $S(Z)$ be a Szász space and $0 < a < 1$. Let T_a be the set of restrictions to $[a, 1]$ of functions in $T(Z) := L^2(0, 1) \ominus S(Z)$. Then T_a is dense in $L^2(a, 1)$.*

Proof. Let $g \in L^2(a, 1) \ominus \overline{T_a}$, and f the function defined on $(0, 1)$ by

$$f(x) = \begin{cases} 0 & \text{if } 0 < x \leqslant a, \\ g(x) & \text{if } a < x \leqslant 1. \end{cases}$$

Then $f \in L^2(0, 1) \ominus T(Z) = S(Z)$. But Mf is an entire function and Proposition 5 of the Appendix implies $f = 0$, $g = 0$ and $\overline{T_a} = L^2(a, 1)$. $\quad\square$

3.22   J. Alcántara-Bode, An Algorithm for the Evaluation of Certain Fredholm Determinants, Integral Equations Operator Theory 39 (2001), 153–158

The paper contains a corrected proof of the equality $D^*(z) := e^z T(z)$ claimed in 3.9. There is also a useful review of the properties of compact operators in Hilbert spaces.

3.23   J.-F. Burnol, An Adelic Causality Problem Related to Abelian *L*-Functions, J. Number Theory 87 (2001), 253–269

The author generalizes Nyman's criterion to abelian *L*-functions of global fields, establishing a connexion with the scattering theory of Lax and Phillips. There are also remarks and variants on Nyman's criterion itself.

Appendix: Szász Spaces

All theorems and properties in this Appendix are classical but this summary may be convenient for the reader; for more information see [9] and [13].

(a) Let Z be a multiset of complex numbers in the half-plane $\Re s > \frac{1}{2}$, $\widetilde{Z} := \{1 - \bar{\rho}, \rho \in Z\}$ and $E_Z := Z' \cup \operatorname{clos} \widetilde{Z}$ the set of accumulation points of Z and of adherent points of $\widetilde{Z}$. Assume that

$$\sum_{\rho \in Z} \frac{\Re(\rho - 1/2)}{|\rho|^2} < +\infty, \tag{3.5}$$

where each ρ is counted according to its multiplicity. Then the **Blaschke product**

$$B(s) = B_Z(s) := \prod_{\rho \in Z} \left(\frac{|\rho(1 - \rho)|}{\rho(1 - \rho)} \frac{s - \rho}{s + \bar{\rho} - 1} \right),$$

converges uniformly on compact subsets of $\mathbb{C} \backslash E$ ($\supset \left\{ \Re s > \frac{1}{2} \right\}$) and defines a meromorphic function in $\mathbb{C} \backslash (Z \cup \widetilde{Z})'$, with poles at points of $\widetilde{Z}$. Moreover (even in cases where E contains a set of positive linear measure on $\frac{1}{2} + \mathbb{R}i$), $B_Z(s)$ has an a.e. unimodular non tangential limit on the line $\frac{1}{2} + \mathbb{R}i$.

The set of functions of H^{2+} (resp. H^{2-}) vanishing on Z (resp. $\widetilde{Z}$) is $B_Z H^{2+}$ (resp. $B_Z^{-1} H^{2-}$).

(b) Let Z be a multiset of complex numbers in the half-plane $\Re s > \frac{1}{2}$. We define

$$\mathcal{S}(Z) := \operatorname*{span}_{L^2(0,1)} \left\{ t \mapsto (\log t)^{k-1} t^{\rho-1}, \, \rho \in Z, \, 0 \le k < m(\rho) \right\},$$

where $m(\rho)$ denotes the multiplicity of $\rho \in Z$.

Theorem 17 (Szász 1916 [19]). $S(Z) \neq L^2(0,1)$ *if and only if* (3.5) *holds.*

If (3.5) holds, we call $S(Z)$ the **Szász space** defined by the multiset Z.

Proposition 3. *Let* $S(Z)$ *be a Szász space and* $T(Z) := L^2(0,1) \ominus S(Z)$. *Then* $T(Z)$ *is invariant under all dilations* K_λ, $\lambda \geqslant 1$.

(c) Mellin transforms of elements of a Szász space have specific properties.

Proposition 4. *Let* $S(Z)$ *be a Szász space. Then*

$$MS(Z) = H^{2+} \ominus B_{\overline{Z}} H^{2+} = H^{2+} \cap B_{\overline{Z}} H^{2-},$$

where $B_{\overline{Z}}$ *is the Blaschke product associated with the multiset* $\overline{Z} := \{\overline{\rho}, \rho \in Z\}$.

Thus the Mellin transform of $f \in S(Z)$ should be thought of as a couple (F, G) where $F(s)$ is holomorphic for $\Re s > 1/2$, $F \in H^{2+}$, $G(s)$ is meromorphic for $\Re s < 1/2$ with poles in $\overline{Z}$, $G \in B_{\overline{Z}} H^{2-}$, and boundary values of F and G coincide on the line $\sigma = 1/2$. Note that $f = 0$ if and only if $G \in H^{2-}$. In case $(\overline{Z} \cup \widetilde{\overline{Z}})' \neq \frac{1}{2} + \mathbb{R}i$, the edge-of-the-wedge theorem ([3, Theorem 3.6.23, p. 265]) implies that F and G are the restrictions of a single function that is analytic in the domain $\mathbb{C} \setminus E_{\overline{Z}}$, with poles at points of $\widetilde{\overline{Z}}$, and non-polar singularities at points of $(\overline{Z} \cup \widetilde{\overline{Z}})' \subsetneq \frac{1}{2} + \mathbb{R}i$.

Proposition 5. *Let* $\alpha \in (1/2, +\infty]$, Z *a multiset satisfying* (3.5) *and*

$$Z \subset \{z, \ 1/2 < \Re z < \alpha\}.$$

If $f \in S(Z)$ *is such that* $Mf(s)$ *is holomorphic for* $\Re s > 1 - \alpha$, *then* $f = 0$.

Proof. Let $G \in B_{\overline{Z}} H^{2-}$ be associated with f as above, $G = B_{\overline{Z}} G_1$, $G_1 \in H^{2-}$. Let G_2 be the analytic continuation of Mf to $\Re s > 1 - \alpha$. Since G_1 and $B_{\overline{Z}}^{-1} G_2$ are analytic for $1 - \alpha < \Re s < 1/2$, with slow growth near $\sigma = 1/2$ and the same boundary values on $\sigma = 1/2$, they coincide in $1 - \alpha < \Re s < 1/2$, and hence on $\overline{Z}$. Therefore, G_1 vanishes on $\overline{Z}$, which implies the factorization $G_1 = B_{\overline{Z}}^{-1} G_3$, with $G_3 \in H^{2-}$. Then $G = G_3 \in H^{2-}$ and $f = 0$. $\qquad\square$

Proposition 6. *Let* $\alpha \in (1/2, +\infty)$, Z *a multiset in the half-plane,* $\Re s > 1/2$ *satisfying* (3.5). *The two following assertions are equivalent.*
  (i) $Z \subset \{z, \ 1/2 < \Re z \leqslant \alpha\}$;
  (ii) $f \in S(Z)$ *and* $\int_0^1 t^{-\alpha} |f(t)| \, dt < +\infty \Rightarrow f = 0$.

Proof. First, if $\rho \in Z$ with $\Re\rho > \alpha$, then $f(t) = t^{\rho-1}$ is in $S(Z)$, $f \neq 0$ and $\int_0^1 t^{-\alpha}|f(t)|\,dt < +\infty$. Hence (ii) $\Rightarrow$ (i).

If $f \in S(Z)$ and $\int_0^1 t^{-\alpha}|f(t)|\,dt < +\infty$, then $Mf(s)$ is holomorphic for $\Re s > 1 - \alpha$. Assuming $Z \subset \{z, 1/2 < \Re z \leqslant \alpha\}$, the same proof as for Proposition 5 shows that the only possible poles of $G(s)$ are on the line $\sigma = 1 - \alpha$. Since analytic continuation provides $G(s) = \int_0^1 t^{s-1}f(t)\,dt$ for $\Re s > 1 - \alpha$, the convergence of $\int_0^1 t^{-\alpha}|f(t)|\,dt$ proves that there are in fact no poles at all, and the same argument as for Proposition 5 leads to $f = 0$. □

(d) If $S(Z)$ is a Szász space, the family of functions

$$t \mapsto f_{k,\rho}(t) := (\log t)^k t^{\rho-1}, \quad \rho \in Z, 0 \leqslant k < m(\rho),$$

is a topological basis of $S(Z)$. The coordinates of $f \in S(Z)$ in this basis are given by a formal application of Mellin inversion and the residue theorem:

$$f(t) \sim \sum_{\rho \in Z} \mathrm{Res}\left(G(s)t^{-s}, s = 1 - \rho\right).$$

(e) If the elements of Z are numbered $\rho_1, \rho_2, \ldots$, each $\rho \in Z$ being repeated with its multiplicity, the Gram-Schmidt orthonormalization process produces an orthonormal basis of $S(Z)$, $(g_n)_{n \geqslant 1}$, where

$$Mg_n(s) = \frac{(2\Re\rho_n - 1)^{1/2}}{s + \rho_n - 1} \prod_{k=1}^{n-1} \left(\frac{|\rho_k(1 - \rho_k)|}{\bar\rho_k(1 - \bar\rho_k)} \frac{s - \bar\rho_k}{s + \rho_k - 1}\right).$$

(f) The theory of Szász spaces is by no means complete. We conclude with an open question.

Question 8. *What is a necessary and sufficient condition on the multiset Z for the existence of a positive constant $c(Z)$ such that*

$$\int_0^{1/2} |f(t)|^2\,dt \geqslant c(Z) \int_0^1 |f(t)|^2\,dt$$

for all $f \in S(Z)$?

Note added in proof (February 2002). L. Báez-Duarte recently gave a positive answer to **Question 7**.

References

[1] D. E. Alspach and A. Matheson, *Projections onto translation-invariant subspaces of $L_1(R)$*, Trans. Amer. Math. Soc. **277** (1983), 815–823.

[2] L. Báez-Duarte and G. A. Cámera, *Horizontal growth of harmonic functions*, Stud. Appl. Math. **98** (1997), 195–202.

[3] C. A. Berenstein and R. Gay, *Complex variables. An introduction*, Grad. Texts in Math., vol. 125, Springer-Verlag, New York, 1991.

[4] A. Beurling, *On two problems concerning linear transformations in Hilbert space*, Acta Math. **81** (1949), 239–255.

[5] ――――, *On a closure problem*, Ark. Mat. **1** (1951), 301–303.

[6] ――――, *A theorem on functions defined on a semigroup*, Math. Scand. **1** (1953), 127–130.

[7] A. Borichev and H. Hedenmalm, *Completeness of translates in weighted spaces on the half-line*, Acta Math. **174** (1995), 1–84.

[8] W. F. Donoghue, Jr., *Distributions and Fourier transforms*, Pure Appl. Math., vol. 32, Academic Press, 1969.

[9] M. M. Džrbašjan, *A characterization of the closed linear spans of two families of incomplete systems of analytic functions*, Mat. Sb. **42** (1982), 1–70.

[10] V. P. Gurarii, *Group methods in commutative harmonic analysis*, Encyclopaedia Math. Sci. (V. P. Havin and N. K. Nikolski, eds.), Commutative harmonic analysis II, vol. 25, Springer-Verlag, 1998.

[11] B. Landreau and F. Richard, *Le critère de Beurling et Nyman pour l'hypothèse de Riemann. Aspects numériques*, submitted.

[12] M. Newman, C. Ryavec, and B. N. Shure, *The use of integral operators in number theory*, J. Funct. Anal. **32** (1979), 123–130.

[13] N. Nikolski, *Lectures on the shift operator*, Grundlehren Math. Wiss., vol. 273, Springer-Verlag, 1986.

[14] W. Rudin, *Real and complex analysis*, McGraw-Hill, 1974.

[15] C. Ryavec, *Zero-free regions for $\zeta(s)$*, Mathematika **22** (1975), 92–96.

[16] ――――, *Inequalities for zeros of $\zeta(s)$*, Acta Arith. **33** (1977), 255–260.

[17] ———, *Inequalities for the zeros of Dirichlet L-functions*, Linear Algebra Appl. **8** (1979), 205–211.

[18] B. Sz.-Nagy and C. Foias, *Harmonic analysis of operators on Hilbert space*, North-Holland, Elsevier, Akadémiai Kiadó, 1970.

[19] O. Szász, *Über die Approximation stetiger Funktionen durch lineare Aggregate von Potenzen*, Math. Ann. **77** (1916), 482–496.

[20] E. C. Titchmarsh, *The theory of functions*, 2nd ed., Oxford University Press, 1939.

[21] ———, *The theory of the Riemann zeta-function*, 2nd ed., Clarendon Press, Oxford, 1986, Revised by D. R. Heath-Brown.

A Survey on Pure and Mixed Exponential Sums Modulo Prime Powers

Todd Cochrane[1] and Zhiyong Zheng[1]

1 Introduction

In this paper we give a survey of recent work by the authors and others on pure and mixed exponential sums of the type

$$\sum_{x=1}^{p^m} e_{p^m}(f(x)), \qquad \sum_{x=1}^{p^m} \chi(g(x)) e_{p^m}(f(x)),$$

where p^m is a prime power, $e_{p^m}(\cdot)$ is the additive character $e_{p^m}(x) = e^{2\pi i x / p^m}$ and χ is a multiplicative character $\pmod{p^m}$. The goals of this paper are threefold; first, to point out the similarity between exponential sums over finite fields and exponential sums over residue class rings $\pmod{p^m}$ with $m \geq 2$; second, to show how mixed exponential sums can be reduced to pure exponential sums when $m \geq 2$; and third, to make a thorough review of the formulae and upper bounds that are available for such sums. Included are some new observations and consequences of the methods we have developed as well as a number of open questions, some very deep and some readily accessible, inviting the reader to a further investigation of these sums.

2 Pure Exponential Sums

We start with a discussion of pure exponential sums of the type

$$S(f, q) = \sum_{x=1}^{q} e_q(f(x)) \tag{2.1}$$

[1]The first author wishes to thank Tsinghua University, Beijing, P.R.C., and the National Center for Theoretical Sciences, Hsinchu, Taiwan, for hosting his visits and supporting this work. The second author was supported by the N.S.F. of the P.R.C. for distinguished young scholars.

where $f(x)$ is a polynomial over $\mathbb{Z}$, q is any positive integer and $e_q(f(x)) = e^{2\pi i f(x)/q}$. These sums enjoy the multiplicative property

$$S(f,q) = \prod_{i=1}^{k} S(\lambda_i f, p_i^{m_i})$$

where $q = \prod_{i=1}^{k} p_i^{m_i}$ and $\sum_{i=1}^{k} \lambda_i q/p_i^{m_i} = 1$, reducing their evaluation to the case of prime power moduli.

For the case of prime moduli we have the fundamental result of Weil [92] that for any polynomial $f(x)$ of degree d not divisible by p, there exists a set of $d-1$ complex numbers $\omega_1, \ldots \omega_{d-1}$, each of modulus $\sqrt{p}$ such that

$$S(f,p) = -(\omega_1 + \omega_2 + \cdots + \omega_{d-1}). \tag{2.2}$$

Moreover, if the sum is extended to the finite field $\mathbb{F}_{p^j}$ by setting

$$S_j = \sum_{x \in \mathbb{F}_{p^j}} e_p(\mathrm{Tr}_j(f(x))),$$

where Tr_j is the trace mapping from $\mathbb{F}_{p^j}$ into $\mathbb{F}_p$, then $S_j = -(\omega_1^j + \cdots + \omega_{d-1}^j)$. These results follow from the rationality of the L-function associated with the sums S_j, together with the accompanying Riemann Hypothesis. In Weil's original work [92] it is only shown that the number of characteristic values ω_i is less than or equal to $d-1$. Bombieri [3] proved the number actually equals $d-1$. The reader is referred to the works of Schmidt [82] and Stepanov [90] for elementary proofs of these results.

It is perhaps less well known that there is a striking analogy for prime power moduli p^m with $m \geq 2$. To illustrate this let t be the largest power of p dividing all of the coefficients of $f'(x)$. Note $t = 0$ if $p > \deg(f)$. We define a

Critical Point Congruence: $\quad C(x) := p^{-t} f'(x) \equiv 0 \pmod{p}, \quad (2.3)$

together with a set of

Critical Points: $\quad A = \{\alpha \in \mathbb{F}_p : C(\alpha) \equiv 0 \pmod{p}\}. \quad (2.4)$

To evaluate $S(f,p^m)$ it is convenient to break up the sum over the different residue classes $\pmod{p}$, $S(f,p^m) = \sum_{\alpha=1}^{p} S_\alpha(f,p^m)$ where

$$S_\alpha(f,p^m) = \sum_{\substack{x=1 \\ x \equiv \alpha \pmod{p}}}^{p^m} e_{p^m}(f(x)). \tag{2.5}$$

Upon setting $x = u + p^{m-1}v$ with u running from 1 to p^{m-1}, subject to the constraint $u \equiv \alpha \pmod{p}$, and v running from 1 to p, one readily sees that if p is odd and $m \geq t+2$ or $p = 2$ and $m \geq 3$ then $S_\alpha = 0$ unless α is a critical point, and so

$$S(f, p^m) = \sum_{\alpha \in A} S_\alpha(f, p^m). \tag{2.6}$$

In particular, if there are no critical points then the sum is zero.

If α is a critical point of multiplicity one, then for odd p and $m \geq t+2$,

$$S_\alpha(f, p^m) = \begin{cases} e_{p^m}(f(\alpha^*))p^{\frac{m+t}{2}}, & \text{if } m - t \text{ is even,} \\ \left(\frac{A_\alpha}{p}\right) e_{p^m}(f(\alpha^*))p^{\frac{m+t-1}{2}} \mathcal{G}_p, & \text{if } m - t \text{ is odd,} \end{cases} \tag{2.7}$$

where α^* is the unique lifting of α to a solution of the congruence

$$p^{-t}f'(x) \equiv 0 \pmod{p^{[(m-t+1)/2]}},$$

$A_\alpha \equiv 2p^{-t}f''(\alpha) \pmod{p}$, $\left(\frac{\cdot}{p}\right)$ is the Legendre symbol and $\mathcal{G}_p$ is the quadratic Gauss sum

$$\mathcal{G}_p := \sum_{x=0}^{p-1} e_p(x^2) = \begin{cases} \sqrt{p}, & \text{if } p \equiv 1 \pmod{4}, \\ i\sqrt{p}, & \text{if } p \equiv 3 \pmod{4}. \end{cases}$$

A similar formula holds for $p = 2$; see [12], Section 5.

It follows from (2.6) and (2.7) that if p is odd, $m \geq t+2$, and all critical points are of multiplicity one, then

$$\sum_{x=1}^{p^m} e_{p^m}(f(x)) = \begin{cases} \sum_{\alpha \in A} e_{p^m}(f(\alpha^*))p^{\frac{m+t}{2}}, & \text{if } m - t \text{ is even,} \\ \sum_{\alpha \in A} \left(\frac{A_\alpha}{p}\right) e_{p^m}(f(\alpha^*))p^{\frac{m+t-1}{2}} \mathcal{G}_p & \text{if } m - t \text{ is odd.} \end{cases}$$
$$\tag{2.8}$$

It is not known to the authors when this formula may have first appeared in the literature but the technique dates back at least to the work of Salié [80]. Katz [47, p. 110], and Dabrowski and Fisher [17] have stated generalizations of this formula in higher dimensions and refer to such formulae as stationary phase formulae. In the result of Dabrowski and Fisher it is also allowed that the critical points have multiplicity greater than one, but the result obtained in this case is less explicit. The authors derived the formula as stated above in their work [12].

To illustrate the formula consider the quadratic Gauss sum $S(Ax^2, p^m)$, with p odd, $p \nmid A$. The critical point congruence is just $2Ax \equiv 0 \pmod{p}$, and so $t = 0$ and there is a single critical point $\alpha = 0$ of multiplicity one. Thus

$$S(Ax^2, p^m) = \begin{cases} p^{m/2}, & \text{if } m \text{ is even,} \\ \left(\frac{A}{p}\right) \mathcal{G}_p p^{\frac{m-1}{2}}, & \text{if } m \text{ is odd.} \end{cases}$$

This of course is a well known formula due to Gauss.

The formula in (2.7) holds as well for rational functions; see Katz [47], Dabrowski and Fisher [17] or the authors' work [14]. For the Kloosterman sum $S(Ax + Bx^{-1}, p^m)$, with $p \nmid AB$, the critical point congruence is $Ax^2 - B \equiv 0 \pmod{p}$. If $\left(\frac{AB}{p}\right) = -1$ then the sum is zero for $m \geq 2$, while if $\left(\frac{AB}{p}\right) = 1$ then there are two critical points of multiplicity one and (2.8) yields the classical formula of Salié [80], proven also by Whiteman [93], Estermann [23], Carlitz [4] and Williams [94].

The extra factor $t/2$ occurring in the exponent on the right-hand side of (2.8) may be understood by realizing that $S(f, p^m)$ actually degenerates to a complete sum $\pmod{p^{m-t}}$ in the following sense:

$$S(f, p^m) = p^t \sum_{x=1}^{p^{m-t}} e_{p^m}(f(x)). \tag{2.9}$$

The latter sum may be regarded as a nonstandard exponential sum $\pmod{p^{m-t}}$ since $f(x) \equiv f(y) \pmod{p^m}$ if $x \equiv y \pmod{p^{m-t}}$. By (2.8), it may be expressed as a sum of complex numbers of moduli $p^{\frac{m-t}{2}}$.

It is quite straightforward to obtain the formula in (2.7). For instance, if m is even and $t = 0$ then write $x = u + p^{m/2}v$ with u and v running from 1 to $p^{m/2}$, and note that $f(x) \equiv f(u) + f'(u)p^{m/2}v \pmod{p^m}$. Thus

$$S(f, p^m) = \sum_u e_{p^m}(f(u)) \sum_v e_{p^{m/2}}(f'(u)v) = p^{m/2} \sum_{p^{m/2} | f'(u)} e_{p^m}(f(u)),$$

which is the stated result. Since there are at most $d - 1$ critical points, we see from (2.8) that $S(f, p^m)$ can be expressed as a sum of at most $d - 1$ complex numbers of moduli $p^{\frac{m+t}{2}}$, and moreover the values of these numbers are explicit.

Question 1. *Is there a general formula for the values ω_i in (2.2) for* $\pmod{p}$ *exponential sums?*

This is a very deep and unyielding problem. For the case of Gauss sums $S(Ax^d, p)$ formulae (sometimes with sign ambiguities) exist for a number of small values of d. The reader is referred to the book of Berndt, Evans and

Williams [2] for a report on progress that has been made for Gauss sums. The value of the Kloosterman sum also is unknown when $m = 1$, although it is known that it may be expressed as a sum of two complex numbers of moduli $\sqrt{p}$. The formula obtained above for $m \geq 2$ unfortunately cannot be extrapolated to the case $m = 1$. Curiously, if the Kloosterman sum is modified by inserting the Legendre symbol to create a Salié sum, $\sum_{x=1}^{p} \left(\frac{x}{p}\right) e_p(Ax + \frac{B}{x})$, then it can be evaluated, and the formula one obtains is the extrapolation of the formula for odd $m \geq 3$ that we state in (7.7). In general, however, it appears that the set of critical points does not play any role in the evaluation of (mod p) exponential sums.

Question 2. *In what cases does the formula stated here for $m \geq 2$ hold as well for the prime field case $m = 1$? Is this coincidence a sporadic phenomenon?*

Question 3. *Is there a unified theory of exponential sums that yields both the results of Weil for the case $m = 1$ and the formula stated here for $m \geq 2$?*

This question becomes perhaps more compelling when one realizes that the same analogy holds between mixed exponential sums with prime moduli and with prime power moduli; see (7.7).

There are two shortcomings to the formula in (2.8). The first is the case $m = t+1$. (We may trivially assume $m > t$ by reducing the degree of f via the identity $x^{p^m} \equiv x^{p^{m-1}} \pmod{p^m}$.) In this case $S(f, p^m)$ degenerates to a nonstandard (mod p) exponential sum. A well known example here is the Heilbronn sum $\sum_{x=1}^{p^2} e_{p^2}(x^p) = p \sum_{x=1}^{p} e_{p^2}(x^p)$. Nontrivial estimates for this sum have been obtained by Heath-Brown [28] and Heath-Brown and Konyagin [29], the latter being $\left| \sum_{x=1}^{p} e_{p^2}(x^p) \right| \ll p^{7/8}$.

Question 4. *Does there exist in general a nontrivial bound for $S(f, p^m)$ when $m = t+1$?*

The second limitation of the formula occurs when there is a critical point of multiplicity greater than one. Consider for example the Gauss sum $S(Ax^d, p^m)$ with $p > d \geq 3$ and $m \geq 2$. For this sum there is a single critical point $x = 0$ of multiplicity $d - 1 > 1$, and thus the formula above doesn't apply. However, by (2.6) we see that $S(Ax^d, p^m) = S_0(Ax^d, p^m)$, that is the only contribution to the sum comes from values of $x \equiv 0 \pmod{p}$. We obtain $S(Ax^d, p^m) = S(Ap^d x^d, p^{m-1}) = p^{d-1} S(Ax^d, p^{m-d})$. The argument can be repeated provided $m - d \geq 2$. In particular, if $d \mid m$ then we see that

$$S(Ax^d, p^m) = p^{(d-1)\frac{m}{d}} = p^{m(1-\frac{1}{d})},$$

a formula known to Hardy and Littlewood [26], [27], in connection with their work on Waring's problem. Two useful observations come out of this example. First is the idea of a recursion formula for evaluating $S(f, p^m)$, an idea we elucidate in more generality in Section 5, and second is the realization that the magnitude of such a sum can be as large as $p^{m(1-\frac{1}{d})}$.

3   Upper Bounds on Pure Exponential Sums

Hua [30], [31], [32] established the following uniform upper bound on exponential sums: For any nonconstant polynomial f (mod p) of degree d and any prime power p^m with $m \geq 1$,

$$|S(f, p^m)| \leq c_1(d) p^{m(1-\frac{1}{d})}, \tag{3.1}$$

with $c_1(d) = d^3$. In view of the preceding example, the exponent here is best possible, but improvements on $c_1(d)$ were made by Chen [6], [7], Chalk [5], Ding [20], [21], Loh [58], [59], Lu [62], [63], Mit'kin [68], Nečaev [72], [73] and Stečkin [89]. Stečkin showed that $c_1(d)$ could be taken as an absolute constant although he did not indicate how large it must be. In [15] we established that one may take $c_1(d) = 4.41$, including the case $p = 2$. This value for $c_1(d)$ follows readily from (5.5).

Question 5. *What is the best possible value for $c_1(d)$? Is it possible to take $c_1(d) = 1 + o(d)$?*

Nečaev and Topunov [74] determined the best possible constant to be 1.986 and 2.263 for polynomials of degree 3 and 4, respectively. Nečhaev [72], Chen [7] and the authors [15] obtained the optimal upper bound $c_1(d) = 1$, under the additional assumption that $p > (d-1)^{\frac{2d}{d-2}}$, the interval where Weil's upper bound implies that $|S(f, p)| \leq p^{1-\frac{1}{d}}$.

Using the multiplicative property of exponential sums one obtains corresponding upper bounds for $S(f, q)$ with q arbitrary, but there still remains the basic question:

Question 6. *What is the best possible constant $c_2(d)$ such that for an arbitrary modulus q, $|S(f, q)| \leq c_2(d) q^{1-\frac{1}{d}}$, for any polynomial f that is nonconstant (as a function) (mod p) for each prime divisor p of q?*

Currently, the best values available for a general modulus are $c_2(d) = e^{d + O(d/\log d)}$ due to Stečkin [89] and $c_2(d) = e^{1.74d}$, due to Qi and Ding [79]; see also Chen [6], [7], Lu [62], [63], Nečaev [72], [73], Qi and Ding [77], [78], and Zhang and Hong [96]. As noted by these authors the biggest obstacle to improving the value of $c_2(d)$ is the task of improving the Weil upper bound for (mod p) exponential sums when p is small relative to d.

4 On Improving the Weil Upper Bound

From (2.2) we have the basic upper bound of Weil for (mod p) exponential sums,

$$|S(f,p)| \le (d-1)\sqrt{p}. \tag{4.1}$$

The bound is valid for any prime p and polynomial f of degree d over $\mathbb{Z}$ that is nonconstant (mod p), with the exception $(p,d) = (2,2)$. In one sense this bound is best possible. Mit'kin [67] proved that for any d and $\epsilon > 0$ there exists an infinite family of pairs (p,λ) with $p \nmid \lambda$ such that

$$\left|S(\lambda x^d, p)\right| \ge (d-1-\epsilon)\sqrt{p}.$$

In this example d is fixed and the prime p grows arbitrarily large. The real interest is in improving the Weil upper bound when p is small relative to d, in particular when $p < d^2$, where the bound is trivial. The only significant progress in this direction appears to be for the case of sparse polynomials, polynomials having relatively few nonzero terms in comparison with the degree. Mordell [70] showed that for the polynomial $f = \sum_{i=1}^{n} a_i x^{d_i}$, with $p \nmid a_i$ for all i and $1 \le d_1 < d_2 < \cdots < d_n$,

$$|S(f,p)| \le (d_1 d_2 \ldots d_n (p-1, d_1, \ldots, d_n))^{\frac{1}{2n}} p^{1-\frac{1}{2n}}. \tag{4.2}$$

This is actually a slightly refined version of Mordell's result as given by Shparlinski [83]. Thus, for instance, if $p \nmid AB$, then

$$\left|S(Ax^d + Bx, p)\right| \le d^{1/4} p^{3/4}, \tag{4.3}$$

which is sharper than (4.1) for $p < (d-1)^4/d$.

The case of Gauss sums $S(Ax^d, p)$ has received much attention. Heath-Brown and Konyagin [29], sharpening earlier bounds of Konyagin and Shparlinski [52], Mullen and Shparlinski [71] and Shparlinski [84] established the following: For $p \nmid a$,

$$\left|S(Ax^d, p)\right| \ll \begin{cases} d^{5/8} p^{5/8}, \\ d^{3/8} p^{3/4}. \end{cases} \tag{4.4}$$

These bounds are sharper than Weil and nontrivial on the interval $d^{\frac{3}{2}} \ll p \ll d^3$. Konyagin [51] has recently sharpened (4.4) to obtain nontrivial bounds for $p > d^{\frac{4}{3}+\epsilon}$. Konyagin [50] and Konyagin and Shparlinski [52] give upper bounds of a much weaker type that are nontrivial for much smaller values of p relative to d. The reader is referred to the recent book of Konyagin and Shparlinski [53] for further discussion of this problem.

A long-standing and very deep question is the following.

Question 7. *For any $\epsilon > 0$ is there a constant $C(\epsilon)$ such that $|S(f,p)| \leq C(\epsilon)(dp)^{\frac{1}{2}+\epsilon}$ for any nonconstant polynomial f (mod p)?*

This type of upper bound is plausible if one believes that the $d-1$ values ω_i in (2.2) are randomly distributed on the circle of radius $\sqrt{p}$. Montgomery, Vaughan and Wooley [69] gave a heuristic argument to suggest that for Gauss sums one has the bound $|S(f,p)| \ll (dp\log(p))^{1/2}$; the bound they state is even more precise. As pointed out to the authors by Igor Shparlinski, results of Karatsuba [44, Theorem 4] and [45] show that for Gauss sums such a bound would be best possible; see also Levenstein [54]. In particular, it is not possible to obtain a uniform upper bound of the type $|S(f,p^m)| \leq Cd^{1/2}\sqrt{p}$ with C an absolute constant. (Mit'kin's example above also prohibits such a bound.)

A more modest question is the following.

Question 8. *Obtain any nontrivial bound for $S(f,p)$ when $p < (d-1)^2$, under the appropriate assumptions on f. For instance, under what restrictions on f does one have $|S(f,p)| \leq \frac{p}{2}$ or $|S(f,p)| \leq p - \sqrt{p}$?*

It is important that certain restrictions be placed on f in view of examples such as $S(x^p - x, p) = p$ and $S(x^{\frac{p-1}{2}}, p) = 1 + (p-1)\cos(2\pi/p)$. For the two term polynomial $Ax^d + Bx$ progress can be made. Combining the bound in (4.3) with the work of Akulinichev [1] we are able to show that if $p \nmid AB$ and $d \not\equiv 1$ (mod $p-1$) then

$$|S(Ax^d + Bx, p)| \leq \frac{7}{8}p. \tag{4.5}$$

5   Local Upper Bounds on Pure Exponential Sums

Chalk [5] established the following upper bound for any nonconstant polynomial f (mod p) and any $m \geq 2$:

$$|S(f,p^m)| \leq d \left(\sum_{\alpha \in A} \nu_\alpha \right) p^{t/(M+1)} p^{m(1-\frac{1}{M+1})}, \tag{5.1}$$

where ν_α is the multiplicity of α as a zero of the critical point congruence and $M = \max_{\alpha \in A}(\nu_\alpha)$. He suggested that one may be able to eliminate the value d altogether from the right-hand side, and this was proven independently by Ding [21], Loh [59] and the authors [12], under the assumption p is odd and $m \geq t + 2$. In [15] we went one step further and proved that for p odd and $m \geq t + 2$ or $p = 2$ and $m \geq t + 3$,

$$|S_\alpha(f,p^m)| \leq \lambda_\alpha p^{\frac{t}{\nu_\alpha+1}} p^{m(1-\frac{1}{\nu_\alpha+1})}, \tag{5.2}$$

with $\lambda_\alpha = \min\{\nu_\alpha, 3.06\}$, and thus

$$|S(f, p^m)| \le \left(\sum_{\alpha \in \mathcal{A}} \lambda_\alpha\right) p^{\frac{t}{M+1}} p^{m\left(1 - \frac{1}{M+1}\right)} = \left(\sum_{\alpha \in \mathcal{A}} \lambda_\alpha\right) p^t p^{(m-t)\left(1 - \frac{1}{M+1}\right)}.$$

$$(5.3)$$

Moreover, if $p > (d-1)^{2d/(d-1)}$ then we can take $\lambda_\alpha = 1$. The second expression on the right-hand side gives a better reflection of the fact that the sum degenerates to a (mod p^{m-t}) exponential sum in the manner of (2.9).

Question 9. *Is it possible to replace $\sum_\alpha \lambda_\alpha$ in (5.3) with an absolute constant?*

A related upper bound in which the value M is replaced by the maximum multiplicity of the zeros of $f'(x)$ over $\mathbb{C}$ was considered by Smith [88], Loxton and Smith [60] and Loxton and Vaughan [61], the latter obtaining

$$|S(f, p^m)| \le (d-1) p^{(\delta + \tau)/(M'+1)} p^{m\left(1 - \frac{1}{M'+1}\right)}, \qquad (5.4)$$

where M' is the maximum multiplicity of any of the complex zeros of f', $\tau = 0$ if $d < p$, $\tau = 1$ if $d \ge p$, and $\delta = \operatorname{ord}_p(\mathcal{D}(f'))$, where $\mathcal{D}(f')$ is the different of f'. In many cases one will have $M' < M$ and so this upper bound will be stronger (as m grows large) than (5.3). Loxton and Vaughan give several examples for which the bound in (5.4) is essentially best possible.

In applications it is sometimes more useful to have upper bounds stated in terms of $d_p(f)$ and $d_p(f_1)$, the degrees of f and f_1 (mod p), where $f_1 = p^{-t} f'$. In [15], Theorem 2.1, we proved that if p is odd and $m \ge t + 2$ or $p = 2$ and $m \ge t + 3$, and $d_p(f) \ge 1$, then

$$|S(f, p^m)| \le 3.06 \, p^{\frac{t}{d_p(f_1)+1}} p^{m\left(1 - \frac{1}{d_p(f_1)+1}\right)}. \qquad (5.5)$$

This essentially follows from (5.2). Under the same hypotheses it was shown in [14], inequality (5.1), that

$$|S(f, p^m)| \le d_p(f_1) p^{m\left(1 - \frac{1}{d_p(f)}\right)}, \qquad (5.6)$$

with an extra factor of $\sqrt{2}$ in case $p = 2$. If $p > \deg(f)$ then $d_p(f_1) = d_p(f) - 1$ and so these two bounds are essentially of the same strength, but for smaller p it is possible to have $d_p(f_1) > d_p(f)$ in which case the latter bound is stronger; consider, e.g., $f = x^p + px^d$. The upper bound in (5.6) follows from the stronger bound given in (5.10).

The proofs of all of the upper bounds stated in this section follow the same general inductive line of argument. For any polynomial $h(X)$ over $\mathbb{Z}$, let $\mathrm{ord}_p(h(X))$ denote the largest power of p dividing all of the coefficients of $h(X)$. For any critical point α define

$$\sigma = \sigma_\alpha = \mathrm{ord}_p(f(pY + \alpha) - f(\alpha)), \qquad (5.7)$$

and

$$g_\alpha(Y) = p^{-\sigma}(f(pY + \alpha) - f(\alpha)). \qquad (5.8)$$

Setting $x = \alpha + py$ with y running from 1 to p^{m-1} one readily obtains the following relationship for $m \geq \sigma$.

Recursion Relationship: $\quad S_\alpha(f, p^m) = e_{p^m}(f(\alpha))p^{\sigma-1}S(g_\alpha, p^{m-\sigma}).$ (5.9)

If $m < \sigma$ then we just have $S_\alpha(f, p^m) = e_{p^m}(f(\alpha))p^{m-1}$. The upper bounds can then be obtained by induction on m. By successive applications of the recursion relationship one can often succeed in either explicitly evaluating $S(f, p^m)$ or at least expressing $S(f, p^m)$ in terms of (mod p) exponential sums. In either case one obtains $S(f, p^m)$ as a sum of complex numbers having moduli that are powers of $\sqrt{p}$. As an example, suppose that we wish to evaluate $S(f, p^4)$, with $f = 3x^4 - 4x^3$ and $p > 3$. There are two critical points, $\alpha = 0$ of multiplicity 2 and $\alpha = 1$ of multiplicity one. For $\alpha = 0$ we have $\sigma = 3$, $g_\alpha(y) = y^3(3py - 4)$, and so $S_\alpha(f, p^4) = p^2 S(-4x^3, p)$. We are left with a cubic Gauss sum (mod p). For $\alpha = 1$, $S_\alpha(f, p^4) = e_{p^4}(-1)p^2$. Thus altogether, $S(f, p^4)$ is a sum of two complex numbers of modulus $p^{5/2}$ and one of modulus p^2.

Another local type of upper bound, in terms of the parameter σ_α, was established in [14], again using the recursion relationship above: For p odd $m \geq t + 2$ and f nonconstant (mod p) we have for any critical point α,

$$|S_\alpha(f, p^m)| \leq \nu_\alpha p^{m(1-\frac{1}{\sigma_\alpha})}. \qquad (5.10)$$

Since $\sigma_\alpha \leq \nu_\alpha + 1 + t$ the upper bound here is often sharper than (5.2).

6   Pure Exponential Sums with Rational Function Entries

Let $f = f_1/f_2$ be a nonconstant (mod p) rational function over $\mathbb{Z}$ with $d = \deg(f_1) + \deg(f_2)$, and $d^* = \max\{\deg(f_1), \deg(f_2)\}$. We define

$$S(f, p^m) = \sum_{\substack{x=1 \\ p \nmid f_2(x)}}^{p^m} e_{p^m}(f_1(x)\overline{f_2(x)})$$

where the overline denotes multiplicative inverse (mod p^m). The value t, the critical point congruence and the set of critical points are defined in the same way as in (2.3) and (2.4) above. In [14] we established that the equality in (2.8) holds for any rational f. Also, by the method of [15] it is not hard to show that the upper bound in (5.2) holds also for any nonconstant (mod p) rational f, and any $m \geq t + 2$. In [14] we obtained the following analogue of the Hua upper bound: For any odd prime p and nonconstant (mod p) rational function f and $m \geq 2$,

$$|S(f, p^m)| \leq dp^{m(1 - \frac{1}{d^*})}. \tag{6.1}$$

For $p = 2$ one needs an extra factor of $\sqrt{2}$ on the right-hand side. One cannot obtain the fraction $1/d^*$ in the exponent directly from (5.3) since it is possible for the multiplicity of a critical point can be as large as $d - 1$; consider $f = x^p/(1 + x^k)$. However, over $\mathbb{C}$ the maximum multiplicity of any zero of f' is at most $d^* - 1$, and so one might expect the bound in (6.1) in view of the result of Loxton and Vaughan (5.4). The upper bound here sharpens earlier results of Ismoilov [41] and Stepanov and Shparlinski [86]. It is deduced from (5.10) and the fact that $\sigma \leq d_p^*(f)$.

Question 10. *Is it possible to replace the value d on the right-hand side of (6.1) with an absolute constant?*

Question 11. *Does an upper bound of the type (5.4) hold for rational functions?*

7 Mixed Exponential Sums

Now let χ be a multiplicative character (mod p^m) and let $f(x), g(x)$ be rational functions over $\mathbb{Z}$. We consider the general "mixed" or "hybrid" sum

$$S(\chi, g, f, p^m) = \sum_{x=1}^{p^m} \chi(g(x)) e_{p^m}(f(x)).$$

It is understood that in the sum x runs only through values for which $f(x)$ and $g(x)$ are both defined (mod p^m) and $g(x)$ is nonzero (mod p). For $m = 1$ it follows from the work of Weil [92] (see also Perel'muter [76]) that if the sum is nondegenerate (that is, either $f \neq f_1^p - f_1$ for any $f_1 \in \overline{\mathbb{F}}_p[x]$ or $g \neq g_1^k$ for any $g_1 \in \overline{\mathbb{F}}_p[x]$ where k is the order of χ) then

$$|S(\chi, g, f, p)| \leq (n_1 + n_2 - 2 + \deg(f)_\infty) \sqrt{p}, \tag{7.1}$$

where n_1 is the number of poles or zeros of g, n_2 is the number of poles of f and $(f)_\infty$ is the divisor $(f)_\infty = \sum_{i=1}^{n_2} m_i P_i$ with m_i the multiplicity of the pole P_i.

For $m \geq 2$ our strategy for estimating the mixed exponential sum is to convert it into a pure exponential sum. Suppose that p is odd and let a be a primitive root $\pmod{p^m}$. In order to evaluate the mixed exponential sum we work over the ring of p-adic integers $\mathbb{Z}_p$ and make use of the p-adic logarithm

$$\log(1 + px) = \sum_{i=1}^{\infty} \frac{(-1)^{i-1}(px)^i}{i},$$

which takes on p-adic integer values for $x \in \mathbb{Z}_p$. Define r by $a^{p-1} = 1 + rp$ and let R be the p-adic integer $R = p^{-1}\log(1 + rp)$. Let c be the unique integer with $1 \leq c \leq p^{m-1}(p-1)$ such that for any integer k,

$$\chi(a^k) = e^{\frac{2\pi i c k}{p^{m-1}(p-1)}}.$$

We extend the additive character $e_{p^m}(\cdot)$ to $\mathbb{Z}_p$ by setting for $x \in \mathbb{Z}_p$, $e_{p^m}(x) = e_{p^m}(\tilde{x})$ where $\tilde{x}$ is the image of x in $\mathbb{Z}_p/(p^m) \simeq \mathbb{Z}/(p^m)$. The key to untwisting the mixed exponential sum is to observe that the multiplicative character χ acts on the subgroup of residues congruent to 1 $\pmod p$ in the following manner; see [9, Lemma 2.1]. For any p-adic integer y,

$$\chi(1 + py) = e_{p^m}(\overline{R}c \log(1 + py)). \tag{7.2}$$

In particular,

$$\chi(1 + p^{m/2}y) = e_{p^{m/2}}(\overline{R}cy). \tag{7.3}$$

To illustrate the argument, suppose that m is even. Writing $x = u + p^{m/2}v$ with u and v running from 1 to $p^{m/2}$ one obtains from (7.3)

$$\begin{aligned}
S(\chi, g, f, p^m) &= \sum_u \sum_v \chi(g(u + p^{m/2}v))e_{p^m}(f(u + p^{m/2}v)) \\
&= \sum_u \sum_v \chi(g(u) + g'(u)p^{m/2}v)e_{p^m}(f(u) + f'(u)p^{m/2}v) \\
&= \sum_u \chi(g(u))e_{p^m}(f(u)) \sum_v e_{p^{m/2}}(\overline{Rcg(u)}g'(u)v + f'(u)v) \\
&= p^{m/2} \sum_{cg'(u) + Rgf'(u) \equiv 0 \pmod{p^{m/2}}} \chi(g(u))e_{p^m}(f(u)).
\end{aligned}$$

This leads us to define a *critical point congruence*

$$C(x) := p^{-t}(Rg(x)f'(x) + cg'(x)) \equiv 0 \pmod{p}, \tag{7.4}$$

where $t = \mathrm{ord}_p(Rgf' + cg')$, together with a set of *critical points*

$$\mathcal{A} = \{\alpha \in \mathbb{F}_p : C(\alpha) \equiv 0 \pmod{p} \quad \text{and} \quad g(\alpha) \not\equiv 0 \pmod{p}\}. \tag{7.5}$$

We again have the basic decomposition,

$$S(\chi, g, f, p^m) = \sum_{\alpha \in A} S_\alpha(\chi, g, f, p^m),$$

for p odd and $m \geq t+2$ or $p = 2$ and $m \geq t+3$. Also, if p is odd, $m \geq t+2$ and α is a critical point of multiplicity one, then

$$S_\alpha(\chi, g, f, p^m) = \begin{cases} \chi(g(\alpha^*))e_{p^m}(f(\alpha^*))p^{\frac{m+t}{2}}, & m-t \text{ even}, \\ \chi(g(\alpha^*))e_{p^m}(f(\alpha^*))\left(\frac{A_\alpha}{p}\right)\mathcal{G}_p p^{\frac{m+t-1}{2}}, & m-t \text{ odd}, \end{cases}$$

$$(7.6)$$

where α^* is the unique lifting of α to a solution of the congruence

$$C(x) \equiv 0 \pmod{p^{[(m-t+1)/2]}},$$

and $A_\alpha \equiv 2r(C/g)'(\alpha) \pmod p$. The details of this derivation are worked out in [9], but the essence of the argument is as given above.

If each critical point has multiplicity one, then for p odd and $m \geq t+2$ we obtain the formula

$$S(\chi, g, f, p^m)$$
$$= \begin{cases} \sum_{\alpha \in A} \chi(g(\alpha^*))e_{p^m}(f(\alpha^*))p^{\frac{m+t}{2}}, & \text{if } m-t \text{ even}, \\ \sum_{\alpha \in A} \chi(g(\alpha^*))e_{p^m}(f(\alpha^*))\left(\frac{A_\alpha}{p}\right)\mathcal{G}_p p^{\frac{m+t-1}{2}}, & \text{if } m-t \text{ odd}, \end{cases} \quad (7.7)$$

which is a sum of complex numbers of moduli $p^{\frac{m+t}{2}}$.

Suppose now that f and g are polynomials. It is clear that there are at most $\deg(f) + \deg(g) - 1$ critical points, but we can be even more precise. Let n be the number of distinct roots of g over $\overline{\mathbb{F}}_p$. Then writing the critical point congruence in the manner

$$p^{-t}(Rf' + c\frac{g'}{g}) \equiv 0 \pmod p, \quad (7.8)$$

we see that the number of critical points is at most $\deg(f) + n - 1$. Interestingly, this is the same value that appears in the corresponding (mod p) exponential sum. To be precise, if f, g are polynomials such that $p \nmid \deg(f)$ and $(\mathrm{ord}(\chi), \deg(g)) = 1$ then there exist complex numbers ω_i, $1 \leq i \leq \deg(f) + n - 1$, of moduli $\sqrt{p}$ such that

$$S(\chi, g, f, p) = -(\omega_1 + \omega_2 + \cdots + \omega_{\deg(f)+n-1});$$

see [82] or [90]. We again ask Questions 2 and 3 in this more general setting.

8   Upper Bounds on Mixed Exponential Sums

For critical points of multiplicity greater than one, we obtain [9] an analogue
of the recursion formula stated for pure exponential sums. Suppose that f
and g are rational functions and that α is a critical point of multiplicity ν.
Let

$$F_\alpha(Y) := c\overline{R}\log\left(\overline{g(\alpha)}g(\alpha + pY)\right) + f(\alpha + pY) - f(\alpha). \quad (8.1)$$

$F_\alpha(Y)$ may be expanded into a formal power series of the type

$$F_\alpha(Y) = \sum_{j=1}^{\infty} a_j Y^j, \quad (8.2)$$

with p-adic integer coefficients a_j. Define

$$\sigma := \mathrm{ord}_p(F_\alpha(Y)) = \min_{j \geq 1}(\mathrm{ord}_p(a_j)), \quad (8.3)$$

$$G_\alpha(Y) := p^{-\sigma}F_\alpha(Y). \quad (8.4)$$

Note, for the case of pure exponential sums $G_\alpha(Y)$ and σ coincide with the
definitions given earlier in (5.8), (5.7). We have for $m \geq \sigma$ the

Conversion Formula:

$$S_\alpha(\chi, g, f, p^m) = p^{\sigma-1}\chi(g(\alpha))e_{p^m}(f(\alpha))S(G_\alpha, p^{m-\sigma}), \quad (8.5)$$

where $S(G_\alpha, p^{m-\sigma}) = \sum_{y=1}^{p^{m-\sigma}} e_{p^{m-\sigma}}(G_\alpha(y))$. The function G_α, defined
a priori as an infinite series with p-adic integer coefficients, may be viewed
as a polynomial over $\mathbb{Z}$ in the exponential sum $S(G_\alpha, p^{m-\sigma})$, since its coefficients are p-adic integers and the high order coefficients all vanish modulo
$p^{m-\sigma}$. Thus we have succeeded in reducing the mixed exponential sum to
the pure exponential sum $S(G_\alpha, p^{m-\sigma})$.

We proceed now to obtain a relationship between G_α and the critical
point function $\mathcal{C}$ defined in (7.4). Note that

$$F'_\alpha(Y) = c\overline{R}p\frac{g'(\alpha + pY)}{g(\alpha + pY)} + pf'(\alpha + pY) = p^{t+1}\overline{R}\frac{\mathcal{C}(\alpha + pY)}{g(\alpha + pY)}. \quad (8.6)$$

Develop $\overline{R}\mathcal{C}/g$ into a Taylor expansion about α,

$$\overline{R}\frac{\mathcal{C}(X)}{g(X)} = \sum_{j=0}^{\infty} c_j(X - \alpha)^j, \quad (8.7)$$

with p-adic integer coefficients c_j, and note that since α is a zero of $\mathcal{C}$ of multiplicity ν,

$$\text{ord}_p(c_j) > 0 \quad \text{for } 0 \leq j < \nu \text{ and ord}_p(c_\nu) = 0. \tag{8.8}$$

It follows from (8.6) that

$$F_\alpha(Y) = p^{t+1} \sum_{j=0}^{\infty} c_j p^j \frac{Y^{j+1}}{j+1}, \tag{8.9}$$

and that

$$G_\alpha(Y) = p^{-\sigma} F_\alpha(Y) = p^{-\sigma} \sum_{j=1}^{\infty} a_j Y^j = p^{t-\sigma} \sum_{j=1}^{\infty} \frac{c_{j-1}}{j} p^j Y^j. \tag{8.10}$$

Using local upper bounds of the type (5.5) and (5.6) we are then able to establish [9] that (5.2) and (5.3) hold identically for any mixed exponential sum under the assumption $m \geq t + 2$ (p odd). In particular, if p is odd, f, g are rational functions over $\mathbb{Z}$ and $m \geq t + 2$ then

$$|S(\chi, g, f, p^m)| \leq \left(\sum_{\alpha \in \mathcal{A}} \lambda_\alpha \right) p^{\frac{t}{M+1}} p^{m(1 - \frac{1}{M+1})}, \tag{8.11}$$

where M is the maximum multiplicity of any critical point. A similar upper bound holds for $p = 2$; see [9], Section 7.

If χ is not a primitive character (mod p^m) and f' is not identically zero (mod p) then $M \leq d - 1$, where $d = \deg(f_1) + \deg(f_2)$ is the total degree of $f = f_1/f_2$. We recover a Hua-type upper bound,

$$|S(\chi, g, f, p^m)| \leq 3.06 \, p^{\frac{t}{d}} p^{m(1 - \frac{1}{d})}, \tag{8.12}$$

for p odd and $m \geq t + 2$.

Suppose now that χ is a primitive character. Writing $f = f_1/f_2$, $g = g_1/g_2$, $d = \deg(f_1) + \deg(f_2)$ and letting n being the number of distinct zeros of $g_1 g_2$ over $\overline{\mathbb{F}}_p$, it follows from (7.8) that the maximum multiplicity M of any critical point satisfies,

$$M \leq \max\{\deg(f) + n - 1, 2\deg(f_2) + n - 1\}. \tag{8.13}$$

If $\deg(f_2) \leq \deg(f_1)$ then we deduce from (8.11) the upper bound

$$|S(\chi, g, f, p^m)| \leq \left(\sum_{\alpha \in \mathcal{A}} \lambda_\alpha \right) p^{\frac{t}{d+n}} p^{m(1 - \frac{1}{d+n})}, \tag{8.14}$$

for p odd and $m \geq t + 2$.

Question 12. *Can the upper bound on M in (8.13) be improved for certain values of* $\deg(f_1)$, $\deg(f_2)$ *and* n?

If f, g are polynomials of degrees d_1, d_2, we can state a more precise uniform upper bound. If p is odd, $m \geq 2$, and $S(\chi, g, f, p^m)$ does not degenerate to a sum of smaller modulus, then we have [9, Corollary 1.1],

$$|S(\chi, g, f, p^m)| \leq 4.41\, p^{m\left(1 - \frac{1}{d_1 + d_2}\right)}. \tag{8.15}$$

For $p = 2$ the same holds with constant 8.82 in place of 4.41. The sum degenerates to one of smaller modulus if $d_p(f) = 0$ and either χ is not primitive or $d_p(g) = 0$. This bound includes as a special case the upper bound of Hua for pure exponential sums, and the upper bound in Section 11 for character sums. Questions 5 and 6 may again be asked for this more general sum.

Perhaps the most important question to ask in regard to mixed exponential sums, especially in view of the important role pure exponential sums have played in number theory and elsewhere, is the following.

Question 13. *Are there any applications of mixed exponential sums and of the conversion formula and estimates stated in this section?*

In the next five sections we consider a number of special types of pure and mixed exponential sums.

9   The Twisted Exponential Sum $S(\chi, x, f, p^m)$

The authors' exploration of mixed exponential sums started with the consideration of the twisted exponential sum $S(\chi, x, f, p^m)$ with f a polynomial of degree d over $\mathbb{Z}$. It had been conjectured by E. Bombieri, M. Liu, and W.M. Schmidt that an upper bound analogous to that of Hua should be available for this sum. One might even expect a stronger uniform upper bound than that of Hua due to extra cancellation coming from the twisted terms, but just the opposite occurs. If f and χ are chosen so that the critical point congruence

$$p^{-t}(Rxf'(x) + c) \equiv 0 \pmod{p},$$

has a single zero of multiplicity d, then the upper bound of Hua can fail to hold. In [12], Example 9.2, we give an example of a polynomial f with $S(\chi, x, f, p^m) = p^{m(1 - \frac{1}{d+1})}$ for infinitely many values of m. From (8.15) we have the uniform upper bound

$$\left| \sum_{x=1}^{p^m} \chi(x) e_{p^m}(f(x)) \right| \leq 4.41\, p^{m(1 - \frac{1}{d+1})}, \tag{9.1}$$

valid for any odd p, any nonconstant (mod p) polynomial f and any $m \geq 1$. Again, for $p = 2$ one needs an extra factor of 2 on the right-hand side. In most cases there will not be a critical point of multiplicity d and so sharper bounds are available.

Question 14. *Can an analogue of the upper bound of Loxton and Vaughan [61] be stated for the twisted exponential sum $S(\chi, x, f, p^m)$?*

For rational functions $f = f_1/f_2$ with $d_p(f_1) + d_p(f_2) \geq 1$ and $m \geq 2$ we established in [14] that

$$\left| \sum_{x=1}^{p^m} \chi(x) e_{p^m}(f(x)) \right| \leq 4Dp^{m(1 - \frac{1}{D+1})}, \tag{9.2}$$

where $D = \max(\deg(f), 2\deg(f_2))$. Also, in Example 6.1 of [14] we give examples where $\deg(f_1) = \deg(f_2)$ and $|S(\chi, x, f, p^m)| = p^{m(1 - \frac{1}{D+1})}$ for infinitely many m.

10   Gauss Sums

For the Gauss sum $S(\chi, x, x, p^m)$ the critical point congruence is $Rx + c \equiv 0$ (mod p). If χ is imprimitive, that is $p \mid c$, then there is no critical point, and the sum is zero for $m \geq 2$. If χ is primitive, then there is a single critical point $\alpha \equiv -c\overline{R}$ (mod p) of multiplicity one, and by (7.7)

$$\sum_{x=1}^{p^m} \chi(x) e_{p^m}(x) = \begin{cases} \chi(\alpha^*) e_{p^m}(\alpha^*) p^{m/2}, & \text{if } m \text{ is even,} \\ \chi(\alpha^*) e_{p^m}(\alpha^*) \left(\frac{-2Rc}{p} \right) \mathcal{G}_p p^{\frac{m-1}{2}}, & \text{if } m > 1 \text{ is odd,} \end{cases} \tag{10.1}$$

where $\alpha^* \equiv -c\overline{R}$ (mod $p^{\lfloor (m+1)/2 \rfloor}$). A variation of this formula was obtained by Odoni [75] and Mauclaire [65], [66]; see also the book by Berndt, Evans and Williams [2].

11   Two Term Exponential Sums

The sum $S(Ax^d + Bx, p^m)$ arises in the application of the circle method to Waring's problem; see [91]. For this sum Hua [31] obtained the upper bound

$$\left| \sum_{x=1}^{p^m} e_{p^m}(Ax^d + Bx) \right| \leq c(d) p^{m/2}(p^m, B),$$

for $d \geq 2$ and $p \nmid A$. Loxton and Smith [60], Smith [88], Dabrowski and Fisher [17] and Ye [95] made further refinements. In [13], as an application of the method presented here, the authors obtained

$$\left| \sum_{x=1}^{p^m} e_{p^m} \left(A x^d + B x \right) \right| \leq d p^{m/2} (p^m, B)^{1/2}. \tag{11.1}$$

A more precise version of our upper bound may be stated in terms of the parameters β, δ and h, defined by

$$p^\beta \| d, \qquad p^\delta \| B, \qquad p^h \| (d-1).$$

For $m \geq 2$ we have

$$\left| \sum_{x=1}^{p^m} e_{p^m} \left(A x^d + B x \right) \right| \leq (d-1, p-1) p^{\delta/2} p^{h/2} p^{\frac{1}{2} \min(1,\beta)} p^{m/2}. \tag{11.2}$$

This upper bound is nearly best possible but there is still room for improvement.

Question 15. *Can the term $p^{\frac{1}{2}\min(1,\beta)}$ be eliminated from the right-hand side? Can the value $(d-1, p-1)$ be improved?*

For the mixed exponential sum $S(\chi, x, Ax^d + Bx, p^m)$ we established in [16] the upper bound

$$\left| \sum_{x=1}^{p^m} \chi(x) e_{p^m} \left(A x^d + B x \right) \right| \leq 2 d p^{2m/3}, \tag{11.3}$$

and showed that the exponent $2m/3$ is best possible in general. This upper bound is valid for any prime p, $m \geq 1$, $d \geq 2$ and $p \nmid (A, B)$. If χ is a character with conductor p (e.g., the principal character or Legendre symbol), $p \nmid (A, B)$, $d \geq 2$, $m \geq 1$ then we obtain [16] the stronger bound

$$\left| \sum_{x=1}^{p^m} \chi(x) e_{p^m} \left(A x^d + B x \right) \right| \leq d p^{m/2}. \tag{11.4}$$

A more precise statement analogous to (11.2) is given in Theorem 1.3 of [16]. It is of interest to note that for the mixed exponential sum in (11.4) we are able to eliminate the factor $(p^m, B)^{1/2}$, required for the upper bound in (11.1).

Question 16. *Can the value d in (11.4) be replaced by $O(\sqrt{d})$ or $O(d^{\frac{1}{2}+\epsilon})$?*

12 Sparse Polynomials

Let f be a sparse polynomial of the type $f(x) = a_1 x^{d_1} + a_2 x^{d_2} + \cdots + a_n x^{d_n}$, with $1 \le d_1 < d_2 < \cdots < d_n = d$ and the a_i nonzero integers. For the case of $(\bmod\ p)$ exponential sums we already noted in (4.2) that an upper bound exists that is sometimes sharper than the Weil bound. For $m \ge 2$ upper bounds exist that are sharper than the Hua upper bound provided that we restrict x to nonzero residues $(\bmod\ p)$. If $p > d$ then one can show that the maximum possible multiplicity of any nonzero critical point associated with the sum $S(f, p^m)$ is $n - 1$. Thus by (5.3) it follows that if $d_p(f) \ge 1$, then

$$\left| \sum_{\substack{x=1 \\ p \nmid x}}^{p^m} e_{p^m}\left(f(x)\right) \right| \le d^{3/2} p^{m\left(1 - \frac{1}{n}\right)}. \tag{12.1}$$

If $p \le d$ then there may exist zeros of larger multiplicity $(\bmod\ p)$. However, over $\mathbb{C}$, the multiplicity will still be at most $n - 1$. Thus one can use the upper bound of Loxton and Vaughan (5.4) to obtain an upper bound with the same exponent on p (but different constant). General upper bounds of the same order of magnitude have been obtained by Loh [59] and Shparlinski [85]. Shparlinski dealt also with Laurent polynomials and showed a connection between these sums and exponential sums with linear recurrence sequences.

13 Character Sums

By a character sum we mean a sum of the type

$$S(\chi, g, 0, p^m) = \sum_{x=1}^{p^m} \chi(g(x)),$$

with $g = g_1/g_2$ a nonconstant rational function over $\mathbb{Z}$ and χ a multiplicative character $(\bmod\ p^m)$. We may assume that χ is a primitive character for otherwise the sum degenerates to one of smaller modulus. Such sums were first studied by Ismoilov [33], [34], [35], [37], [36], [38], [39], [40] and then by Liu [55], [56], [57]. The critical point congruence for the sum is

$$p^{-t} g'(x) \equiv 0 \pmod{p}.$$

It is plain that the maximum possible multiplicity of any critical point is $d - 1$ where d is the total degree of g and so we obtain from (8.11), for p

odd and $m \geq t + 2$,

$$\left| \sum_{x=1}^{p^m} \chi(g(x)) \right| \leq d \, p^{m(1-\frac{1}{d})}. \tag{13.1}$$

Ismoilov and Liu stated bounds of comparable strength, some being more precise. Ismoilov noted that the exponent in (13.1) is best possible in general. Indeed, if $g = Ax^d + B$, $d \geq 2$, $p \nmid ABd$ and $d \mid m$ then for any primitive character χ, $|S(\chi, g, 0, p^m)| = p^{m(1-\frac{1}{d})}$. Ismoilov [40] also indicated that in certain cases the value d in the exponent of (13.1) may be replaced by $d^* := \max(\deg(g_1), \deg(g_2))$. In [10] Cochrane, Liu and Zheng established the more general result that if $d_p(g) \geq 1$ and χ is a primitive character (mod p^m) then for any $m \geq 2$ we have

$$|S(\chi, g, p^m)| \leq (d-1)p^{m(1-\frac{1}{d^*})}, \tag{13.2}$$

with an extra factor of 2 on the right-hand side in case $p = 2$.

Question 17. *Can the constant $(d-1)$ in (13.2) be replaced by an absolute constant?*

For the case of polynomial g the answer is yes. From (8.15) we have the upper bound

$$\left| \sum_{x=1}^{p^m} \chi(g(x)) \right| \leq 4.41 \, p^{m(1-\frac{1}{d})}, \tag{13.3}$$

for any odd p with $d_p(g) \geq 1$, $m \geq 2$ and any primitive character χ (mod p^m).

The analogues of Questions 6,7 and 8 for character sums are all important here. For a general modulus q, Liu [56], [57] obtained the upper bound

$$\left| \sum_{x=1}^{q} \chi(g(x)) \right| \leq e^{1.8d}q^{1-\frac{1}{d}},$$

for any rational g that is nonconstant (mod p) for each $p \mid q$, and any primitive character χ (mod q). The authors are not aware of any progress that has been made towards sharpening the basic upper bound of Weil for prime modulus, when d is large relative to p.

14   Kloosterman and Salié Sums

Kloosterman and Salié sums of the type

$$S(Ax + Bx^{-1}, p^m) \qquad \text{and} \qquad S(\chi, x, Ax + Bx^{-1}, p^m),$$

respectively, with $p \nmid AB$, occur as Fourier coefficients of certain automorphic forms; see Iwaniec [43], p. 78. It was Kloosterman [48] who first noted this connection. The Kloosterman sum for $m \geq 2$ was evaluated by Salié [80], and later by Whiteman [93], Estermann [23], Carlitz [4] and Williams [94]. For $m = 1$ we have the upper bound $2\sqrt{p}$ due to Weil [92], but the evaluation is unknown. For the case of the quadratic character χ, odd p and $m \geq 1$, the sum $S(\chi, x, Ax + Bx^{-1}, p^m)$ was evaluated by Iwaniec [42], [43], p. 68, and Sarnak [81], p. 90. For $p = 2$ an evaluation for quadratic χ was given by Dedeo [18], chap. 4. Duke, Friedlander and Iwaniec [22] have found applications of such estimates in their work on bilinear forms with Kloosterman fractions.

Suppose that p is odd. The critical point congruence for such a sum is

$$RAx^2 + cx - RB \equiv 0 \pmod{p}.$$

Let $\Delta = c^2 + 4R^2 AB$. It follows that if Δ is a quadratic nonresidue $\pmod{p}$ then the sum is zero while if Δ is a quadratic residue $\pmod{p}$ then there are two critical points of multiplicity one, and we obtain the classical formula for the Kloosterman and Salié sum. If $\Delta \equiv 0 \pmod{p}$ then there is a single a critical point of multiplicity two. In any case,

$$\left| \sum_{x=1}^{p^m} \chi(x) e_{p^m}(Ax + Bx^{-1}) \right| \leq \begin{cases} 2p^{m/2}, & p \nmid \Delta, \\ 2p^{2m/3}, & p \mid \Delta. \end{cases}$$

A more detailed discussion may be found in [14], including the case $p = 2$.

A multidimensional (or hyper) Kloosterman sum is defined by

$$K_n(A, p^m) := \sum_{x_1=1}^{p^m} \cdots \sum_{x_n=1}^{p^m} e_{p^m}(x_1 + x_2 + \cdots + x_n + A(\overline{x_1 x_2 \ldots x_n}))$$

with $p \nmid A$. Smith [87] established that for p odd and $m \geq 1$,

$$|K_n(A, p^m)| \leq (n+1)p^{nm/2},$$

the case $m = 1$ being due to Deligne [19]. Dabrowski and Fisher [17, Example 1.17] sharpened the estimate of Smith for certain values of n, to which Ye [95] made an application. In [11] the authors, with M.-C. Liu, made a further sharpening to obtain

$$|K_n(A, p^m)| \leq (n+1, p-1)p^{\frac{1}{2}\min(\gamma, m-2)}p^{mn/2}, \qquad (14.1)$$

for any odd prime p and $m \geq 2$, where $p^\gamma \parallel (n+1)$. For $p = 2$ the same bound is obtained with an extra factor of 2 on the right-hand side. Katz [46] studied general twisted hyper-Kloosterman sums over finite fields. Evans [24], [25] has recently made evaluations and estimates for twisted hyper-Kloosterman sums modulo p^m with $m \geq 2$, and for Kloosterman sums over rings of algebraic integers.

Question 18. *Can an upper bound of the type (14.1) be established for twisted hyper-Kloosterman sums?*

References

[1] N. M. Akulinichev, *Estimates for rational trigonometric sums of a special type*, Doklady Acad. Sci. USSR **161** (1965), 743–745, Transl. in Soviet Math. Dokl. **161** (1965), 480–482.

[2] B. C. Berndt, R. J. Evans, and K. W. Williams, *Gauss and Jacobi sums*, Canadian Math. Soc. series of monographs and advanced texts, vol. 21, Wiley, New York, 1998.

[3] E. Bombieri, *On exponential sums in finite fields*, Amer. J. Math. **88** (1966), 71–105.

[4] L. Carlitz, *A note on multiple Kloosterman sums*, J. Indian Math. Soc. **29** (1965), 197–200.

[5] J. H. H. Chalk, *On Hua's estimate for exponential sums*, Mathematika **34** (1987), 115–123.

[6] J. R. Chen, *On the representation of natural numbers as a sum of terms of the form $x(x+1)\ldots(x+k-1)/k!$*, Acta Math. Sin. **8** (1958), 253–257.

[7] ———, *On Professor Hua's estimate of exponential sums*, Sci. Sinica **20** (1977), 711–719.

[8] J. R. Chen and C. Pan, *Analytic number theory in China I*, Number Theory and Its Applications in China, Contemp. Math., vol. 77, Amer. Math. Soc., 1988, pp. 1–17.

[9] T. Cochrane, *Exponential sums modulo prime powers*, To appear in Acta Arith.

[10] T. Cochrane, C. Liu, and Z. Zheng, *Upper bounds on character sums with rational function entries*, Preprint, 2000.

[11] T. Cochrane, M.-C. Liu, and Z. Zheng, *Upper bounds on n-dimensional Kloosterman sums*, Preprint, 2000.

[12] T. Cochrane and Z. Zheng, *Pure and mixed exponential sums*, Acta Arith. **91** (1999), 249–278.

[13] _____, *Bounds for certain exponential sums*, Asian J. Math. **4** (2000), 757–774.

[14] _____, *Exponential sums with rational function entries*, Acta Arith. **95** (2000), 67–95.

[15] _____, *On upper bounds of Chalk and Hua for exponential sums*, Proc. Amer. Math. Soc. **29** (2001), 2505–2516.

[16] _____, *Upper bounds on a two-term exponential sum*, Sci. China Ser. A **44** (2001), 1003–1015.

[17] R. Dabrowski and B. Fisher, *A stationary phase formula for exponential sums over $\mathbb{Z}/p^m\mathbb{Z}$ and applications to GL(3)-Kloosterman sums*, Acta Arith. **80** (1997), 1–48.

[18] M. Dedeo, *Graphs over the ring of integers modulo 2^r*, Ph.D. thesis, UCSD, 1998.

[19] P. Deligne, *Applications de la formula des traces aux sommes trigonométric*, Cohomologie Etale SGA $4\frac{1}{2}$, Lecture Notes in Math., vol. 569, Springer Verlag, New York, 1977, pp. 168–232.

[20] P. Ding, *An improvement to Chalk's estimation of exponential sums*, Acta Arith. **59** (1991), 149–155.

[21] _____, *On a conjecture of Chalk*, J. Number Theory **65** (1997), 116–129.

[22] W. Duke, J. Friedlander, and H. Iwaniec, *Bilinear forms with Kloosterman fractions*, Invent. Math. **128** (1997), 23–43.

[23] T. Estermann, *On Kloosterman's sum*, Mathematika **8** (1961), 83–86.

[24] R. Evans, *Twisted hyper-Kloosterman sums over finite rings of integers*, Preprint, 2000.

[25] _____, *Gauss sums and Kloosterman sums over residue rings of algebraic integers*, Trans. Amer. Math. Soc. **353** (2001), 4429–4445.

[26] G. H. Hardy and J. E. Littlewood, *Some problems of "Partitio Numerorum" 1: A new solution of Waring's problem*, Nachrichten von der K. Gesellschaft der Wissenschaften zu Göttingen, Math.-Phys. Klasses (1920), 33–54.

[27] ———, *Some problems of "Partitio Numerorum"*, Math. Zeit. **23** (1925), 1–37.

[28] D. R. Heath-Brown, *An estimate for Heilbronn's exponential sum*, Analytic number theory, Vol. 2 (Allerton Park, IL, 1995), Birkhäuser Boston, Boston, MA, 1996, pp. 451–463.

[29] D. R. Heath-Brown and S. Konyagin, *New bounds for Gauss sums derived from kth powers, and for Heilbronn's exponential sum*, Quart. J. Math. **51** (2000), 221–235.

[30] L. K. Hua, *On exponential sums*, J. Chinese Math. Soc. **20** (1940), 301–312.

[31] ———, *On exponential sums*, Sci. Record (N.S.) **1** (1957), 1–4.

[32] ———, *Additive Primzahltheorie*, B. G. Teubner Verlagsgesellschaft, Leipzig, 1959.

[33] D. Ismoilov, *Estimate of a character sum of polynomials*, Dokl. Acad. Nauk Tadzhik SSR **29** (1986), 567–571.

[34] ———, *Estimate of a character sum of rational functions*, Dokl. Acad. Nauk Tadzhik SSR **29** (1986), 635–639.

[35] ———, *Estimates for character sums of polynomials and rational functions*, 1989, Construction Methods and Algorithms in Number Theory, Abstracts of Reports, Minsk.

[36] ———, *Lower bounds on character sums of polynomials with respect to a composite modulus*, Dokl. Acad. Nauk Tadzhik SSR **33** (1990), 501–505.

[37] ———, *On lower bounds of character sums of rational functions with respect to a composite modulus*, Vestnik Tadzhik. Gos. Univ. Mat. **5** (1990), 27–32.

[38] ———, *Estimates of complete character sums of polynomials*, Trudy Mat. Inst. Steklov **200** (1991), 171–186, Transl. in Proc. Steklov Inst. Math. **200** (1993), 189-203.

[39] ———, *A lower bound estimate for complete sums of characters of polynomials and rational functions*, Acta Math. Sinica, New Series **9** (1993), 90–99.

[40] _____, *Estimates for complete trigonometric sums*, Trudy Mat. Inst. Steklov **207** (1994), 153–171, Transl. in Proc. Steklov Inst. Math. **207** (1995), 137-153.

[41] _____, *On a method of Hua Loo-Keng of estimating complete trigonometric sums*, Adv. in Math. (China) **23** (1994), 31–49.

[42] H. Iwaniec, *Fourier coefficients of modular forms of half-integral weight*, Invent. Math. **87** (1987), 385–401.

[43] _____, *Topics in classical automorphic forms*, Graduate Studies in Mathematics, vol. 17, American Math. Soc., Providence, 1997.

[44] A. A. Karatsuba, *On estimates of complete trigonometric sums*, Mat. Zametki **1** (1967), 199–208, Transl. in Math. Notes **1** (1968), 133–139.

[45] _____, *Lower bounds for sums of the characters of polynomials*, Mat. Zametki **14** (1973), 67–72, Transl. in Math. Notes **14** (1973), 593-596.

[46] N. Katz, *Gauss sums, Kloosterman sums and Monodromy Groups*, Ann. of Math. Stud., vol. 116, Princeton University Press, Princeton, 1988.

[47] _____, *Travaux de Laumon*, Astérisque **161–162** (1988), 105–132, Séminaire Bourbaki, Exp. 691.

[48] H. D. Kloosterman, *Asymptotische formeln für die Fourierkoeffizienten ganzer Modulformen*, Abh. Math. Sem. Univ. Hamburg, vol. 5, 1927, pp. 337–352.

[49] N. Koblitz, *p-adic Numbers, p-adic Analysis, and Zeta-Functions*, second ed., Springer-Verlag, New York, 1984.

[50] S. V. Konyagin, *Estimates for Gaussian sums and Waring's problem modulo a prime*, Trudy Mat. Inst. Steklov **198** (1992), 111–124, Transl. in Proc. Steklov Inst. Math. **198** (1994), 105–107.

[51] _____, *Exponential sums over multiplicative groups of residues*, Preprint, 2000.

[52] S. V. Konyagin and I. E. Shparlinski, *On the distribution of residues of finitely generated multiplicative groups and their applications*, Macquarie mathematics reports, Macquarie University, 1995.

[53] _____, *Character sums with exponential functions and their applications*, Cambridge Univ. Press, Cambridge, 1999.

[54] V. I. Levenstein, *Bounds for packing in metric spaces and certain applications*, Probl. Kibernetiki (Problems of Cybernetics) **40** (1983), 44–110.

[55] C. Liu, *Character sums of rational functions*, Sci. China Ser. A **38** (1995), 182–187.

[56] ———, *Dirichlet character sums*, Acta. Arith. **88** (1999), 299–309.

[57] ———, *Dirichlet character sums of rational functions*, Preprint, 2000.

[58] W. K. A. Loh, *Hua's Lemma*, Bull. Austral. Math. Soc. **50** (1994), 451–458.

[59] ———, *Exponential sums on reduced residue systems*, Canad. Math. Bull. **41** (1998), 187–195.

[60] J. H. Loxton and R. A. Smith, *On Hua's estimate for exponential sums*, J. London Math. Soc. **26** (1982), 15–20.

[61] J. H. Loxton and R. C. Vaughan, *The estimation of complete exponential sums*, Canad. Math. Bull. **28** (1985), 442–454.

[62] M. Lu, *A note on the estimate of a complete rational trigonometric sum*, Acta Math. Sin. **27** (1984), 817–823.

[63] ———, *The estimate of complete trigonometric sums*, Sci. Sin. **28** (1985), 561–578.

[64] ———, *A note on complete trigonometric sums for prime powers*, Sichuan Daxue Xuebao **26** (1989), 156–159.

[65] J.-L. Mauclaire, *Sommes de Gauss modulo p^α I*, Proc. Japan Acad. Ser. A **59** (1983), 109–112.

[66] ———, *Sommes de Gauss modulo p^α II*, Proc. Japan Acad. Ser. A **59** (1983), 161–163.

[67] D. A. Mit'kin, *The size of character sums of polynomials*, Mat. Zametki **31** (1982), 827–835, Transl. in Math. Notes **31** (1982), 418–422. 1982.

[68] ———, *Estimates and asymptotic formulas for rational exponential sums that are nearly complete*, Matem. Sborn. **122** (1983), 527–545.

[69] H. L. Montgomery, R. C. Vaughan, and T. D. Wooley, *Some remarks on Gauss sums associated with kth powers*, Math. Proc. Cambridge Philos. Soc. **118** (1995), 21–33.

[70] L. J. Mordell, *On a sum analogous to a Gauss's sum*, Q. J. Math. **3** (1932), 161–167.

[71] G. L. Mullen and I. E. Shparlinski, *Open problems and conjectures in finite fields*, Finite Fields and Applications (Glasgow, 1995) (S. Cohen and H. Niederreiter, eds.), London Math. Soc. Lecture Note Series, vol. 233, Cambridge University Press, Cambridge, 1996, pp. 243–268.

[72] V. I. Nečaev, *An estimate of a complete rational trigonometric sum*, Mat. Zametki **17** (1975), 839–849, Transl. in Math. Notes **17** (1975), 504–511.

[73] ———, *On the least upper bound on the modulus of complete trigonometric sums of degrees three and four*, Investigations in number theory, Saratov. Gos. Univ., Saratov, 1988, pp. 71–76.

[74] V. I. Nečaev and V. L. Topunov, *Estimation of the modulus of complete rational trigonometric sums of degree three and four*, Trudy Mat. Inst. Steklov **158** (1981), 125–129, Transl. in Proc. Steklov Inst. Math. Analytic number theory, mathematical analysis and their applications, Amer. Math. Soc., Providence, 1983, 135–140.

[75] R. Odoni, *On Gauss sums* (mod p^n), $n \geq 2$, Bull. London Math. Soc. **5** (1973), 325–327.

[76] G. I. Perel'muter, *Estimate of a sum along analgebraic curve*, Mat. Zametki **5** (1969), 373–380.

[77] M. Qi and P. Ding, *Estimate of complete trigonometric sums*, Kexue Tongbao **29** (1984), 1567–1569.

[78] ———, *On estimate of complete trigonometric sums*, China Ann. Math. **B6** (1985), 110–120.

[79] ———, *Further estimate of complete trigonometric sums*, J. Tsinghua Univ. **29** (1989), 74–85.

[80] H. Salié, *Über die Kloostermanschen Summen $S(u, v; q)$*, Math. Zeit. **34** (1931), 91–109.

[81] P. Sarnak, *Some applications of modular forms*, Cambridge Tracts in Mathematics, vol. 99, Cambridge Univ. Press, Cambridge, 1990.

[82] W. M. Schmidt, *Equations over finite fields*, Lecture notes in mathematics, vol. 536, Springer-Verlag, Berlin, 1976.

[83] I. E. Shparlinski, *Computational and algorithmic problems in finite fields*, Kluwer Academic Pub., Boston, 1992.

[84] _____, *On Gaussian sums for finite fields and elliptic curves*, Proc. 1-st French-Soviet Workshop on Algebraic Coding, Paris, 1991, Lect. Notes in Computer Sci., vol. 537, 1992, pp. 5–15.

[85] _____, *On exponential sums with sparse polynomials and rational functions*, J. Number Theory **60** (1996), 233–244.

[86] I. E. Shparlinski and S.A. Stepanov, *Estimation of trigonometric sums with rational and algebraic functions*, Automorphic Functions and Number Theory, Part I, II, vol. 253, Akad. Nauk SSSR Dalćprime nevostochn. Otdel., Vladivostok, 1989, pp. 5–18.

[87] R. A. Smith, *On n-dimensional Kloosterman sums*, J. Number Theory **11** (1979), 324–343.

[88] _____, *Estimate for exponential sums*, Proc. Amer. Math. Soc. **79** (1980), 365–368.

[89] S. B. Stečkin, *Estimate of a complete rational trigonometric sum*, Proc. Steklov Inst. **143** (1977), 188–220, Transl. in Proc. Steklov Inst. Math., Analytic number theory, mathematical analysis and their applications, American Math. Soc., Providence, 1980, pp. 201–220.

[90] S. A. Stepanov, *Arithmetic of algebraic curves*, Monographs in Contemporary Math., Consultants Bureau, New York, 1994.

[91] R. C. Vaughan, *The Hardy Littlewood Method*, 2nd ed., Cambridge Tracts in Math., vol. 125, Cambridge Univ. Press, Cambridge, 1997.

[92] A. Weil, *On some exponential sums*, Proc. Nat. Acad. Sci. U.S.A. **34** (1948), 204–207.

[93] A. L. Whiteman, *A note on Kloostermans sums*, Bull. Amer. Math. Soc. **51** (1945), 373–377.

[94] K. S. Williams, *Note on the Kloosterman sum*, Proc. Amer. Math. Soc. **30** (1971), 61–62.

[95] Y. Ye, *Kloosterman sums and estimation of exponential sums of polynomials of higher degree*, Acta Arith. **86** (1998), 255–267.

[96] M. Zhang and Y. Hong, *On the maximum modulus of complete trigonometric sums*, Acta Math. Sinica, New Series **3** (1987), 341–350.

One Hundred Years of Normal Numbers

Glyn Harman

Prologue

In this paper we survey the development of metric number theory in the twentieth century with an emphasis on Borel's notion of a *normal number*. The early work in this subject is best described as "analysis" or "probability", rather than "number theory". However, as the subject has developed through the twentieth century, arithmetical questions and methods have come to the fore. We will demonstrate this by placing Borel's normal number theorem in a wider context of the study of the arithmetical properties of the sequence $F(\alpha, n)$, $n = 1, 2, \ldots$, for families of functions F. We shall also survey other properties of a "typical" number and lay down a few challenges for the next century (or next millennium?).

1 Paris in 1900

Back in 1874, in Germany, Cantor had demonstrated that the continuum was uncountable. His ideas were resisted by some leading mathematicians at the time, notably Kronecker, but they found a more sympathetic audience in France where several influential figures had come to accept them by the turn of the century. Cantor's demonstration had set a limit on mankind's knowledge of numbers. It could not be possible to list all the real numbers with a description, even allowing for an infinite list. To embark on any analysis of the properties of real numbers would therefore require a system of classification: one would have to fit the reals into classes and describe the nature of those classes. As the 19th century drew to a close and the 20th century dawned, Borel and Lebesgue were laying the foundations of measure theory and a new definition of the integral. Measure theory gave researchers an excellent tool for an analysis of the continuum. A natural question to ask was: What is true for a typical number—in the sense that the exceptional set has measure zero? If you pick a number at random, what properties might you expect it to enjoy? The simplest analysis of the continuum is its division into rationals and irrationals. A "typical" number is irrational since the set of rationals has measure zero. We pause

57

to remark that it was Borel who first demonstrated that a countable set has measure zero; indeed, this was the motivating point in his foundation of measure theory (see [11, pp. 499–501]).

In 1898 Borel wrote his influential book "Leçons sur la théorie des fonctions" [6] in which he introduces the idea of measure and considers examples of uncountable sets of measure zero. On pages 41–42 he essentially proves the following result.

Let A be the set of real α for which there are infinitely many solutions in integers a and q to

$$\left| \alpha - \frac{a}{q} \right| < \frac{1}{q^4}.$$

Then A is an uncountable set with measure zero.

Putting this another way: A "typical number" satisfies

$$\left| \alpha - \frac{a}{q} \right| > \frac{C(\alpha)}{q^4} \tag{1.1}$$

for all a and q, although there are uncountably many exceptions.

This appears to be the first result in metric number theory in the literature. Borel's original proof from our vantage point may seem unnecessarily longwinded, until we remember that he was working on the foundation of a completely new theory. In 1898 he could easily have shown the best possible form of the above result with the right hand side of (1.1) replaced by $\psi(q)/q$ for any non-negative function $\psi(q)$ such that

$$\sum_{q=1}^{\infty} \psi(q)$$

converges. It looks obvious to us now, but Borel had not yet stated the result which is known to us at present (after Cantelli's removal of the hypothesis of independence in the convergence case) as the Borel–Cantelli lemma!

Many count the birth of metric number theory from Borel's paper [7] in which he considered the properties of 'typical' numbers from the viewpoint of their decimal or continued fraction expansion. Write a real number α to base b ($b \in \mathbb{Z}, b \geq 2$) as

$$\alpha = [\alpha] + \sum_{n=1}^{\infty} a_n b^{-n},$$

with

$$0 \leq a_n \leq b - 1, \qquad a_n < b - 1 \text{ infinitely often.}$$

Let
$$A(d, b, N) = |\{n \leq N : a_n = d\}| \,.$$
Then α is said to be *simply normal* to base b if, for $0 \leq d \leq b - 1$,
$$\lim_{N \to \infty} \frac{A(d, b, N)}{N} = \frac{1}{b},$$
and *entirely normal* to base b if α is simply normal to each base $b, b^2, b^3, \ldots$. Finally, call a number *absolutely normal* if it is simply normal to every base $b \geq 2$. Such a number is usually referred to as a *normal number*. Borel's original definitions were equivalent to those we have given here (see [17, 1–4]). He established the following result.

Theorem (Borel 1909). *Almost all real numbers are normal.*

Since a countable union of sets with measure zero itself has measure zero, it suffices to prove that, given any base b, almost all real numbers are simply normal to base b.

From Borel's result we see that a normal number is indeed a 'typical' number. We note that an equivalent definition of 'entirely normal to base b' is that
$$\lim_{N \to \infty} \frac{A(B_k, b, N)}{N} = \frac{1}{b^k}$$
for all $k \geq 1, B_k$, where B_k is a block of k digits to base b, and $A(B_k, b, N)$ denotes the number of times B_k occurs in the first N digits of α (see [17, 5–7]). This can be described more dramatically by pointing out that if your telephone number has 11 digits, say, and you pick a number α at random, then in the first N digits (N very large!) of the decimal expansion of α your telephone number should occur $N \cdot 10^{-11}$ times. Alternatively, since an individual's genetic code can be considered as a block of digits to base 4, a normal number contains the genetic code of every human being who has ever lived, or will live, with the correct asymptotic frequency. A similar argument shows that a normal number contains the sum of human knowledge: past, present and future. Of course, as with similar remarks (the large number of monkeys typing for many millennia to come up with the works of Shakespeare, for example), it would take a finite rational mind longer than the lifetime of the universe to find even one individual's genetic code in any given normal number!

2 Normal Numbers Through the Twentieth Century

The immediate interest after Borel's result was either to give constructions of normal numbers [26], or to give a quantitative estimate for the discrepancy

$$D(d, b, N) = A(d, b, N) - \frac{N}{b}$$

for almost all α. After earlier work by Hausdorff and Hardy and Littlewood, the definitive answer to this problem was supplied by Khintchine [23].

Theorem. *Suppose $b \in \mathbb{Z}, b \geq 2$. For almost all α, for $0 \leq d \leq b - 1$, we have*

$$\limsup_{N\to\infty} \frac{D(d, b, N)}{\sigma f(N)} = 1, \qquad \liminf_{N\to\infty} \frac{D(d, b, N)}{\sigma f(N)} = -1,$$

with

$$f(N) = (2N \log\log N)^{\frac{1}{2}}, \qquad \sigma^2 = b^{-1}(1 - b^{-1}).$$

Remark. The referee pointed out to the author that it follows that the set of limit points of $\frac{D(d,b,N)}{\sigma f(N)}$ is, with probability 1, precisely the interval $[-1, 1]$.

From our vantage point in the year 2000 the results of Borel and Khintchine could be considered "obvious". The a_n are independent variables taking the values 0 to $b - 1$ with equal probability. The quantity σ^2 represents the variance. However, it must be remembered that Borel's result is the first occurrence of what is now referred to as "the law of large numbers" after the work of Kolmogorov [25], and Khintchine's result is the first mention of the law of the iterated logarithm (see [5] for a survey of this topic).

Another more modern perspective on Borel's result is afforded by Ergodic Theory (see [4]). If $\{x\}$ represents the fractional part of x, then the mapping $T: \{x\} \to \{bx\}$ is measure preserving and ergodic on the interval $[0, 1)$ with respect to Lebesgue measure. The (pointwise) Ergodic Theorem (see [4, p. 13]) states that, for integrable functions f,

$$\frac{1}{N} \sum_{n=1}^{N} f(T^{n-1}\alpha) \longrightarrow \int_0^1 f(x)\, dx$$

for almost all α. Borel's theorem follows on taking

$$f(x) = \begin{cases} 1 & \text{if } \frac{d}{b} \leq x < \frac{d+1}{b} \\ 0 & \text{otherwise.} \end{cases}$$

One natural question arising from Borel's normal number theorem is the analysis of the set of non-normal numbers. Clearly no rational can be normal (although a rational can be simply normal to a given base, or finite set of bases, it cannot be entirely normal to any base). By modifying

Cantor's argument it is straightforward to show that the set of non-normal numbers is uncountable. With the introduction of Hausdorff dimension the technology became available to give a more precise definition of the exceptional set—it has dimension 1. From one point of view it is thus as large as it could be and still have measure zero. We pause briefly here to contrast the measure-theorist's and topologist's perspectives. The set of normal numbers is a set of the first Baire category (it is a countable union of nowhere dense sets), while the non-normal numbers form a set of the second category—and thus appear larger from this topological point of view.

One challenge has been to write down numbers which are entirely normal to a given base. Taking 10 as the natural example base, Champernowne [10] demonstrated that 0.1234567891011121314151617181920212... is entirely normal to base 10. He conjectured that one could replace the naturals in order with the primes to show that 0.23571113171923293137414347... is entirely normal to base 10. Copeland and Erdös [12] proved this as a corollary to a more general result on decimals built up from a sufficiently dense sequence of integers. Davenport and Erdös [13] replaced the naturals with the values of a non-constant polynomial taking only non-negative values. The particular case that 0.149162536496481100121144... is entirely normal to base 10 had been dealt with earlier by Besicovitch. We note that Davenport and Erdös used Weyl's inequality applied to

$$\frac{f(n)}{10^m}, \qquad f(x) \in \mathbb{Z}[x],$$

and so genuinely number-theoretic ideas were involved. These last three examples may at first sight appear to be rather surprising. It is clear that there must be large discrepancies for these numbers. However, as the length of the number added each time increases, the effect of these irregularities is masked to the extent that $D(10^n, b, N)/N \to 0$ as $N \to \infty$, even though infinitely often $D(10, b, N) \gg N/\log N$ (compare [35]). In the opposite direction see [27] for the construction of numbers with minimal discrepancy.

Various authors have investigated sets of numbers which are entirely normal to one given base (or set of bases), but not to another base or set of bases. Note that if α is entirely normal to base b_1, it is entirely normal to base b_2 where $b_2^r = b_1^s, r, s \in \mathbb{Z}^+$. For results on the dimension of such sets see [29], [32] (and the earlier work of Cassels [9] and Schmidt [36]). Brown, Moran and Pearce [8] considered writing every real β as $\beta_1 + \beta_2$, where each β_j is entirely normal to base b_1, but not to base b_2 (where $\log b_1 / \log b_2 \notin \mathbb{Q}$). Wagner [39] constructed rings of numbers which, with 0 excepted, are normal to one base, but not to another. Schmidt [37] and

Pollington [31] established hybrid results concerning α not normal to any given base and satisfying Diophantine approximation conditions.

Other authors have checked the decimal expansions of the best known irrationals such as $\sqrt{2}, e, \pi$, to see how closely they resemble normal numbers. This leads to a great challenge for mathematicians of the coming century (or millennium!):

Challenge. *Determine whether or not $\sqrt{2}, e$ or π are normal.*

Here are some weaker questions that still seem far too hard at present.

Challenge. *Find one base to which $\sqrt{2}$ is simply normal, or prove that no such base exists.*

Challenge. *Find a base $b > 2$ for which each of $0, \ldots, b-1$ occurs infinitely often in the base b expansion of $\sqrt{2}, e, \pi$.*

The result of this last challenge is trivial, of course, if $b = 2$. We note that Mahler [28] gave an approximation to the above:

For any given irrational α, positive integer N, and base $b > 2$ there exists an integer M such that every block of length N occurs infinitely often in the expansion of $M\alpha$ to base b.

We close this section with just one simple case which illustrates the difficulty of the problem. Suppose there is no 1 in the base 3 expansion of $\sqrt{2}$ from some point onwards. Then there exists m such that, upon writing $\beta = 3^m \sqrt{2}$, we have either

$$\{3^n\beta\} < \frac{1}{3} \quad \text{or} \quad \{3^n\beta\} > \frac{2}{3}$$

for all $n \geq 1$. Could this be possible? The set of β for which the above holds has Hausdorff dimension $\log 2/\log 3$, of course, for it is essentially Cantor's ternary set.

3   Normal Numbers Generalized

In this section we reformulate the normal number problem in a way that admits generalization, yet which still has a satisfactory solution. Let $f(n, \alpha)$ be a function of a positive integer variable n and a real variable α. The question is to investigate the arithmetical properties of $F(n, \alpha) = [f(n, \alpha)]$ for almost all α. Normal numbers correspond to taking $f(n, \alpha) = b^n \alpha$ and investigating the property $P(\alpha, b)$, that $F(n, \alpha)$ is equidistributed in the residue classes (mod b) as $n \to \infty$. Cast in this light, normal numbers are those for which $P(\alpha, b)$ is true for all $b \geq 2$. Now let $f(n, \alpha) = g^n \alpha$ ($g \in \mathbb{Z}$,

$g \geq 2$), and let $P(\alpha, b, g)$ be the property that $F(n, \alpha)$ is equidistributed in the residue classes (mod b) as $n \to \infty$. Using the fact that a countable union of sets with measure zero itself has measure zero, we need only investigate $P(\alpha, b, g)$ for any given pair (b, g). It is not difficult to show that $P(\alpha, b, g)$ is true for almost all α. This could be proved directly, or one could use

$$\left\{\frac{\alpha g^n}{b}\right\} \text{ uniformly distributed (mod 1)} \Rightarrow P(\alpha, b, g) \text{ true.}$$

Thus almost all α satisfy a more stringent condition than normality:

$$\lim_{N \to \infty} \max_{0 \leq a \leq b-1} \left| \frac{1}{N} \sum_{\substack{[g^n \alpha] \equiv a (\bmod \, b) \\ n \leq N}} 1 - \frac{1}{b} \right| = 0,$$

for every pair b, g.

It should be noted that all we have said so far could be described as *hitting a fixed target*. Given an infinite sequence $\mathcal{C} = c_1, \ldots, c_n, \ldots$ of integers from 0 to $b-1$ write $P(\mathcal{C}, g, b, \alpha)$ for the property that

$$\frac{1}{N} |\{n \leq N : [g^n \alpha] \equiv c_n \pmod{b}\}| \to \frac{1}{b}.$$

This could be described as a *moving target problem* (the terminology was suggested by [21]). Now we cannot have a result true for all possible $\mathcal{C}$ (since there are uncountably many!), but if we are given one sequence $\mathcal{C}(b, g)$ for each pair (b, g), then, for almost all α, $P(\mathcal{C}, g, b, \alpha)$ is true for every pair (b, g) with $b, g \geq 2$. This follows since we can give an asymptotic formula, valid for almost all α, for the number of solutions to

$$\left\{\frac{g^n \alpha - c_n}{b}\right\} < \frac{1}{b}$$

(see [30]).

We now give a much wider generalization of the normal number concept. Let a_n be any increasing sequence of positive reals with $a_{n+1} > a_n + n^{-1}(\log n)^{1+\delta}$ for some $\delta > 0$. Let $f(n, \alpha) = a_n \alpha$ with $P(\alpha, b)$ the property that

$$\lim_{N \to \infty} \max_{0 \leq a \leq b-1} \left| \frac{1}{N} \sum_{\substack{n \leq N \\ F(n, \alpha) \equiv a \pmod{b}}} 1 - \frac{1}{b} \right| = 0.$$

Then $P(\alpha, b)$ is true for almost all α. This follows from [17, Theorem 5.8*]. Given a countable set of sequences a_n we could thus let $P(\alpha)$ be the property that $P(\alpha, b)$ is true for all $b \geq 2$, and for each of the given sequences. We then have that $P(\alpha, b)$ is true for almost all α. One can turn this into a "moving target" problem also, but at present the slightly faster growth criterion $a_{n+1} > a_n + n^\sigma$ with $\sigma > -1$ is required (see [17, Theorem 7.5]).

Since $\{b^n\alpha\}$ being uniformly distributed (mod 1) is equivalent to α being entirely normal to base b, it has become customary to define normality to a non-integer base $b > 1$ by using the criterion that $\{b^n\alpha\}$ should be uniformly distributed (mod 1). It follows from our previous discussion that given any countable set S of bases (for example $\mathbb{Q} \cap (1, \infty)$) almost all α are normal to every given base in S. The subject of uniform distribution gives many other characterisations of what should be expected of a 'typical' number (see [17, Chapter 5]). Write

$$D_N(a_n) = \sup_{\mathcal{I} \subset [0,1)} \left| \sum_{\substack{\{a_n\} \in \mathcal{I} \\ n \leq N}} 1 - \lambda(\mathcal{I})N \right|,$$

where $\mathcal{I}$ denotes an interval, and λ is used for Lebesgue measure. Khintchine [22] established the following result using the metric theory of continued fractions.

Theorem. *For almost all α we have*

$$D_N(n\alpha) \ll (\log N)G(\log \log 9N)$$

for any positive non-decreasing function G such that

$$\sum_{n=1}^{\infty} \frac{1}{G(n)}$$

converges. On the other hand, for almost all α we have

$$\limsup_{N \to \infty} \frac{D_N(n\alpha)}{(\log N)H(\log \log N)} > 0$$

where $H(x)$ is any positive non-decreasing function such that

$$\sum_{n=1}^{\infty} \frac{1}{H(n)}$$

diverges.

This gives a very precise result on the discrepancy of $n\alpha$ for almost all α. If a_n is an increasing sequence of integers then it is known that

$$D_N(a_n\alpha) = O\left(N^{\frac{1}{2}}(\log N)^{\frac{3}{2}+\epsilon}\right)$$

for almost all α [1], and that the exponent of the logarithm cannot be reduced below $\frac{1}{2}$ [2]. It is very frustrating that we are so close to knowing the exact order of magnitude for this problem, but there remains the $\log N$ gap.

Challenge. *Characterise the slowest growing functions $F(N)$ such that, for any increasing sequence of integers a_n,*

$$D_N(a_n\alpha) \ll F(N) \text{ for almost all } \alpha.$$

4 Prime Values of $\mathbf{F(n, \alpha)}$

In the previous section we considered the distribution of $F(n, \alpha)$ in arithmetic progressions. The property of integer sequences of greatest interest to number-theorists is usually whether they infinitely often take prime values. The author considered this question in [16] (see also [17, Chapter 8]). Among the results we proved was the following.

Theorem. *Let a_n be an increasing sequence of positive reals. Then $[a_n\alpha]$ is infinitely often prime, or only finitely often prime, for almost all α according to whether*

$$\sum_{n=1}^{\infty} \frac{R(n)}{\log a_n}$$

diverges or converges, respectively. Here

$$R(n) = \left(\sum_{\substack{m\leq n \\ |a_m-a_n|\leq 1}} 1\right)^{-1}.$$

A limited quantitative version of this result can be proved [18]. If $a_{n+1} \geq a_n + 1$, $P(\alpha, N) = |\{n \leq N : [a_n\alpha] \text{ prime}\}|$, and

$$V(N) = \sum_{n=1}^{N} \frac{1}{\log a_n},$$

then

$$\limsup_{N\to\infty} \frac{P(\alpha, N)}{V(N)} \geq 1, \qquad \liminf_{N\to\infty} \frac{P(\alpha, N)}{V(N)} \leq 1.$$

We might expect

$$\lim_{N\to\infty} \frac{P(\alpha, N)}{V(N)} = 1,$$

and this can be proved if a_n is a lacunary sequence. In particular, we called a number α *prime normal* to base b if

$$\frac{1}{\log N} |\{n \le N \colon [b^n\alpha] \text{ prime}\}| \to \frac{1}{\log b} \text{ as } N \to \infty.$$

Our result showed that almost all $\alpha > 0$ are prime normal to every integer base $b \ge 2$. Hence, not only do all blocks of integers show up with the expected frequency in the base b expansion of a "typical" number, but also the number of primes occurring in the first N places obeys the expected asymptotic law. As would be expected, much more number-theoretic machinery is required to prove a result of this type.

5   Diophantine Approximation to a Typical Number

We now return to Borel's other main result in his 1909 paper which gave a characterisation of a typical number through the growth of the partial quotients in its continued fraction expansion. We write

$$\alpha = [a_0; a_1, \ldots, a_n, \ldots]$$

in the usual continued fraction notation [15], [34].

Theorem (Borel 1909). *Let $\psi(n) > 0$ be a decreasing function of n and let*

$$S = \sum_{n=1}^{\infty} \psi(n).$$

Then, if S diverges, for almost all α

$$\limsup_{n\to\infty} a_n\psi(n) = \infty,$$

while, if S converges, for almost all α,

$$\lim_{n\to\infty} a_n\psi(n) = 0.$$

This completely describes the behaviour of the large values of the partial quotients of a typical number. The behaviour of the average value of a_n was given later by the following result.

Theorem (Khintchine 1935). *For almost all α we have*

$$(a_1 \ldots a_n)^{\frac{1}{n}} \to \prod_{r=1}^{\infty} \left(1 + \frac{1}{r(r+2)}\right)^{\frac{\log r}{\log 2}} \quad as \ n \to \infty.$$

Indeed, Khintchine proved a more general result which also gives

$$\lim_{N \to \infty} \frac{1}{N} \sum_{\substack{n \leq N \\ a_n = r}} 1 = \frac{\log\left(1 + \frac{1}{r(r+2)}\right)}{\log 2}$$

for almost all α. It can be shown [20] that, given a finite set of positive integers $\{r_1, \ldots, r_k\}$, the limit

$$\lim_{N \to \infty} \frac{1}{N} |\{n \leq N : a_{n+j} = r_j, 1 \leq j \leq k\}|$$

exists and is constant for almost all α. This is a consequence of the pointwise ergodic theorem. Thus, for almost all α, not only does a given block of digits (to base b) occur in its base b expansion with a fixed frequency, the same block also occurs as consecutive numbers among the partial quotients with a certain fixed frequency also. The referee remarks that most limit theorems, such as the Law of the Iterated Logarithm, or the Central Limit Theorem, carry over from the independent case as the random variables involved are ψ-mixing with an exponential mixing rate.

The above results lead naturally on to a discussion of Diophantine approximation. We are now concerned with the fractional rather than integer parts of a sequence. Let

$$f_1(n, \alpha) = \min_{m \in \mathbb{Z}} |n\alpha - m|,$$

$$f_2(n, \alpha) = \min_{\substack{m \in \mathbb{Z} \\ (m,n)=1}} |n\alpha - m|,$$

and suppose that $\psi(n)$ is a sequence of positive reals. Let $P_j(\alpha, \psi)$ denote the property that $f_j(n, \alpha) < \psi(n)$ infinitely often. Then Khintchine proved the following result.

Theorem (Khintchine 1924). *If $n\psi(n)$ is monotonically decreasing then $P_1(\alpha, \psi)$ holds for almost all α or almost no α according to whether the series*

$$\sum_{n=1}^{\infty} \psi(n)$$

diverges or converges, respectively.

The convergence case is very easy: it could have been proved by Borel in 1898. The divergence case requires more work and has been the starting point for much further work (see [17, Chapters 2–4]). Khintchine's original proof depended on Borel's work on continued fractions. The condition that $n\psi(n)$ should be decreasing can be relaxed to $\psi(n)$ decreasing [14]. We note that if $\psi(n)$ is decreasing then

$$P_2(\alpha, \psi) \text{ true} \Rightarrow P_1(\alpha, \psi) \quad \text{true},$$

and, if $\alpha \notin \mathbb{Q}$,

$$P_1(\alpha, \psi) \text{ true} \Rightarrow P_2(\alpha, \psi) \quad \text{true}.$$

If our earlier problems were 'fixed target' or 'moving target', then results of the above type can be considered 'shrinking target'. One can, of course, consider targets which are moving and shrinking at once:

$$|n\alpha - m - \beta_n| < \psi(n)$$

(see [17, Chapter 3]). Having said this though, questions involving $P_2(\alpha, \psi)$ are the more natural, and the most important unsolved problem in this area is as follows.

Conjecture (Duffin and Schaeffer 1941). *Let $\psi(n)$ be any non-negative function of an integer variable. Then*

$$\sum_{n=1}^{\infty} \psi(n)\frac{\phi(n)}{n} = \infty \Rightarrow P_2(\alpha, \psi) \text{ true}.$$

We note that if the series converges then $P_2(\alpha, \psi)$ is false for almost all α as a simple consequence of the Borel–Cantelli lemma. This conjecture has stood for nearly 60 years and has been proved subject to various conditions on ψ (see [17, Chapters 2–4]). Will the next 100 years see a proof (or counterexample)?

The author has considered Khintchine's theorem with both variables restricted to sets of number theoretic interest [17, Chapter 6]. As an example of what can be obtained we cite the following.

Theorem. *Let $\psi(n)$ be monotonic decreasing. Then the inequality*

$$|\alpha p - q| < \psi(p), \quad p, q \text{ primes},$$

has infinitely many solutions or only finitely many solutions for almost all α according to whether the series

$$\sum_{n=2}^{\infty} \frac{\psi(n)}{\log^2 n}$$

diverges or converges, respectively.

6   Higher Dimensions

All the problems considered so far can be extended to $\alpha \in \mathbb{R}^k$. Usually the extra freedom of more variables makes the problem easier. For example, the Duffin and Schaeffer conjecture is known to be true in two or more dimensions [33]. The more interesting—and more demanding!—problems in higher dimensions concern restricting attention to a submanifold and working with the appropriate measure. Diophantine approximation problems from this perspective are covered in the recent tract of Bernik and Dodson [3]. This is a very active area of research. One goal likely to be achieved in the next few years is an analogue of Khintchine's theorem for all 'reasonable' manifolds. It should be noted (see Sprindzuk's theorem, for example [38]) that the convergence case is no longer trivial for such problems.

Very recently the author [19] has considered the arithmetical structure of the set of the integer parts of points on a curve. Among the results that can be proved we mention the following (where $R(N)$ has the same meaning as in §4.).

Theorem. *Let $b_1 = 1$, $b_j > (20/19)b_{j-1}$ $(j = 2, \ldots, k)$. Let a_n be an increasing sequence of positive reals. Then $[(\alpha a_n)^{b_1}], \ldots, [(\alpha a_n)^{b_k}]$ are infinitely often simultaneously prime for almost all positive α or almost no α according to whether the series*

$$\sum_{n=1}^{\infty} \frac{R(n)}{(\log a_n)^k}$$

diverges or converges respectively.

Of course, many of the classical conjectures concerning prime numbers (for example, prime twins) can be expressed in terms of finding prime points on lines or curves.

7   The Crucial Feature of the Proofs

We finish by outlining the most significant step in the proof of many of the results outlined above. Say $P(\alpha, n)$ denotes a property as discussed above, while $P(\alpha)$ indicates the property asymptotically holds the expected number of times (or alternatively at least that it holds infinitely often). To prove $P(\alpha)$ for almost all α we usually need first to restrict attention to an interval of finite length, say $\mathcal{I}$, and to consider

$$\lambda(\mathcal{B}_n \cap \mathcal{B}_m),$$

where
$$\mathcal{B}_n = \{\alpha \in \mathcal{I} \colon P(\alpha, n) \text{ true}\}.$$

For Borel's normal number theorem we can take $\mathcal{I} = [0, 1)$ and the sets $\mathcal{B}_n$ are independent, so that

$$\lambda(\mathcal{B}_m \cap \mathcal{B}_n) = \lambda(\mathcal{B}_m)\lambda(\mathcal{B}_n).$$

In general, to get results involving an asymptotic formula, we take $\mathcal{I}$ to have unit length and must show that

$$\lambda(\mathcal{B}_m \cap \mathcal{B}_n) = \lambda(\mathcal{B}_n)\lambda(\mathcal{B}_m) + g(m, n)$$

where

$$\sum_{1 \leq m, n \leq N} g(m, n) = O\left(\frac{S(N)^2}{\log^{3+\epsilon} S(N)}\right) \quad \text{with} \quad S(N) = \sum_{n=1}^{N} \lambda(\mathcal{B}_n);$$

see [17, Theorem 1.5]. In some problems one can show that

$$\lambda(\mathcal{B}_m \cap \mathcal{B}_n) = \lambda(\mathcal{B}_m)\lambda(\mathcal{B}_n)\left(1 + O\left(q^{-|m-n|}\right)\right).$$

[34, p. 159] where $0 < q < 1$, which suffices. In general, it is the 'm near n' terms which are problematical ('near' meaning 'in absolute value' or '$\gcd(m, n)$ large', depending on the context).

In order to show only that a property holds infinitely often for almost all α it suffices to prove that

$$\lambda(\mathcal{I})\lambda(\mathcal{B}_m \cap \mathcal{B}_n) \leq K\lambda(\mathcal{B}_m)\lambda(\mathcal{B}_n)$$

for *all* open intervals $\mathcal{I}$ with K an absolute constant. In fact, in many cases it can be shown that a "zero-one" law holds, in which case the above inequality need only be proved for *one* open interval $\mathcal{I}$.

These overlap estimates often involve obtaining an upper bound for the number of solutions to inequalities of the form

$$\left|\frac{m}{n} - \frac{r}{s}\right| < \eta,$$

with various restrictions on the integer variables m, n, r and s. This is the part of the proof which requires the most number-theoretic input. The reader will see its relation to the equation

$$ms - nr = 1$$

which has been at the heart of number theory from Euclid onwards. Sometimes the inequalities take the form

$$|\theta r - s| < A$$

where $\theta = a_n/a_m$, with a_n a given sequence. The proof of the appropriate upper bound then branches in different directions depending on Diophantine approximations to θ. It may be that there are "too many" potential solutions for certain θ: the idea is to show that this happens for few pairs m, n.

8 Conclusion

In the last one hundred years we have come a long way in characterising the arithmetical properties of almost all real numbers, yet several difficult challenges remain. Those presented at the end of Section 2 seem especially difficult! After presenting this talk Andrew Granville asked if I could exhibit an irrational absolutely abnormal number: that is a number which is not entirely normal to any base (clearly no rational can be entirely normal to any base). Greg Martin quickly provided one nice construction. Carl Pomerance thought that

$$\sum_{n=1}^{\infty} \frac{1}{(n!)^{n!}}$$

should work—here is another challenge for the reader!

Acknowledgements. I would like to thank the referee and my colleagues at Royal Holloway—Eira Scourfield and Gar de Barra—for their comments.

References

[1] R. C. Baker, *Metric number theory and the large sieve*, J. London Math. Soc. **24** (1981), 34–40.

[2] I. Berkes and W. Philipp, *The size of trigonometric and Walsh series and uniform distribution mod* 1, J. London Math. Soc. **50** (1994), 454–463.

[3] V. I. Bernik and M. M Dodson, *Metric Diophantine approximation on manifolds*, Cambridge Tracts in Mathematics, vol. 137, Cambridge University Press, Cambridge, 1999.

[4] P. Billingsley, *Ergodic theory and information*, Wiley, New York, 1965.

[5]   N. H. Bingham, *Variants on the law of the iterated logarithm*, Bull. London Math. Soc. **18** (1986), 433–467.

[6]   E. Borel, *Leçons sur la théorie des fonctions*, Gauthier-Villars, Paris, 1898.

[7]   ———, *Les probabilités dénombrables et leurs applications arithmétiques*, Rend. Circ. Math. Palermo **27** (1909), 247–271.

[8]   W. Brown, W. Moran, and C. E. M. Pearce, *Riesz products and normal numbers*, J. London Math. Soc. **32** (1985), 12–18.

[9]   J. W. S. Cassels, *On a problem of Steinhaus about normal numbers*, Collect. Math. **7** (1959), 95–101.

[10] D. G. Champernowne, *The construction of decimals normal in the scale of ten*, J. London Math. Soc. **8** (1933), 254–260.

[11] E. F. Collingwood, *Emile Borel*, J. London Math. Soc. **34** (1959), 488–512.

[12] A. H. Copeland and P. Erdös, *Note on normal numbers*, Bull. Amer. Math. Soc. **52** (1946), 857–860.

[13] H. Davenport and P. Erdös, *Note on normal decimals*, Canad. J. Math. **4** (1952), 58–63.

[14] R. J. Duffin and A. C. Schaeffer, *Khintchine's problem in metric Diophantine approximation*, Duke Math. J. **8** (1941), 243–255.

[15] G. H. Hardy and E. M. Wright, *An introduction to the theory of numbers*, Clarendon Press, Oxford, 1979.

[16] G. Harman, *Metrical theorems on prime values of the integer parts of real sequences*, Proc. London Math. Soc. **75** (1997), 481–496.

[17] ———, *Metric number theory*, Clarendon Press, Oxford, 1998.

[18] ———, *Variants of the second Borel–Cantelli lemma and their applications in metric number theory*, Number Theory, HBA, New Delhi, 2000, pp. 121–140.

[19] ———, *Metrical theorems on prime values of the integer parts of real sequences. II*, J. London Math. Soc. (2) **64** (2001), 287–298.

[20] G. Harman and K. C. Wong, *A note on the metric theory of continued fractions*, Amer. Math. Monthly **107** (2000), 834–837.

[21] R. Hill and S. L. Velani, *The shrinking target problem for matrix transformations of tori*, J. London Math. Soc. **60** (1999), 381–398.

[22] A. Khintchine, *Einige Sätze über Kettenbrüche mit Anwendungen auf die Theorie der Diophantischen Approximationen*, Math. Ann. **92** (1924), 115–125.

[23] _____, *Über einen Satz der Wahrscheinlichkeitsrechnung*, Fund. Math. **6** (1924), 9–20.

[24] _____, *Metrische Kettenbrücheprobleme*, Comp. Math. **1** (1935), 361–382.

[25] A. N. Kolmogorov, *Über die Summen durch den Zufall bestimmten unabhängiger Grössen*, Math. Ann. **99** (1928), 309–319.

[26] H. Lebesgue, *Sur certains démonstrations d'existence*, Bull. Soc. Math. France **45** (1917), 132–144.

[27] M. B. Levin, *On the discrepancy estimate of normal numbers*, Acta Arith. **88** (1999), 99–111.

[28] K. Mahler, *Arithmetical properties of the digits of the multiples of an irrational number*, Bull. Austral. Math. Soc. **8** (1973), 191–203.

[29] K. Nagasaka, *La dimension de Hausdorff de certaines ensembles dans* [0, 1], Proc. Japan Acad. Ser. A Math. Sci. **54** (1979), 109–112.

[30] W. Philipp, *Some metrical theorems in number theory II*, Duke Math. J. **37** (1970), 447–458.

[31] A. D. Pollington, *The Hausdorff dimension of a set of nonnormal well approximable numbers*, Number Theory, Carbondale (1979), Lecture Notes in Math. 751, Springer-Verlag, Berlin, 1979, pp. 256–264.

[32] _____, *The Hausdorff dimension of a set of normal numbers*, Pacific J. Math. **95** (1981), 193–204.

[33] A. D. Pollington and R. C. Vaughan, *The k-dimensional Duffin and Schaeffer conjecture*, Mathematika **37** (1990), 190–200.

[34] A. M. Rockett and P. Szüsz, *Continued fractions*, World Scientific, Singapore, 1992.

[35] J. Schiffer, *Discrepancy of normal numbers*, Acta Arith. **47** (1986), 175–186.

[36] W. M. Schmidt, *On normal numbers*, Pacific J. Math. **10** (1960), 661–672.

[37] _____, *On badly approximable numbers*, Mathematika **12** (1965), 10–20.

[38] V. G. Sprindzuk, *A proof of Mahler's conjecture on the measure of the set of S-numbers (in Russian)*, Izv. Akad. Nauk SSSR, Ser. Mat. **29** (1965), 379–436.

[39] G. Wagner, *On rings of numbers which are normal to one base but non-normal to another*, J. Number Theory **54** (1995), 211–231.

On Theorems of
Barban-Davenport-Halberstam Type

C. Hooley

1 Foreword

In this survey all equations and notation should be interpreted in the manner usual in the analytical theory of numbers. Beyond this encompassing comment it would be pedantic in a work of this type to expand, save to say that the precise meaning of each statement is easily supplied by the reader in the knowledge that x is usually regarded as tending to infinity, that the constants implied by the O-notation are independent of k and Q, and that A denotes a positive number, often arbitrarily large, that is not necessarily the same at each occurrence.

2 Introduction

The subject of primes in arithmetical progression is of great interest both because it forms an important facet in the theory of prime numbers and because of its potential applications to a host of problems in additive and multiplicative number theory. Undoubtedly primarily initiated for the former reason, the study of this subject especially involves the behaviour of entities that are associated with the sum[1]

$$\theta(x; a, k) = \sum_{\substack{p \le x \\ p \equiv a \bmod k}} \log p \qquad ((a, k) = 1) \qquad (2.1)$$

and the formal remainder term

$$E(x; a, k) = \theta(x; a, k) - x/\phi(k). \qquad (2.2)$$

---

[1] Note that we persevere in our practice of using a notation in which a, k appear in θ and other functional symbols in the reverse of the usual order. This idiosyncrasy, started more by accident than design, has the advantage that k always follows a as in the underlying residue class $a \bmod k$.

Confining descriptions to (2.1) and (2.2) for technical reasons and appreciating they apply equally well to cognate items such as

$$\psi(x; a, k) = \sum_{\substack{n \le x \\ n \equiv a \bmod k}} \Lambda(n),$$

we can summarize the main state of the subject reached before the Second World War by enunciating the unconditional estimate

$$E(x; a, k) = O(x \log^{-A} x) \qquad ((a, k) = 1) \qquad (2.3)$$

for any positive constant A, which was subject to the improved

$$E(x; a, k) = O\left(x e^{-A\sqrt{\log x}}\right)$$

provided that k were not the modulus of a Dirichlet's L-function having an exceptional 'Siegel-zero'.

Although both the fruit and inspiration of substantial development in the understanding of the zeta function and the L-functions, these results were disappointing in that they only gave substantial information when the common difference of the arithmetical progression was small compared with its length x—so much so that many relevant problems in number theory remained out of reach with the welcome exception of the Goldbach three primes conjecture settled by Vinogradov. In compensation, however, Titchmarsh [37] showed in 1930 on the extended Riemann hypothesis (E.R.H.) that

$$E(x; a, k) = O\left(x^{1/2} \log^2 x\right) \qquad ((a, k) = 1) \qquad (2.4)$$

with the implication that we are provided with substantial conditional information for moduli k almost, but not quite, as large as $x^{1/2}$. The impact of this was that a conditional mechanism was supplied that brought about the resolution of some important conjectures when it was accompanied, if necessary, by an ancillary supporting apparatus. For example, in the latter part of his paper, Titchmarsh settled the divisor sum problem associated with

$$\sum_{p < n} d(n - p)$$

with the aid of Brun's sieve method, while later in 1957 the author conditionally proved the asymptotic formula conjectured by Hardy and Littlewood for the number of representations of a large n as the sum of two squares and a prime by creating an altogether more elaborate structure for coping with the situations where (2.4) did not suffice.

Even on E.R.H. the results yet mentioned are seriously deficient because they do not reflect our expectation that $\theta(x; a, k)$ should be asymptotic to $x/\phi(k)$ for values of k greater than $x^{1/2}$. But this shortcoming was partially ameliorated on E.R.H. by Turán in 1937, whose result was later slightly improved by Montgomery [31] to yield

$$\sum_{\substack{0 \leq a \leq k \\ (a,k)=1}} E^2(x; a, k) = O(x \log^4 x) \qquad (k \leq x) \qquad (2.5)$$

by a simplification in the method; thus, in mean-square average,

$$E(x; a, k) = O\left(\frac{x^{1/2} \log^2 x}{\phi^{1/2}(k)}\right), \qquad (2.6)$$

which conclusion not only accords with what might have been forecast on probability grounds but also indicates in some sense that an asymptotic formula for $\theta(x; a, k)$ remains true for k as large as $x^{1-\epsilon}$. Somewhat surprisingly no applications of this potentially rewarding theorem seem to have been published, though we should perhaps mention that while studying the Hardy-Littlewood problem on small values of $p - p'$ we ascertained in 1956 with the aid of a close analogue that infinitely often on E.R.H. we have

$$0 < p - p' < (.466 \cdots + \epsilon) \log p \qquad (2.7)$$

and thus improved on previous work of Rankin. We should observe also that (2.5) is so strong that it actually implies the Titchmarsh estimate (2.4) for individual $E(x; a, k)$.

In post-war years the theory received a vital transfusion that had its genesis in Linnik's important 1941 paper on his large sieve. Conceived as its name suggests to deal with certain sieving problems in which many residues for each appropriate congruential modulus are removed from sequences, Linnik's method was developed and reinterpreted to such an extent that powerful new procedures emerged that far outran its original purpose and that were seen to have applications to prime number theory. Thence, following the work of Linnik and Renyi in particular, the progress of the method accelerated in the 1960s and culminated after the work of Roth in Bombieri's [3] version of the so called Bombieri-Vinogradov theorem[2] to the effect that, for any positive constant A_1, there exists another positive constant A_2 such that

$$\sum_{k \leq Q} \max_{(a,k)=1} \max_{1 \leq y \leq x} |E(y; a, k)| = O\left(\frac{x}{\log^{A_1} x}\right) \qquad (2.8)$$

---

[2]Vinogradov's result was rather weaker and did not depend on the large sieve.

when

$$Q = x^{1/2} \log^{-A_2} x.$$

But, although denominated through the title of large sieve, the procedures behind these developments do not constitute sieve methods in themselves but depend on sums such as

$$S(\theta) = S(x, \theta) = \sum_{n \le x} e^{2\pi i n \theta},$$

and their moments

$$\sum_{k \le Q} \sum_{\substack{0 < h \le k \\ (h,k)=1}} |S(h/k)|^2, \qquad (2.9)$$

inequalities for which have implications for sieving problems as well as the wider theory described.

The Bombieri-Vinogradov theorem, as a moment's reflection will reveal, shows that almost always up to values of k nearly as large as $x^{1/2}$ we have $\theta(x; a, k) \sim x/\phi(k)$ for $(a, k) = 1$ and that we therefore have a workable unconditional surrogate for the conditional bound (2.4) in the problems that had previously depended on it. In particular, superseding Linnik's unconditional treatment of the Hardy-Littlewood problem on sums of two squares and prime, we can simply substitute the use of (2.8) for (2.4) in our earlier paper referred to above, as was shown by Elliott and Halberstam or by us in our tract [17]; also, in somewhat parallel manner, Bombieri and Davenport obtained the bound (2.7) without any hypothesis.

This is as far as we take the story about individual values of $\theta(x; a, k)$ or their involvement in the sums in (2.8) because the first part of the review we now reach will concern the parallel developments in our knowledge of the sum in (2.5) that stem from, or have been motivated by, estimates for the sum (2.9). We could not have veered from this direction earlier because we shall need the results already quoted and because we wished to set the survey in a proper context.

So far we have deliberately confined our focus on the particular sequence of prime numbers, since this is the subject on which we weight our concentration and by which we are then moved to widen our thoughts to other sequences. Such sequences may either be familiar ones such as that of the square-free numbers for which analogues of (2.3) are known for small moduli or, more generally, simply ones for which such analogues are postulated, the emphasis of the enquiries being what can then be discovered about the distribution on average of the sequences among arithmetical progressions with large common differences.

3 The Barban-Davenport-Halberstam Theorem

At about the time of the promulgations of the Bombieri-Vinogradov theorem, Barban ([1], [2]) on the one hand and Davenport and Halberstam [6] on the other initiated a complementary advance in prime number theory by proving versions of a theorem[3] that stands in the same relationship to (2.8) as did on E.R.H. the conditional (2.5) to the conditional (2.4). Although the demonstrations—also based on bounds for (2.9)—are much simpler even than the later more compact derivations of the Bombieri-Vinogradov theorem, we shall see that this result in its own sphere of relevance is almost as influential in its applications as the theorem it parallels.

In sketching the principles used by Barban, Davenport, and Halberstam, we follow Gallagher's [9] improved treatment and statement of their work. We begin with his form of the large sieve inequality for (2.9) that states that

$$\sum_{\substack{k \le Q}} \sum_{\substack{0 < h \le k \\ (h,k)=1}} |S(h/k)|^2 \le \left(\pi x + Q^2\right) \sum_{n \le x} |a_n|^2, \tag{3.1}$$

which is roughly tantamount to certain earlier versions and which was derived in his characteristically elegant manner (this is the most accessible form of the inequality but is not best possible; Montgomery and Vaughan have shown that the multiplier π can be removed, although this is of no significance here). From this, since for any primitive character χ_q^* mod q, we have

$$\chi_q^*(m) = \frac{1}{\tau\left(\overline{\chi}_q^*\right)} \sum_{0 < h \le q} \overline{\chi}_q^*(h) e^{2\pi i m h/q}$$

with the usual definition of $\tau(\chi_q^*)$, it follows that the sums

$$T(\chi_q^*) = \sum_{n \le x} a_n \chi_q^*(n) = \frac{1}{\tau\left(\overline{\chi}_q^*\right)} \sum_{n \le x} a_n \sum_{0 < h \le q} \overline{\chi}_q^*(h) e^{2\pi i n h/q}$$

$$= \frac{1}{\tau\left(\overline{\chi}_q^*\right)} \sum_{0 < h \le q} \overline{\chi}_q^*(h) S(h/q)$$

---

[3]But note that this will not imply the truth of (2.8), whereas, as already stated, (2.5) implied (2.4).

are subject to the inequality

$$\sum_{q\leq Q}\frac{q}{\phi(q)}\sum_{\chi^*}|T(\chi_q^*)|^2 = \sum_{q\leq Q}\frac{q}{|\tau\,(\overline{\chi}_q^*)|^2\,\phi(q)}\sum_{\chi_q^*}\left|\sum_{0<h\leq q}\overline{\chi}_q^*(h)S(h/q)\right|^2$$

$$\leq \sum_{q\leq Q}\frac{1}{\phi(q)}\sum_{\chi_q}\left|\sum_{0<h\leq q}\overline{\chi}_q(n)S(h/q)\right|^2$$

$$= \sum_{q\leq Q}\sum_{\substack{0<h\leq q\\(h,q)=1}}|S(h/q)|^2$$

$$\leq \left(\pi x + Q^2\right)\sum_{n\leq x}|a_n|^2.$$

Hence, on choosing a_n to be $\log n$ or 0 according as n is a prime or otherwise, we obtain,

$$\sum_{q\leq Q}\frac{q}{\phi(q)}\sum_{\chi_q^*}|\theta(x,\chi_q^*)|^2 = O\left((x+Q^2)x\log x\right) \tag{3.2}$$

with the usual meaning attached to the sums $\theta(x,\chi)$.

To benefit from this inequality, let now X_k denote the principal character mod k and χ_k a general non-principal character mod k so that for $(a,k)=1$

$$\theta(x;a,k) = \frac{\theta(x,X_k)}{\phi(k)} + \frac{1}{\phi(k)}\sum_{\chi_k}\overline{\chi}_k(a)\theta(x,\chi_k)$$

with the implication that

$$H(x,k) = \sum_{\substack{0<a\leq k\\(a,k)=1}}E^2(x;a,k)$$

$$= \sum_{\substack{0<a\leq k\\(a,k)=1}}\left(\theta(x;a,k)-\frac{\theta(x,X_k)}{\phi(k)}\right)^2 + \frac{(x-\theta(x,X_k))^2}{\phi(k)}$$

$$= \frac{1}{\phi(k)}\sum_{\chi_k}|\theta(x,\chi_k)|^2 + O\left(\frac{x^2e^{-A\sqrt{\log x}}}{\phi(k)}\right)$$

because $\theta(x,X_k)$ is the mean value of $\theta(x;a,k)$ for given k. Thence, having consequently expressed the moment

$$G(x,Q) = \sum_{k\leq Q}H(x,k) = \sum_{k\leq Q}\sum_{\substack{0<a\leq k\\(a,k)=1}}E^2(x;a,k) \tag{3.3}$$

as

$$\sum_{k \leq Q} \frac{1}{\phi(k)} \sum_{\chi_k} |\theta(x, \chi_k)|^2 + O\left(x^2 e^{-A\sqrt{\log x}}\right)$$

when $Q \leq x$, we proceed to consider the contribution to the first term in this due to those values of k for which χ_k is associated with a given primitive character χ_q^* mod q, where therefore $q|k$. Thus, as obviously

$$\theta(x, \chi_k) = \theta(x, \chi_q^*) + O\left(\sum_{p|k} \log p\right) = \theta(x, \chi_q^*) + O(\log 2k),$$

we deduce that

$$G(x, Q) = O\left(\sum_{q \leq Q} \sum_{\chi_q^*} |\theta\left(x, \chi_q^*\right)|^2 \sum_{\substack{k \leq Q \\ k=0 \bmod q}} \frac{1}{\phi(k)}\right)$$

$$+ O\left(\sum_{k \leq Q} \log^2 2k\right) + O\left(x^2 e^{-A\sqrt{\log x}}\right)$$

$$= O\left(\sum_{q \leq Q} \frac{1}{\phi(q)} \log \frac{2Q}{q} \sum_{\chi_q^*} |\theta(x, \chi_q^*)|^2\right) + O\left(x^2 e^{-A\sqrt{\log x}}\right)$$

$$= O\left(G_1(x, Q)\right) + O\left(x^2 e^{-A\sqrt{\log x}}\right), \quad \text{say.} \tag{3.4}$$

The proof is completed by using (3.2) in tandem with the estimate

$$\theta(x, \chi_q^*) = O\left(x \log^{-A} x\right),$$

from which in fact (2.3) above flows. Then, if we set $Q_1 = \log^{A+2} x$, we have[4]

$$G_1(x, Q) \leq \sum_{q \leq Q_1} + \sum_{Q_1 < q \leq Q} = G_1(x, Q_1) + G_2(x, Q), \quad \text{say,} \tag{3.5}$$

in which first

$$G_1(x, Q_1) = O\left(\frac{x^2 \log x}{\log^{2A+3} x} \sum_{q \leq Q_1} 1\right) = O\left(x^2 \log^{-A} x\right). \tag{3.6}$$

---

[4]The inequality sign is needed to take care of the case $Q < Q_1$, which in fact is of little significance.

Also, by (3.2) and partial summation, it is evident that

$$G_2(x, Q) = O\left(\frac{x^2 \log^2 x}{Q_1}\right) + O(Qx \log x)$$

$$= O\left(x^2 \log^{-A} x\right) + O(Qx \log x). \tag{3.7}$$

We conclude from (3.4), (3.5), (3.6), and (3.7) the truth of

Theorem 1 (the Barban-Davenport-Halberstam theorem; Gallagher). *Let $G(x, Q)$ be defined through* (2.2) *and* (3.3)*. Then, for $Q \leq x$ and any positive constant A, we have*

$$G(x, Q) = O(Qx \log x) + O\left(x^2 \log^{-A} x\right).$$

Although we associate this proposition with the names of Barban, Davenport, and Halberstam, we should stress that in its present form it is due to Gallagher and is stronger then its precursors obtained by the earlier writers. As here enunciated, it is certainly best possible in two of its aspects, while the earlier results only asserted that

$$G(x, Q) = O\left(x^2 \log^{-A} x\right),$$

for $Q \leq x \log^{-B} x$ where $B = B(A)$ in Barban's work and $B = A + 5$ in that of Davenport and Halberstam. Thus our theorem asserts one thing the others cannot, namely, that for $x \log^{-A} x < Q \leq x$ and any A, we have

$$G(x, Q) = O(Qx \log x), \tag{3.8}$$

which as we shall soon see is an echo of two aspects of affairs and which is consistent with the expectation that (2.6) can be almost improved to

$$E(x; a, k) = O\left((x/\phi(k))^{1/2} \log^{1/2} x\right)$$

for most values of k bounded away from 1. Similarly, no betterment in the second term in the estimate for $G(x, Q)$ is possible at present because otherwise by considering $\sqrt{G(x, Q)}$ for smallish values of Q we could infer a better version of the classical estimate (2.3) that is constrained by our lack of knowledge of 'Siegel' zeros; this is a comment that has been attributed to Montgomery.

Apart from their basically different structures, one other distinction between the Bombieri-Vinogradov theorem and the Barban-Davenport-Halberstam theorem (shortened now to 'B.D.H. theorem' for convenience) is that the upper bound symbol before E only appears in (2.8). It is indeed tempting to think that this symbol could also be affixed to the enunciation of the latter because Montgomery [32], forming what he terms a maximal large sieve inequality, has shown that (3.1) may be generalized to

$$\sum_{k \leq Q} \sum_{0 < h \leq k} \max_{1 \leq y \leq x} \left| \sum_{n \leq y} a_n e^{2\pi i n h/k} \right|^2 = O \left((x + Q^2) \sum_{h \leq x} |a_n|^2 \right)$$

by appealing to deep work by Hunt on Fourier series. However, it is naïve to think that this aspiration can be easily fulfilled since the connecting link between (3.1) and Theorem 1 is insufficiently direct. Nevertheless, even before the appearance of Montgomery's paper, Uchiyama [38] had used an ingenious argument to get an estimate that implies that

$$G^\dagger(x, Q) = \sum_{k \leq Q} \sum_{\substack{0 < a \leq k \\ (a,k)=1}} \max_{1 \leq y \leq x} \left| E^2(y; a, k) \right| = O \left(Qx \log^3 x \right) \qquad (3.9)$$

for $x \log^{-A} x < Q \leq x$ and that is a useful tool for some applications. Uchiyama's result, whose proof associates a combinational idea with minor adjustments to Theorem 1, is probably imperfect to the extent of a super-fluous $\log^2 x$ in the estimate, as is rendered probable by the conclusion of the author (unpublished) to the effect that, if

$$E_1(x; a, k) = \sum_{\substack{p \leq x \\ p \equiv a \bmod k}} \left(1 - \frac{p}{x} \right) \log p - \frac{x}{2\phi(k)} \qquad ((a, k) = 1),$$

then

$$\sum_{k \leq Q} \sum_{\substack{0 < a \leq k \\ (a,k)=1}} \max_{1 \leq y \leq x} \left| E_1^2(y; a, k) \right| = O(Qx \log x)$$

for $x \log^{-A} x < Q \leq x$. Furthermore, unfolding the implications of this by means of a Tauberian argument, we were led to the relations

$$G^\dagger(x, Q) = \begin{cases} O(Qx \log x), & \text{if } x \log^{-1} x < Q \leq x, \\ O \left(Qx \log x (\log \log x)^2 \right), & \text{if } x \log^{-A} x < Q \leq x \log^{-1} x, \end{cases}$$

that were announced at our address to the I.C.M. in 1974 [12]. This improvement on (3.9), a proof of which may appear in due course, is a little disappointing because the refinement wrought by the technique would become nugatory as soon as we used methods, conditional or unconditional, for invading territory belonging to smaller values of Q. The elucidation of the status of $G^\dagger(x, Q)$ therefore remains a not unimportant aspect of the theory.

Theorem 1 and the earlier estimates, whether or not accompanied by variations in detail, are usually equally effective for the sort of applications we have in mind. But from the aspect of pure prime number theory the establishment and development of (3.8) for smaller values of Q is a desideratum. This is a matter to which we shall constantly return but, in the meanwhile, mention that in accord with Theorem 0 of our paper [17] equation (3.8) is seen to be valid on E.R.H. for $Q > x^{1/2} \log^3 x$ by substituting the use of (2.5) for (2.3) in our proof of Theorem 1 above.

4 Applications of the Barban-Davenport-Halberstam Theorem

We first illustrate the way in which Theorem 1 or its variants can be applied to problems in number theory by establishing a theorem of Bombieri-Vinogradov type for the set of almost primes s of the form $p_1 p_2$ affected with a weight

$$\psi(s) = \begin{cases} 2 \log p_1 \log p_2, & \text{if } s = p_1 p_2 \text{ (or } s = p_2 p_1\text{) and } p_1 \neq p_2, \\ \log^2 p, & \text{if } s = p^2, \end{cases}$$

where p_1, p_2 do not exceed a large number $x^{1/2}$. Then, the analogues of $\theta(x)$, $\theta(x; a, k)$, and $E(x; a, k)$ in the prime number problem being

$$\Theta(x) = \sum_{s \leq x} \psi(s) = \sum_{p_1, p_2 \leq x^{1/2}} \log p_1 \log p_2 = \theta^2 \left(x^{1/2} \right) \sim x,$$

$$\Theta(x; a, k) = \sum_{\substack{s \leq x \\ s \equiv a \bmod k}} \psi(s),$$

and

$$E(x; a, k) = \Theta(x; a, k) - \frac{x}{\phi(k)},$$

the appropriate object of study in this instance is the sum

$$\Gamma(x, Q) = \sum_{\substack{k \leq Q \\ (a,k)=1}} \max_{0 < a \leq k} |E(x; a, k)|.$$

First, letting $\bar{b}$ denote a solution of $bu \equiv 1 \bmod k$ when $(b, k) = 1$, and supposing that $(a, k) = 1$, we have

$$\Theta(x; a, k) = \sum_{\substack{p_1, p_2 \leq x^{1/2} \\ p_1 p_2 \equiv a \bmod k}} \log p_1 \log p_2$$

$$= \sum_{\substack{0 < b \leq k \\ (b,k)=1}} \sum_{\substack{p_1 \leq x^{1/2} \\ p_1 \equiv \bar{b} \bmod k}} \log p_1 \sum_{\substack{p_2 \leq x^{1/2} \\ p_2 \equiv a\bar{b} \bmod k}} \log p_2$$

$$= \sum_{\substack{0 < b \leq k \\ (b,k)=1}} \theta\left(x^{1/2}; b, k\right) \theta\left(x^{1/2}; a\bar{b}, k\right)$$

$$= \sum_{\substack{0 < b \leq k \\ (b,k)=1}} \left(\frac{x^{1/2}}{\phi(k)} + E\left(x^{1/2}; b, k\right)\right) \left(\frac{x^{1/2}}{\phi(k)} + E\left(x^{1/2}; a\bar{b}, k\right)\right)$$

$$= \frac{x}{\phi(k)} + \frac{x^{1/2}}{\phi(k)} \sum_{\substack{0 < b \leq k \\ (b,k)=1}} \left(E\left(x^{1/2}; b, k\right) + E\left(x^{1/2} a\bar{b}, k\right)\right)$$

$$+ \sum_{\substack{0 < b \leq k \\ (b,k)=1}} E\left(x^{1/2}; b, k\right) E\left(x^{1/2}; a\bar{b}, k\right). \tag{4.1}$$

Since here, for $k \leq x^{1/2}$,

$$\sum_{\substack{0 < b \leq k \\ (b,k)=1}} E\left(x^{1/2}; b, k\right) = \sum_{\substack{p \leq x^{1/2} \\ p \nmid k}} \log p - x^{1/2}$$

$$= \theta\left(x^{1/2}\right) - x^{1/2} + O(\log k) = O\left(x^{1/2} e^{-A\sqrt{\log x}}\right)$$

with a similar estimate for the other sum in the second term on the last line of (4.1),

$$\Theta(x; a, k) - \frac{x}{\phi(k)} = O\left(\frac{x e^{-A\sqrt{\log x}}}{\phi(k)}\right) + \sum_{\substack{0 < b \leq k \\ (b,k)=1}} E\left(x^{1/2}; b, k\right) E\left(x^{1/2}; a\bar{b}; k\right)$$

so that

$$E(x; a, k) = O\left(\frac{x e^{-A\sqrt{\log x}}}{\phi(k)}\right) + O\left(\sum_{\substack{0 < c \leq k \\ (c,k)=1}} E^2\left(x^{1/2}; c, k\right)\right)$$

by the Cauchy-Schwarz inequality. Thus, through an argument that has recalled a well-known relation between covariances, Theorem 1 implies for $Q \leq x^{1/2}$ the estimate

$$\Gamma(x, q) = O\left(Qx^{1/2} \log x\right) + O\left(x \log^{-A} x\right),$$

which is of the type we sought. Alternatively, we may gain an almost equally useful estimate by replacing $E^{1/2}(x; a\bar{b}, k)$ in the last term of (4.1) by $O(x^{1/2} \log x / \phi(k))$ and then using the Cauchy-Schwarz inequality.

The latter reasoning serves as a model for the partial treatment of other problems in which two of the entities appearing are restricted to be prime numbers. Such a situation arises, for example, when we try to enlarge the scope of Lagrange's four square theorem by restricting two of the squares in the representation to be squares of primes. One avenue to this is to follow Greaves [11] by considering the set of positive numbers of the type $n - p_1^2 - p_2^2$ for $p_1, p_2 \leq (1/2)n$ that are congruent to 1 mod 4 and to try to use a sieve method to find a positive lower estimate for the cardinality of such numbers that are indivisible by primes congruent to 3 mod 4. To this end, as is familiar, good estimates for the number of p_1, p_2 satisfying the condition.

$$n - p_1^2 - p_2^2 \equiv 1 \bmod 4; \quad n - p_1^2 - p_2^2 \equiv 0 \bmod d; \quad p_1, p_2 \leq \frac{1}{2} n^{1/2}$$

will be needed either for individual square-free odd d or on average, this requirement being supplied through a variant of Theorem 1 by a modification in previous reasoning for values of d up to $n^{1/2} \log^{-A} n$. But, since it is only when d can be taken as large as $n^{1/2+\epsilon}$ that the $\frac{1}{2}$ residue sieve will show that a non-empty portion of the sequence has members with at most one prime divisor congruent to 3 mod 4, and hence none, the argument in this simple form fails and it was necessary for Greaves to combine it with other ideas in order to secure a satisfactory outcome.

Another approach, if one aims at the slightly more ambitious goal of finding an asymptotic formula for the number $\nu(n)$ of representations of n in the form $X^2 + Y^2 + p_1^2 + p_2^2$, is to use the equation

$$\nu(n) = \sum_{p_1^2 + p_2^2 < n} r(n - p_1^2 - p_2^2) + O(1),$$

where

$$r(\mu) = 4 \sum_{lm=\mu} \chi(\ell)$$

is the number of representations of a positive number μ as the sum of two squares and $\chi(\ell)$ is the non-principal character mod 4. Incorporating the second expression in the first, we obtain a double sum in which either ℓ or m does not exceed $n^{1/2}$ and which, by changes in the order of summation, consists of a sum over such values of ℓ and another over such values of m. The former, for example, can be taken to be

$$4 \sum_{\ell \leq n^{1/2}} \chi(\ell) \sum_{\substack{p_1^2 + p_2^2 < n \\ n - p_1^2 - p_2^2 \equiv 0 \bmod \ell}} 1$$

whose inner sum on average can be calculated adequately through our programmes for values of ℓ as large as $n^{1/2} \log^{-A} n$. The latter having similar attributes, there remains a narrow band of ℓ and m surrounding $n^{1/2}$ for which it is possible to extend in a lengthy way the author's operations in [12] (see §2). In such a manner the asymptotic formula for $\nu(n)$ was found by Shields [37] and independently and later by Plaksin [36].

These examples by no means exhaust the potentialities of the method. We need only point to the recent work of Daniel [5] on sums of squares of primes to confirm that theorems of B.D.H. type continue to play a useful rôle in the theory of numbers.

5   The Barban-Montgomery Theorem

The theme we now pursue is about developments of the B.D.H. theorem that pertain more to the structure of primes in arithmetical progressions than to applications elsewhere.

The first significant advance in this direction was made by Montgomery ([30]; see also his monograph [31]) when he converted the bound for $G(x, Q)$ into an asymptotic formula and incidentally showed that the treatment of Theorem 1 need not depend on the large sieve. In sketching the ideas behind his proof, we mention that he could first approximate closely to $H(x, k)$ in (3.3) by

$$\sum_{\substack{0 < a \leq k \\ (a,k)=1}} \theta^2(x; a, k) - \frac{x^2}{\phi(k)} \tag{5.1}$$

since $H(x, k)$ is almost a dispersion to which the usual rules of evaluation apply. Then, summing over k, he was left with the problem of assessing the sum

$$\sum_{k \leq Q} \sum_{\substack{0 < a \leq k \\ (a,k)=1}} \theta^2(x; a, k) \tag{5.2}$$

because of known asymptotic formulae for the sum

$$\sum_{k \leq Q} \frac{1}{\phi(k)}.$$

In continuation, he acted as though he expressed the inner sum in (5.2) for $k \leq x$ as

$$\sum_{\substack{0<a\leq k \\ (a,k)=1}} \sum_{\substack{p,p'\leq x \\ p\equiv p'\equiv a \bmod k}} \log p \log p' = \sum_{\substack{p,p'\leq x \\ p-p'\equiv 0 \bmod k \\ (pp',k)=1}} \log p \log p'$$

$$= \sum_{\substack{p\leq x \\ p\nmid k}} \log^2 p + 2 \sum_{\substack{p-p'=\ell k \\ p\leq x;(pp',k)=1}} \log p \log p'$$

$$= x \log x - x + O\left(xe^{-A\sqrt{\log x}}\right)$$

$$+ 2 \sum_{\substack{p-p'=\ell k \\ p\leq x}} \log p \log p' + O\left(\log^2 k\right)$$

$$= x \log x - x + O\left(xe^{-A\sqrt{\log x}}\right) + 2\theta_2(x,\ell k), \quad \text{say}, \qquad (5.3)$$

and, having summed over k to obtain (5.2), met the sum

$$\sum_{k\leq Q} \sum_{\ell\leq x/Q} \theta_2(x,\ell k). \qquad (5.4)$$

There being as yet no formula for the prime-twins counting function $\theta_2(x,m)$ available for insertion in this, Montgomery circumvented the apparent obstacle by using a theorem by Lavrik that asserts there is such a formula that is valid on average in a mean-square sense, namely, if $\Psi(h) = 0$ when h is odd but

$$\Psi(h) = 2 \prod_{p>2} \left(1 - \frac{1}{(p-1)^2}\right) \prod_{\substack{p|h \\ p>2}} \left(\frac{p-1}{p-2}\right)$$

when h is even and if

$$F(x,h) = \theta_2(x,h) - (x-h)\Psi(h),$$

then

$$\sum_{0<h\leq x} F^2(x,h) = O(x^3 \log^{-A} x);$$

note here that we have departed slightly from Montgomery's notation and that there are a few minor misprints towards the end of his exposition on this topic in [31]. The sum (5.4) is therefore

$$\sum_{k \leq Q} k \sum_{\ell \leq x/k} \left(\frac{x}{k} - \ell \right) \Psi(\ell k) + \sum_{\substack{k \leq Q \\ \ell k \leq x}} F(x, \ell k),$$

the second term in which is

$$O \left(\sum_{m \leq x} d(m) |F(x, m)| \right) = O \left(\left(\sum_{m \leq x} d^2(m) \right)^{1/2} \left(\sum_{m \leq x} F^2(x, m) \right)^{1/2} \right)$$

$$= O \left(\left(x \log^3 x \cdot x^3 \log^{-2A-3} x \right)^{1/2} \right)$$

$$= O \left(x^2 \log^{-A} x \right). \tag{5.5}$$

Since the first term can be accurately calculated by routine methods, all the constituents in the broken-down form of $G(x, Q)$ have been evaluated when $Q \leq x$ and lead after appropriate cancellations to an asymptotic formula, which in the form originally found by Montgomery is

$$G(x, Q) = \begin{cases} Qx \log Q + O \left(Qx \log \frac{2x}{Q} \right) + O \left(x^2 \log^{-A} x \right) & \text{if } Q \leq x, \\ x^2 \log x + Bx^2 + O \left(x^2 \log^{-A} x \right) & \text{if } Q = x, \end{cases} \tag{5.6}$$

but which he later slightly improved by the replacement of the first remainder term in the first line by $O(Qx)$ (see Croft [4]). Here, as almost always, we have curtailed attention to the most significant case where $Q \leq x$, although we should note that Montgomery makes no such restriction in his work.

In the designation of theorems stating formulæ of the general type (5.6) Montgomery's name has been wont to be linked with that of Barban [2] because the latter had pre-empted the former to the extent of asserting the truth of (5.6) for the special case $Q = x$. Although Montgomery's contribution is undeniably the greater, this practice has at least the virtue of identifying unequivocally the theorems to which the nomenclature appertains. We are therefore ready to state the basic result in the second subject area as

Theorem 2 (Barban-Montgomery theorem). *Let $G(x, Q)$ be defined through (2.2) and (3.3). Then $G(x, Q)$ satisfies the asymptotic formulae in (5.6) for any positive constant A.*

This should be contrasted with Theorem 1 before we assay further developments. Having chosen any large A to establish a point of reference, we easily see that the formulae of these theorems are equivalent for $Q \leq x \log^{-A-1} x$ because each is then equivalent to $G(x, Q) = O\left(x^2 \log^{-A} x\right)$; on the other hand, for $Q > x \log^{-A-1+\epsilon} x$, the formula of Theorem 2 converts the formally dominant term $O(Qx \log x)$ in Theorem 1 into $Qx \log Q + O(Qx)$ and indicates an average value for the dispersion of $E(x; a, k)$ taken over values of a coprime to k, this being the origin of one of our later discussions about the distribution of $E(x; a, k)$.

As mentioned in §2, a constant thread throughout our work is our need to lower the values of Q—and hence indirectly of k—for which our results shed useful information on $E(x; a, k)$. To our comment that on E.R.H. the range of validity of (3.8) can be stretched to $Q > x^{1/2} \log^3 x$ we therefore add that on the same supposition we may replace the last remainder term in Theorem 2 by $O\left(x^{7/4+\epsilon}\right)$, thus obtaining an informative asymptotic formula for $Q > x^{3/4+\epsilon}$. This conclusion, of which Montgomery was most certainly aware, is deduced by ousting Lavrik's theorem in favour of a conditional version that is obtained along the lines of Hardy and Littlewood's work on Partitio Numerorum.

A new and easier way of dealing with theorems of the Barban-Montgomery type ('B.M. type' for brevity) was brought in by the author [17] in 1975 even though the analysis subsumed a prior acquaintance with Theorem 1. To expose the ideas involved, we should mention that the case $Q \leq x \log^{-A-1} x$ was first dismissed in the spirit of the last but one paragraph, whereupon, after setting

$$Q_1 = x \log^{-A-1} x, \quad Q_1 < Q \leq x,$$

and

$$G(x; Q_1, Q_2) = \sum_{Q_1 < k \leq Q_2} \sum_{\substack{0 < a \leq k \\ (a,k)=1}} E^2(x; a, k),$$

we worked with the equations

$$G(x, Q_2) = G(x; Q_1, Q_2) + O(x \log^{-A} x)$$

and

$$G(x; Q_1, Q_2) = G(x; Q_1, x) - G(x; Q_2, x).$$

Then, in analysing the sum $G(x; Q, x)$ for Q equal to Q_1 or Q_2, we followed the previous treatment of $G(x, Q)$ up to (5.3) and, summing over the revised range $Q < k \leq x$, arrived at the sum

$$\sum_{\substack{p-p'=\ell k \\ p \leq x; k > Q}} \log p \log p', \qquad (5.7)$$

the condition $k > Q$ in which permits one to dispense with the use of Lavrik's theorem. Indeed, since now the dummy variable ℓ is less than $x/Q \leq \log^{A+1} x$, a change in the order of summation permits one to use effectively the prime number theorem for arithmetical progressions for the small common difference ℓ in the revised formulation of (5.7) as

$$\sum_{\ell < x/Q} \sum_{\substack{p \equiv p' \bmod \ell \\ p \leq x; p-p' > \ell Q}} \log p \log p_1 = \sum_{\ell < x/Q} \sum_{0 < b \leq \ell} \sum_{\substack{p' < x - \ell Q \\ p' \equiv b \bmod \ell}} \log p' \sum_{\substack{\ell Q + p' < p \leq x \\ p \equiv b \bmod \ell}} \log p.$$

After this and some consequential calculations there emerges a satisfactory expression for (5.7), which when combined with other items in the analysis recoups Theorem 2 in the slightly sharper form

$$G(x, Q) = Qx \log Q + B_1 Q x + O\left(Q^{5/4} x^{3/4}\right) + O\left(x^2 \log^{-A} x\right)$$

for $Q \leq x$.

As well as its easy application to Theorem 2 and its wide-ranging generalizations that we later mention, this method has a number of other advantages over Montgomery's. Not only does it yield a more accurate form of the basic theorem but it also empowers us on the basis of E.R.H. to increase the range of usefulness of the theorem beyond the limit of $Q = x^{3/4+\epsilon}$ imposed by the earlier method. In fact, having already achieved the sharper $Q = x^{4/7+\epsilon}$ in the original paper, we went on to gain the yet better $Q = x^{1/2+\epsilon}$ in the second paper [13] by augmenting the technique through the use of primes in small intervals and a Tauberian argument. On the other hand, as we shall see in the next Section, the advantage does not lie wholly on one side because where applicable the Montgomery method can often yield more information than ours when suitably interpreted.

In continuing the story to the present day, we should remark that some clearly felt that the Hardy-Littlewood circle method was the most fitting instrument for the study of Theorem 2 along lines parallel to ours— a view in which we do not necessarily acquiesce for a reason we shortly give. Prompted by this outlook, Friedlander and Goldston [7] improved our bound of utility on E.R.H. from $x^{1/2+\epsilon}$ to $x^{1/2} \log^{5+\epsilon} x$ although they were partially forestalled by Vaughan, who somewhat similarly had obtained the better $Q = x^{1/2} \log^{3/2+\epsilon} x$ at about the time of our discoveries in [13] but who then laid the matter aside without proceeding to publication. Recently, however, in collaboration with Goldston, Vaughan has returned to the subject and published a proof of his bound with an account of associated topics [38].

We also have revisited earlier work and discovered that we were just a bit too clever in our application to our work in [14] of the properties of primes in small intervals. Had we used less sophisticated bounds in one instance, we would not only have materially simplified the treatment but would have obtained a lower bound for Q commensurate with that of Goldston and Vaughan's. An account of this revision will be given if a suitable opportunity arises.

6   Other Second Moments Containing E(x; a, k)

The information on the remainder $E(x; a, k)$ provided by Theorem 2 and its conditional versions is constrained amongst other things by the very wide domain of summation in $G(x, Q)$ and by the unchanging nature of the limit x for p as a and k vary. Therefore, with the object of obtaining partial confirmation of what the theorem presages, we seek to establish analogues of it in which either x is replaced by a number $x_{a,k}$ or the numbers a, k are subject to extra curtailment.

At one extreme, confining k to a single value in the summation, we wish to examine the sum

$$H(x, k) = \sum_{\substack{0 < a \leq k \\ (a,k)=1}} E^2(x; a, k),$$

for which on E.R.H. we had the Turán-Montgomery bound

$$H(x, k) = O(x \log^4 x) \qquad (k \leq x)$$

already stated in (2.5). This was developed by us [19] into

$$\sum_{\substack{0 < a \leq k \\ (a,k)=1}} \max_{1 \leq u \leq x} E^2(u; a, k) = O(x \log^4 x) \qquad (k \leq x) \qquad (6.1)$$

by an altogether more intricate argument that appealed to Selberg's work on the distribution of the zeros of Dirichlet's L-functions. Having validity for a range of k much wider than for most results of the type to which they belong, these bounds shed light on natural conjectures about $E(x; a, k)$

and are especially useful auxiliary tools in the pursuit of results such as the B.M. theorem in [14] or one to be mentioned. Yet they are probably wanting because by comparison with our bounds for $G(x, Q)$ they seem to contain a superfluous factor $\log^3 x$. We must therefore have recourse to other ideas to investigate whether $H(x, k)$ can be likened to the expected $x \log k$ as k veers away from 1, particularly when E.R.H. is not assumed.

Here Montgomery's ideas score over those used by others in connection with Theorem 2. Improved by minor modifications on which Croft [4] reported, his analysis of $H(x, k)$ gave an equation slightly more precise than

$$
\begin{aligned}
H(x, k) &= x \log k + O\left(x\sigma_{-1/2}(k)\right) + O\left(\frac{x^2}{\phi(k) \log^{2A} x}\right) + O\left(\sum_{\ell \leq x/k} |F(x, \ell k)|\right) \\
&= x \log k + O\left(x\sigma_{-1/2}(k)\right) + O\left(\frac{x^2}{\phi(k) \log^{2A} x}\right) + O(H_1(x, k)), \quad \text{say,}
\end{aligned}
$$

the genesis of which can be discerned from the synopsis at the beginning of §5. Since

$$
\sum_{Q < k \leq 2Q} H_1(x, k) = O\left(x \log^{-3A} x\right)
$$

by (5.5), we infer as in our paper [14] that, if $x \log^{-A} x < Q \leq \frac{1}{2}x$, then

$$
\begin{aligned}
H(x, k) &= x \log k + O\left(x\sigma_{-1/2}(k)\right) + O\left(x \log^{-A} x\right) \\
&= x \log k + O\left(x\sigma_{-1/2}(k)\right)
\end{aligned}
$$

for $Q < k \leq 2Q$ save for at most $O(Q \log^{-A} Q)$ exceptional values of k. Here the remainder term may be replaced by $O(x \log^\epsilon k)$ and actually more ambitiously by an explicit term added to a remainder term of lower order of magnitude; the latter refinement was confirmed and implemented by Friedlander and Goldston [7]. Thus, in summary, we can loosely say that almost always $H(x, k) \sim x \log k$ for values of k closer to x than $x \log^{-A} x$.

A deficiency in this sort of conclusion is that it lacks an absolute character in regard to the moduli k since it does not assert that those k for which $H(x, k) \sim x \log k$ can remain the same when x changes or, in other words, that the exceptional set of k can be chosen to be essentially independent of x. However, in a paper [16] we return to later, we remedied this omission by proving that almost all numbers k have the property that

$$
H(x, k) = x \log k + O\left(x \log^{1/2} k\right) \tag{6.2}
$$

for all x satisfying $k \leq x \leq k \log^A k$ and, indeed, on E.R.H. for all x satisfying $k \leq x \leq k^{5/4-\epsilon}$.

Significant progress was made when Friedlander and Goldston [7], viewing the matter from a different angle, obtained an estimate for *all* k in a given range by showing that

$$H(x,k) > \left(\frac{1}{2} - \epsilon\right) x \log k \qquad (6.3)$$

when

$$x \log^{-A} x < k \leq x. \qquad (6.4)$$

The source of their estimation lay in the properties of a surrogate prime number function $\Lambda_R(n)$, whose definition as later interpreted by us in [26] is tantamount to

$$\Lambda_R(n) = V(R) \sum_{\substack{d|n \\ d \leq R}} \lambda_d,$$

where

$$V(R) = \sum_{d \leq R} \frac{\mu^2(d)}{\phi(d)} = \log R + O(1),$$

and where the real numbers λ_d are chosen for square-free numbers $d \leq R$ as in Selberg's sieve method so that the quadratic form

$$\sum_{d_1, d_2 \leq R} \frac{\lambda_{d_1} \lambda_{d_2}}{[d_1, d_2]}$$

has a conditional minimum subject to the condition $\lambda_1 = 1$. Being likely from its form to mimic $\Lambda(n)$, the function $\Lambda_R(n)$ is used in the formation of sums

$$\psi_R(x; a, k) = \sum_{\substack{n \leq x \\ n \equiv a \bmod k}} \Lambda_R(n)$$

that can be indirectly compared with the sums $\theta(x; a, k)$ through the inequality

$$\theta^2(x; a, k) \geq 2\theta(x; a, k)\psi_R(x; a, k) - \psi_R^2(x; a, k).$$

Accordingly, as the first term in (5.1) is not less that

$$2 \sum_{0 < a \leq k} \theta(x; a, k)\psi_R(x; a, k) - \sum_{0 < a \leq k} \psi_R^2(x; a, k),$$

we meet two substitute sums that are easier to handle than their progenitor. Indeed, using the Bombieri-Vinogrodov theorem in combination with simply derived extensions of familiar properties of the numbers λ_d, we arrive at satisfactory estimates provided that R is not too large, whence (6.3) appears after all terms implicit in the total procedure are taken into account. Also, on adding the force of E.R.H. to this argument, these authors obtained the inequality

$$H(x,k) > \left(2 - \frac{3}{2\alpha} - \epsilon\right) x \log k \qquad (k = x^\alpha) \qquad (6.5)$$

that provides conditional information whenever $x^{3/4+\epsilon} < k \leq x$.

The barrier of the type $k > x \log^{-A} x$ in the applicability of (6.3) was caused by the indirect appeal to (2.3) in the proof. However, by the use of an additional argument that nullified any deleterious effect from Siegel zeros, we showed that this impediment could be relaxed to $k > x e^{-A\sqrt{\log x}}$ and hinted that a further improvement might well be possible [26]. We also on this occasion sharpened the conditional (6.5) to

$$H(x,k) > \left(\frac{3}{2} - \frac{1}{\alpha} - \epsilon\right) x \log k \qquad (k = x^\alpha)$$

by using our estimate (6.1), as did Friedlander and Goldston simultaneously and independently by an identical argument [8].

Going back to theorems that are more recognisable as being of B.M. type, we should report first that our extension of Montgomery's method in [16] also gave the formulae

$$\sum_{k \leq Q} \max_{1 \leq y \leq x} \sum_{\substack{0 < a \leq k \\ (a,k)=1}} E^2(y; a, k) - Qx \log Q$$

$$= \begin{cases} O(Qx) + O\left(x^2 \log^{-A} x\right), & \text{unconditionally,} \\ O(Qx) + O\left(x^{9/5+\epsilon}\right), & \text{on E.R.H.,} \end{cases}$$

which provide a partial answer to the question raised in §3 about Uchiyama's bound (3.9). Also, restricting the domain of the variables of summation in a different way, we [25] have investigated the sum

$$\sum_{k \leq Q} \sum_{\substack{0 < a \leq \lambda k \\ (a,k)=1}} E^2(x; a, k) \qquad (\lambda \leq 1)$$

and established an extension of the B.M. theorem by demonstrating that it equals

$$x\lambda \left(Q \log Q + b_1 Q\right) + O\left(\lambda Q^{5/4} x^{3/4}\right) + O\left(Q^2 \lambda \log x\right) + O\left(x^2 \log^{-A} x\right)$$

for $Q \leq x$ and any $\lambda \leq 1$. In its origin, the method used is more akin to our treatment of Theorem 2 than that of Montgomery's; however, in execution, it involves substantial new procedures.

Before we quit our survey of results in the genre of Theorem 1 and Theorem 2, we should call to mind the reason given after the statement of the former as to why they cannot currently be fully informative in ranges below $Q = x \log^{-A} x$. However, no like consideration debars the verification of useful lower bounds in a wider span of k, since awkward values of $E(x; a, k)$ can be eliminated from $G(x, Q)$ in such a quest; already, in fact, it is easily deduced from our improved version of Friedlander and Goldston's (6.1) that

$$G(x, Q) > \left(\frac{1}{2} - \epsilon\right) Qx \log Q \qquad \left(Q > xe^{-A\sqrt{\log x}}\right). \qquad (6.6)$$

Before the antecedent of this implication was available, Liu [29] had already proved that

$$G(x, Q) > \left(\frac{1}{4} - \epsilon\right) Qx \log Q \qquad \left(Q > x \exp\left(-\log^{3/5-\epsilon} x\right)\right), \qquad (6.7)$$

which is better than (6.6) in terms of range but worse in terms of the multiplier of $Qx \log Q$. Beyond this, Perelli [34] showed that there are certain positive constants A and η with the property that

$$G(x, Q) > AQx \log Q \qquad \left(Q > x^{1-\eta}\right),$$

being the first to breach the limit $Q > x^{1-\epsilon}$ for problems of this sort.

It is of course desirable to replace the coefficient of $Qx \log Q$ in these bounds by $1 - \epsilon$ without weakening the range of k. This we largely achieved in [27] when we improved Liu's result above to

$$G(x, Q) > (1 - \epsilon)Qx \log Q \qquad \left(Q > x \exp\left(-\log^{3/5-\epsilon} x\right)\right), \qquad (6.8)$$

by associating our method of proving Theorem 2 with the use of the function $\Lambda_R(n)$. Yet, the application of $\Lambda_R(n)$ was more radical than before because a reappraisal of part of Selberg's sieve method was needed in order to accommodate the larger values of R involved. Also on the assumption that the Riemann zeta function has no zeros with real part exceeding $3/4$, we proved that

$$G(x, Q) > \left(2 - \frac{1}{\alpha} - \epsilon\right) Qx \log Q \qquad (6.9)$$

when $Q = x^\alpha$ and $1/2 < \alpha \leq 1$; this result, which is interesting if only because it involves a hypothesis weaker than E.R.H., depends on extra ideas including one related to the large sieve.

7   The Sum H(x, k) again and the Distribution of E(x; a, k)

The reader will have noticed the apparent anomaly whereby the known forms of the prime number theorem mentioned in §2 are only valid for smaller values of the common difference k, whereas the results springing from the B.D.H. and B.M. theorems are usually only fully significant for the larger values of k. The most striking illustration of this contrast is perhaps furnished by the conditional theorems available on E.R.H., which are valid for $k \leq x^{1/2-\epsilon}$ in the former instance and essentially $k > x^{1/2+\epsilon}$ in the latter. Indeed, the only important overlap in the two types of ranges of k that has so far occurred has been in connection with the Turán-Montgomery estimate (2.5) and its generalization in (6.1), which as already stated in §5 are almost surely of imperfect sharpness. It is therefore desirable to elicit further theorems that shall illuminate the behaviour of $G(x, Q)$ or $H(x, k)$ for smaller values of Q or k.

Accordingly, emboldened to supply something to bridge this gap, we considered the integral

$$\int_2^x \frac{H(u, k)}{u^2} du$$

in 1975 [18] and showed on E.R.H. that it is $O(\log x \log 2k)$ for $k \leq x$, a result that was certainly consistent with our previously implicitly stated conjecture that $H(x, k) = O(x \log k)$ as soon as k tends to infinity; moreover, on assuming in addition that the zeros of every Dirichlet's L-function formed with a primitive character are simple, we sharpened one aspect of this estimate to

$$\lim_{k \to \infty} \lim_{x \to \infty} \frac{1}{\log x \log k} \int_2^x \frac{H(u, k)}{u^2} du = 1$$

and fortified our expectation of the equivalence

$$H(x, k) \sim x \log k \tag{7.1}$$

that had stemmed from (6.2).

Although the method used is a development of that given by Landau for the mean

$$\int_0^x \frac{(\psi(u) - u)^2}{u^2} du,$$

considerable refinements and modifications in the analysis are needed to achieve uniformity in k and, as for (6.1) above, it is necessary to use Selberg's theory of the distribution of the zeros of L-functions.

Cognate ideas serve to crystallize some notions we have loosely expressed from time to time about $E(x; a, k)$ and its distribution. What we have obtained so far is compatible with the likelihood that $E(x; a, k)$ is usually of size about $(x/\phi(k))^{1/2} \log^{1/2} k$, whatever be the stock of triplets (x, a, k) over which it is taken. Furthermore, since our belief in (7.1) would mean that

$$\frac{E(x; a, k)}{(x/\phi(k))^{1/2} \log^{1/2} k}$$

had a dispersion roughly equal to 1, we are further invited to guess that it has a distribution function of constant form in many underlying circumstances. To progress from these speculations we therefore identified one situation in which a substantial extension of the last method would yield information under the twin assumption of E.R.H. and the linear independence of the imaginary parts of the zeros of the Dedekind zeta-function taken over the (cyclotomic) field $\mathbb{Q}(\sqrt[k]{1})$. On the supposition that a and k were given co-prime integers, the random variable

$$T = \frac{E(e^t; a, k)}{e^{(1/2)t} \log^{1/2} k / \phi^{1/2}(k)} \qquad (t > 0)$$

was considered and shown to have a distribution function $F(y)$ that was independent of a for given k, it then being further demonstrated that

$$\lim_{k \to \infty} F(y) = \Phi(y)$$

where $\Phi(y)$ is the distribution function of the normal distribution $N(0, 1)$ [20].

Let now

$$\mu_r = \begin{cases} 1.3 \ldots (2\nu - 1), & \text{if } r = 2\nu \\ 0, & \text{if } r \text{ is odd,} \end{cases}$$

denote the rth moment of $N(0, 1)$. Then, as stated in [20], the suggested implication that

$$\lim_{k \to \infty} \lim_{v \to \infty} \frac{\phi^{1/2}(k)}{\log v \log^{(1/2)r} k} \int_1^v \frac{E^r(x; a, k) dx}{x^{(1/2)r+1}} = \mu_r$$

is confirmed on the same hypothesis as before by varying the procedure in a less elegant direction. Thus, in the manner of [20], we might be led to believe in the generalized B.M. theorem enunciated in the following

Conjecture. *For some appropriate function X of x such that $X > x \log^{-A} x$, we have, as $x \to \infty$,*

$$\sum_{k \leq Q} \phi^{(1/2)r-1}(k) \sum_{\substack{0 < a \leq k \\ (a,k)=1}} E^r(x; a, k) = (\mu_r + o(1)) Q x^{(1/2)r} \log^{(1/2)r} Q$$

when $X < Q = o(x/\log x)$.

Couched in terms of a factor $\phi^{(1/2)r-1}(k)$ between the inner and outer summation symbols, the conjecture can no doubt be appropriately modified by the insertion of other factors such as $k^{(1/2)r-1}$ or $\phi^{(1/2)r-1/2}(k)$. Anyhow, whatever be the precise form of the conjecture we care to take, the cases $r = 1$ or 2 are trivial or correspond to the B.M. theorem, respectively, so that the real difficulties lying athwart our road begin when $r \geq 3$. For the next case $r = 3$, the author [22] resolved the situation when he proved that

$$\sum_{k \leq Q} \phi(k) \sum_{\substack{0 < a \leq k \\ (a,k)=1}} E^3(x; a, k) = o\left(Q^{3/2} X^{3/2} \log^{3/2} x\right) + O\left(x^3 \log^{-A} x\right)$$

when $Q = o(x/\log x)$ by what turned out to be an elaborate and finely tuned analysis. Beyond this, nothing is yet known although conditional treatments may well have a chance of success.

8   Analogues of Barban-Davenport-Halberstam Type Theorems for Other Sequences

Not surprisingly the promulgation of the B.D.H. and B.M. theorems led to a search for individual sequences that answered to analogues of these results. Being in some ways the closest companions of the primes and having been much studied, the square-free numbers denoted here by s were probably the first to come under this sort of scrutiny, especially as there was already a known asymptotic formula of the type

$$S(x; a, k) = \sum_{\substack{0 < s \leq x \\ s \equiv a \bmod k}} 1 \sim f(a, k)x \tag{8.1}$$

valid in the lower reaches of k. For example, in terms of the notation

$$E(x; a, k) = S(x; a, k) - f(a, k)x \tag{8.2}$$

$$H(x, k) = \sum_{0 < a \leq k} E^2(x; a, k), \qquad G(x, Q) = \sum_{k \leq Q} H(x, k) \tag{8.3}$$

that reflects our previous usage, Orr [33] and Warlimont [41] obtained bounds of the type

$$G(x, Q) = O(Qx)$$

in ranges of the type $x^{1-\alpha} < Q \leq x$, while the latter [42] later improved these to

$$G(x, Q) = O\left(Q^{3/2} x^{1/2}\right)$$

in a similar range. That this was best possible as an upper bound was then substantiated by Croft [4], who obtained the B.M. type asymptotic formula

$$G(x, Q) \sim B_2 Q^{3/2} x^{1/2} \tag{8.4}$$

for $x^{2/3+\epsilon} < Q \leq x$.

The author, however, felt that, instead of considering individual sequences, we would find it more rewarding in the long run to identify the properties a sequence must have in order that it should be the subject of a B.D.H. type theorem. In setting up this investigation we confined our attention to sequences of essentially positive density in the interests of simplicity, believing that any principles established would be applicable to some degree to other sequences. First, if now s denotes a general member of the sequence and if then the meaning of the previous notation in (8.2) and (8.3) is accordingly extended, it is clear there must be an asymptotic formula of type (8.1) since otherwise there could neither be a sensible way of defining $G(x, Q)$ nor any useful estimate for it however $f(a, k)$ were chosen within it. Then, mildly quantifying this condition as

$$S(x; a, k) = f(a, k)x + O\left(\Delta_k(x)\right) \tag{8.5}$$

where $\Delta_k(x) = O(x \log^{-A} x)$ and adding the not unnatural requirement that

$$f(a, k) = g(k, (a, k)) \tag{8.6}$$

(i.e. that $f(a, k)$ depends only on k and the h.c.f. of a and k, to obtain a situation not unlike that afforded by the prime numbers, we showed in [15] that

$$G(x, Q) = O(Qx) + O\left(x^2 \log^{-A} x\right). \tag{8.7}$$

Proved originally through the use of character sums and the large sieve in a manner reminiscent of our description of the proof of Theorem 1, this proposition later received a superior and more illuminating treatment by way of exponential sums under the impetus of our work on certain aspects of Waring's problem [21].

A few years ago we returned to the subject to investigate the circumstances in which (8.7) could be converted to a B.M. type result and proved that [23]

$$G(x, Q) = BQx + O\left(x^2 \log^{-A} x\right) \tag{8.8}$$

where it was further assumed that $g(k, k)$ in (8.6) was a non-zero multiple C of a multiplicative function $\psi(k)$. Of particular importance was the non-negative constant B, the value of which depended only on the form of $f(a, k)$—or, equivalently, C and $\psi(k)$—and was seen to be sometimes positive and sometimes zero. Whereas a genuine asymptotic formula answered to the former situation, a further study of the Dirichlet's series associated with $\psi(k)$ would in the opposite situation sometimes yield a formula similar to the one that was established for square-free numbers (for which our assumed conditions are in place). Nevertheless our work failed to provide a satisfactory criterion for determining the status of B.

A brief word on the condition (8.6) is appropriate before our account unfolds further. Although its presence was felt to be natural, it also seemed hard to dispense with in any of the treatments yet mentioned. This was bound up with a number of associated reasons, the most potent of which is that $H(x, k)$ is no longer generally a dispersion as in the case of the primes and that consequently the sum

$$\sum_{0 < a \leq k} f(a, k) S(x; a, k) \qquad (8.9)$$

occurring in the middle term of its expansion is a covariance that is not necessarily easy to estimate with an adequate remainder term. This difficulty, however, is abated when (8.6) is in place because then (8.9) can be expressed as a sum of dispersions taken over the divisors δ of k.

It was therefore not without some satisfaction that we shortly afterwards demonstrated [24] that all conditions subsidiary to the main (8.5) were in fact superfluous with the corollary that the existence of an asymptotic formula is strictly equivalent to that of a B.D.H. type theorem and even to a B.M. type theorem. The proofs involved were enhancements of the previous ones in which the large sieve method played additional rôles; as before, the B.D.H. type estimate (8.7) was first established as a stepping stone to the more precise (8.8), it therefore being important terminologically to keep both the descriptive titles of B.D.H. and B.M. even though the stronger result is often true in the same circumstances as the weaker.

Alongside the picture painted in [23] and [24], there is the important one portrayed almost simultaneously by Vaughan in two papers [39] and [40]. In the second of these, slightly generalizing the underlying circumstances by attaching weights to the members of the sequence but imposing the condition (8.6) that we removed in [24], he employed the circle method to obtain (8.8) with very accurate remainder terms that depended on a function $\Delta(x)$ appearing in lieu of $\Delta_k(x)$ in the analogue of (8.5). His method and insight also led, amongst other things, to a full understanding of the essence of the constant B in (8.8) that was lacking in our work.

This aspect of his findings we must now briefly describe because, apart from its obvious interest, it lies behind some further developments we shall adumbrate.

Let the formula (8.5) for $k = 1$ be expressed as

$$S(x) = S(x; 0, 1) = Cx + O(\Delta(x))$$

and suppose we attempt formally to calculate $S(x)$ through its representation as

$$\int_0^1 |F(\theta)|^2 \, d\theta \qquad (8.10)$$

where

$$F(\theta) = \sum_{s \leq x} e^{2\pi i \theta s}.$$

To this assignment the circle method of Hardy and Littlewood is partially applicable because (8.5) with (8.6) make possible an accurate rendering of $F(h/k)$ when $(h, k) = 1$ and k is small and hence of $F(\theta)$ when θ is the vicinity of h/k. Regarding such values of h and k as determining the so called major arcs, one can then obtain a contribution to $S(x)$ that is asymptotically equivalent to $\mathfrak{S}x$ because the singular series $\mathfrak{S}$ formally derived is convergent. The remaining portion of $S(x)$ being non-negative because the integrand in (8.10) is non-negative, one can express Vaughan's conclusion in the form[5]

$$B = C - \mathfrak{S},$$

namely, that the main term in a formula of type (8.8) is derived from the minor arcs when the circle method is directed at (8.10). Thus, in the case of the square-free numbers for which Croft's formula (8.4) expresses the situation, the minor arcs make no principal contribution (of order x). For clarity, we should add that this has been a picturesque description, the essential point being the connection between C and the unequivocal meaning of $\mathfrak{S}$. Finally, as shown in the paper at which we now arrive, this interpretation of B remains in place when stipulation (8.6) is removed as in [24].

Our last paper [28] to date stems from ideas similar to those expressed by Professor Montgomery (see [4]) when he stated that Croft's formula (8.4) for square-free numbers could be foreseen from the expected truth of the asymptotic formula

$$S(x) = S(x; 0, 1) = \frac{6x}{\pi^2} + O\left(x^{1/4+\epsilon}\right)$$

---

[5]In our work we obtained the same formula but failed to recognize its connection with a singular series.

for their sum function. We investigate what can be deduced about $H(x,k)$ and $G(x,Q)$ for large k and Q from our knowledge of an estimate $E(x;a,k) = O\left(\Delta_k(x)\right)$ for small values of k. Initially our enquiry might have been whether $G(x,Q)$ for large Q is roughly the same as the sum

$$\sum_{k \leq Q} k \Delta_k^2(x). \tag{8.11}$$

Yet, we would be over ambitious in making such an assertion without qualification because, as we have already implied, the formally dominant term in $G(x,Q)$ depends only on the shape of the function $f(a,k)$. In fact, since the addition of an appropriately thick rogue sequence of zero density to the given sequence would increase the size of $\Delta_k(x)$ without altering $f(a,k)$ and hence the main constituent in $G(x,Q)$, we must reduce our speculations to the possible existence of one sided inequalities between $G(x,Q)$ and (8.11). And we are reinforced in this view by (8.7) that stated that $G(x,Q) = O(Qx)$ for Q close to x even when $\Delta_k(x)$ is merely assured to be not larger than $O\left(x \log^{-A} x\right)$.

We have said enough to indicate that it would be illuminating to study this problem in the case where there is still usually no explicitly stated restriction on the form of $f(a,k)$ and where

$$\Delta_k(x) = O\left((x/k)^\alpha\right) \tag{8.12}$$

for

$$k \leq x^{1/2} \tag{8.13}$$

and $0 < \alpha < 1/2$. In these circumstances, affected by Vaughan's observations on the constant B, we combine in [28] the procedures of [24] with a slightly novel use of the circle method to come close to our conjecture by showing that

$$G(x,Q) = O\left(Q^{2-2\alpha} x^{2\alpha} \log^2 2x/Q\right) \tag{8.14}$$

for largish values of Q up to x; thus $B = 0$ in particular. Much else can also be seen to be true but here we must confine ourselves to the observation that

$$H(x,k) = O\left(k^{1-2\alpha} x^{2\alpha} (x/k)^\epsilon\right)$$

when (8.6) also holds and k is a prime number close to x.

Many consequential matters arise, some of which are taken up in the later part of the paper. There is, for instance, the question as to whether (8.14) is best possible since it would appear to contain a superfluous factor $\log^2 2x/Q$; here we show that this can expunged if we suppose the hypothesis (8.12) to hold in a range of k rather longer than (8.13). Similarly, we would like to know when (8.14) can be converted into an asymptotic formula with a main term of type $B_2 Q^{2-2\alpha} x^{2\alpha}$. Vaughan, indeed, showed in [40] that there is such a formula when the function $f(a,k) = g(k,(a,k))$ is of certain type, although it must be said it is neither clear what sequences conform to such a requirement nor how they can be characterised in terms of the size of $E(x;a,k)$ for small k. This aspect of the theory was therefore followed up in two directions. The first was to consider other circumstances in which an asymptotic formula held, the second being then to construct a sequence for which any such formula failed.

A scrutiny of what we have said with the help of [28] makes it manifest that there is at least a loose connection between the form of $f(a,k)$ and the size of $\Delta_k(x)$. Just what it constitutes is one of several obvious aspects of our work that require further elucidation.

We have been deliberately selective in our provision of references for work cited in the text, confining ourselves in the main to those that relate to the main theme of the survey.

9   Postscript

The value of this report would be diminished if we did not mention some developments that have taken place between the time it was written and its final preparation for the press. These relate to the last two paragraphs of §6 in which lower bounds for $G(x,Q)$ were discussed.

Owing to the history of the subject workers have tended to regard $E(x;a,k)$ and the associated difference

$$E^*(x;a,k) = \theta(x;a,k) - \frac{1}{\phi(k)} \sum_{\substack{p \le x \\ p \nmid k}} \log p$$

as being on a par and have therefore sometimes failed to appreciate the distinction between them. This is that, being a true dispersion, the sum

$$H^*(x,k) = \sum_{\substack{0 < a \le k \\ (a,k)=1}} E^{*2}(x;a,k)$$

bounds $G(x,k)$ from below in accordance with a well known theorem in elementary probability theory. Thus, in regard to problems involving lower bounds, it is the sum

$$G^*(x,Q) = \sum_{k \le Q} H^*(x,k)$$

that is of primary importance even though of course the other sum $G(x,Q)$ still retains its interest.

As a rescrutiny of Liu's paper reveals, it was his failure to remember this point that led him to deduce his lower bound for $G(x,Q)$ in the form (6.7), while actually he could have inferred it for the wider range

$$Q > x^{1-0(1/\log\log x)} \tag{9.1}$$

because of his earlier result for $G^*(x,Q)$. Consequently, both the inequalities

$$G(x,Q),\ G^*(x,Q) > \left(\frac{1}{4} - \epsilon\right) Qx\log Q \tag{9.2}$$

for the range (9.1) should be essentially ascribed to Liu. On the other hand, our inequality (6.8) was derived from the comparable result for $G^*(x,Q)$ by using the minimal property of the latter, which may therefore be added to the left side of the statement.

Recently we developed the method in [27] to show that the inequality (6.9)—previously proved on the strength of a quasi-Riemann hypothesis—was actually true unconditionally; this contains the desired multiplier of $Qx\log Q$ when $\alpha \to 1$ and provides bound of requisite size down to about $Q = x^{\frac{1}{2}}$. Since, however, the method did not extend to $G^*(x,Q)$, this left us confronted by the task of improving the unconditional bounds for this sum implied by our discussion. In response, by a radical reappraisal of Liu's method, we first proved that

$$G^*(x,Q) > \{A(\delta) - \epsilon\}Qx\log Q \qquad (Q = x^{1-\delta}),$$

where $A(\delta)$ is a decreasing function of δ tending to $5\pi^2/6$ as $\delta \to 0$, thus improving (9.2) for a longer range in which the multiplier $1/4$ is increased to a number not far short of the desired 1. Then later, leaning on ideas taken from Heath-Brown's new version of the circle method, we found that $A(\delta)$ could be replaced by a similar function whose limiting value as $\delta \to 0$ was actually 1. There is thus a satisfactory theory for the lower bound for $G^*(x,Q)$ when Q is of the form $x^{1-\delta}$ for small δ.

References

[1] M. B. Barban, *Analogues of the divisor problem of Titchmarsh*, Vestnik Leningrad. Univ. Ser. Mat. Meh. Astronom **18** (1963), 5–13.

[2] ———, *On the average error in the generalized prime number theorem*, Doklady Akademiya Nauk UZ SSR (1964), 5–7.

[3] E. Bombieri, *On the large sieve*, Mathematika **12** (1965), 201–225.

[4] M. J. Croft, *Square-free numbers in arithmetic progressions*, Proc. London Math. Soc. (3) **30** (1975), 143–159.

[5] S. Daniel, *On the sum of a square and a square of a prime*, Math. Proc. Cambridge Philos. Soc. **131** (2001), 1–22.

[6] H. Davenport and H. Halberstam, *Primes in arithmetic progressions*, Michigan Math. J. **13** (1966), 485–489.

[7] J. B. Friedlander and D. A. Goldston, *Variance of distribution of primes in residue classes*, Quart. J. Math. Oxford Ser. (2) **47** (1996), 313–336.

[8] ———, *Note on a variance in the distribution of primes*, Number theory in progress, Vol. 2 (Zakopane-Kościelisko, 1997), de Gruyter, Berlin, 1999, pp. 841–848.

[9] P. X. Gallagher, *The large sieve*, Mathematika **14** (1967), 14–20.

[10] D. A. Goldston and R. C. Vaughan, *On the Montgomery-Hooley asymptotic formula*, Sieve methods, exponential sums, and their applications in number theory (Cardiff, 1995), Cambridge Univ. Press, Cambridge, 1997, pp. 117–142.

[11] G. Greaves, *On the representation of a number in the form $x^2 + y^2 + p^2 + q^2$ where p, q are odd primes*, Acta Arith. **29** (1976), 257–274.

[12] C. Hooley, *The distribution of sequences in arithmetic progressions*, Proceedings of the International Congress of Mathematicians (Vancouver, B.C., 1974), Vol. 1, Canad. Math. Congress, Montreal, Que., 1975, pp. 357–364.

[13] ———, *On the Barban-Davenport-Halberstam theorem: I*, J. Reine Angew. Math. **274/275** (1975), 206–223.

[14] ———, *On the Barban-Davenport-Halberstam theorem: II*, J. London Math. Soc. (2) **9** (1975), 625–636.

[15] ———, *On the Barban-Davenport-Halberstam theorem: III*, J. London Math. Soc. (2) **10** (1975), 249–256.

[16] ———, *On the Barban-Davenport-Halberstam theorem: IV*, J. London Math. Soc. (2) **11** (1975), 399–407.

[17] ———, *Applications of sieve methods to the theory of numbers*, Cambridge University Press, Cambridge, 1976, Cambridge Tracts in Mathematics, No. 70.

[18] ———, *On the Barban-Davenport-Halberstam theorem: V*, Proc. London Math. Soc. (3) **33** (1976), 535–548.

[19] ———, *On the Barban-Davenport-Halberstam theorem: VI*, J. London Math. Soc. (2) **13** (1976), 57–64.

[20] ———, *On the Barban-Davenport-Halberstam theorem: VII*, J. London Math. Soc. (2) **16** (1977), 1–8.

[21] ———, *On a new approach to various problems of Waring's type*, Recent progress in analytic number theory, Vol. 1 (Durham, 1979), Academic Press, London, 1981, pp. 127–191.

[22] ———, *On the Barban-Davenport-Halberstam theorem: VIII*, J. Reine Angew. Math. **499** (1998), 1–46.

[23] ———, *On the Barban-Davenport-Halberstam theorem: IX*, Acta Arith. **83** (1998), 17–30.

[24] ———, *On the Barban-Davenport-Halberstam theorem: X*, Hardy-Ramanujan J. **21** (1998), 9 pp.

[25] ———, *On the Barban-Davenport-Halberstam theorem: XI*, Acta Arith. **91** (1999), 1–41.

[26] ———, *On the Barban-Davenport-Halberstam theorem: XII*, Number theory in progress, Vol. 2 (Zakopane-Kościelisko, 1997), de Gruyter, Berlin, 1999, pp. 893–910.

[27] ———, *On the Barban-Davenport-Halberstam theorem: XIII*, Acta Arith. **94** (2000), 53–86.

[28] ———, *On the Barban-Davenport-Halberstam theorem: XIV*, Acta Arith. **101** (2002), 247–292.

[29] H. Q. Liu, *Lower bounds for sums of Barban-Davenport-Halberstam type (supplement)*, Manuscripta Math. **87** (1995), 159–166.

[30] H. L. Montgomery, *Primes in arithmetic progressions*, Michigan Math. J. **17** (1970), 33–39.

[31] ———, *Topics in multiplicative number theory*, Springer-Verlag, Berlin, 1971, Lecture Notes in Mathematics, Vol. 227.

[32] _____, *Maximal variants of the large sieve*, J. Fac. Sci. Univ. Tokyo Sect. IA Math. **28** (1981), 805–812 (1982).

[33] R. C. Orr, *Remainder estimates for squarefree integers in arithmetic progression*, J. Number Theory **3** (1971), 474–497.

[34] A. Perelli, *The L^1 norm of certain exponential sums in number theory: a survey*, Rend. Sem. Mat. Univ. Politec. Torino **53** (1995), 405–418, Number theory, II (Rome, 1995).

[35] V. A. Plaksin, *Representation of numbers by the sum of four squares of integers, two of which are prime numbers*, Dokl. Akad. Nauk SSSR **257** (1981), 1064–1066.

[36] P. Shields, Ph.D. thesis, Cardiff, 1978.

[37] E. C. Titchmarsh, *A divisor problem*, Rendiconti del Circolo Matematico di Palermo **54** (1930), 414–429.

[38] S. Uchiyama, *The maximal large sieve*, Seminar on Modern Methods in Number Theory (Inst. Statist. Math., Tokyo, 1971), Paper No. 37, Inst. Statist. Math., Tokyo, 1971, p. 5.

[39] R. C. Vaughan, *On a variance associated with the distribution of general sequences in arithmetic progressions. I,*, R. Soc. Lond. Philos. Trans. Ser. A Math. Phys. Eng. Sci. **356** (1998), 781–791.

[40] _____, *On a variance associated with the distribution of general sequences in arithmetic progressions. II*, R. Soc. Lond. Philos. Trans. Ser. A Math. Phys. Eng. Sci. **356** (1998), 793–809.

[41] R. Warlimont, *On squarefree numbers in arthmetic progressions*, Monatsh. Math. **73** (1969), 433–448.

[42] _____, *Über die kleinsten quadratfreien Zahlen in arithmetischen Progressionen mit primen Differenzen*, J. Reine Angew. Math. **253** (1972), 19–23.

Integer Points, Exponential Sums and the Riemann Zeta Function

M. N. Huxley

1 Lattice Points

One of the problems at the origin of mathematics is to define and calculate the area of a plane region D. If parts of the boundary of D are curved, then this problem leads to the integral calculus. Archimedes' construction used two polygons, one inside D, the other containing D. After Descartes' coordinate system, it is natural to consider a zigzag polygon whose vertices are of the form $x = m\delta$, $y = n\delta$, where δ is the size of the smallest square of the printed grid on the graph paper. To find the area of the inner and outer polygons, one counts squares of side δ. This gives upper and lower bounds for the area. A better estimate is given by counting the square bounded by the four points $x = m\delta$, $(m + 1)\delta$, $y = n\delta$, $(n + 1)\delta$ if and only if the point $(m\delta, n\delta)$ lies in D, so the integer lattice point (m, n) lies in $\delta^{-1}D$.

This method links geometry, analysis and number theory. Voronoi saw that an error estimate can be found by approximating the curve by a polygon whose gradients are rational numbers a/q of small height (defined by $h(a/q) = \log \max(|a|, q)$ when $q \geq 1$, $(a, q) = 1$). Voronoi [33] and Sierpiński [31] used these approximating polygons for the divisor and circle problems, respectively, and van der Corput extended the method to a general smooth convex region in his thesis [5]. The error has a contribution from each gradient a/q, which depends on the continued fraction for a/q.

Pfeiffer considered lattice point problems by Fourier methods in the nineteenth century (see van der Corput [5] and Landau [28], [29]). Voronoi [34] gave formal Fourier expansions for the divisor and circle problems. The general convex region had to wait for Kendall [24], followed by Hlawka [14] for a general convex body in higher dimensions. They evaluated the terms of the series to a good approximation; the approximate values correspond to integer vectors normal to the curve or surface, with a small correction from the rest of the curve. These expansions are useful for mean square and omega results, as we heard in Tsang's lecture at this conference.

Van der Corput's less symmetric treatment [6], [7], [8], [9] was more successful. The discrepancy (number of lattice points minus the area of $\delta^{-1}D$) is calculated piecewise on arcs with local coordinates $y = g(x)$. The discrepancy contribution is $\sum \rho(g(m))$, where

$$\rho(t) = [t] - t + \frac{1}{2} \simeq \sum_{h \neq 0} \frac{e(ht)}{2\pi ih}$$

is the usual row-of-teeth function. This leads to the van der Corput sums $\sum e(F(m))$, which he estimated by an iterative method that was purely analytic, with no further input from number theory. There are parameters M, T with

$$m \asymp M, \quad |F^{(r)}(x)| \asymp T/M^r.$$

(The order of magnitude notation used in this paper is '$f = O(g)$' or '$f \ll g$' for '$g > 0$, $|f|/g$ is bounded', '$\asymp g$' for '$f > 0$, $g > 0$, f/g, g/f are both bounded'.)

Step A, Poisson summation, replaces $F(x)$ by its complementary function $G(y)$, for which $F'(x)$, $G'(y)$ are inverse functions, and replaces the size parameter M by T/M. Step B, Weyl differencing, is a mean-to-max step, estimating the maximum of one sum in terms of the mean square of another sum. It introduces a new integer summand $h \ll H$, replaces $F(x)$ by $F(x + h) - F(x) \simeq hF'(x)$, and replaces the parameter T by hT/M. These steps are used in succession to reduce the parameters M and T until the resulting sums are short enough to estimate trivially. Each B step introduces extra integer variables. The multidimensional van der Corput iteration (see Graham and Kolesnik [12]) also uses A and B steps with respect to these new variables. However the errors from each step must be summed trivially over the other variables. This limits the power of the method.

2   Spacing Problems

Although van der Corput's mean-to-max step is purely analytic, other mean-to-max lemmas are useful in analytic number theory. They often lead to a spacing problem, to show that the sums whose mean square is estimated are sufficiently distinct. The sums are labelled by an integer parameter, n say, and we require different values of n to give sums which are non-overlapping or different in some sense. Often, as in Burgess' estimates for character sums [4], some overlapping does occur, and we must count the number of pairs of integers m, n for which the sums overlap. In van der Corput's Step A there is trouble only if $h = m - n$ is zero, giving trivial terms in the upper bound. A useful idea for spacing problems is to show that if the sums labelled by m and n overlap, then there are two integers a and b, constructed from m and n, which satisfy an equation $b = h(a)$ approximately, so that the integer point (a, b) is close to the curve $y = h(x)$. The spacing problem is translated into counting the integer points (a, b) in a certain plane region bounded above and below by arcs of the curves $y = h(x) \pm \delta$, where δ (which can be zero) measures the 'closeness'. If the region for (a, b) is bounded, then there is a nice geometric idea: given $R \geq d + 1$ integer points in d-dimensional space, then either they lie on a $(d-1)$-dimensional hyperplane, or their convex hull has volume at least $(R - d)/d!$.

Estimating the gap between square-free numbers is rapidly reduced to a spacing problem of points-close-to-a-curve type. If no number in $N \leq n \leq N + H$ is square-free, then many of them must be divisible by the square of some large prime. We need a non-trivial bound for the number of solutions of $N \leq p^2 q \leq N + H$, where p runs through large primes. The integer point (p, q) lies close to the curve $x^2 y = N$.

The expected number of lattice points (m, n) with m in an interval of length M, $|n - f(m)| \leq \delta$, is $2\delta M$ (with a possible endpoint correction). For δ large the Fourier method using the series for $\rho(t)$, cut off at $H \asymp 1/\delta$, gives $2\delta M$ plus an error term. The error term increases with H, so there is an optimal value $\delta = \delta_0$ at which both terms are equal. For $\delta < \delta_0$ it is better to count solutions of the weaker inequality $|n - f(m)| \leq \delta_0$. This is partly explained by the possibility of solutions with $\delta = 0$. The curve $y = \sqrt{x}$, $M \leq m \leq 2M$, is suitable for van der Corput's method, but goes through about $(\sqrt{2} - 1)\sqrt{M}$ integer points, so we cannot expect an estimate that tends to zero as $\delta \to 0$.

There is an elementary method that gives upper bounds, not asymptotic formulae, but works better when δ is small. It is built around a determinant formed from the coordinates of r points on a curve $y = f(x)$, which is equal to the Vandermonde determinant of the n values of x, multiplied by $f^{(r-1)}(\xi)/(r - 1)!$. For $r \geq 2$ the derivative is small, but the determinant is approximately an integer when the n points are approximately integer points. In the simplest case $n = 3$, the determinant formed from three integer points P, Q and R is twice the area of the triangle PQR. We call P, Q, R a major triplet if the determinant is zero, so PQR is a straight line, and a minor triplet otherwise. A major arc is a region of the curve where all the integer points close to it lie on the same straight line. Major arcs cannot exist if δ is very small. A minor arc is a region where no three consecutive integer points close to the curve lie on a straight line. The Vandermonde determinant is bounded below, so there cannot be three integer points very close together on a minor arc. Major arcs have rational gradients of small height, and they can be enumerated (Branton and Sargos [3]).

Swinnerton-Dyer's refinement [32] (for the case $\delta = 0$, but the method can be made to work for $\delta \neq 0$, as in Huxley [17]), uses a 4×4 determinant and ingenious divisibility arguments to show that the 3×3 determinants are usually greater than one. The argument is a mean-to-max, considering the mean fourth power of the number of points close to a curve in a short interval.

Bombieri and Pila [2] use high order determinants and intersection theory to give good estimates for the number of points on a curve ($\delta = 0$). The upper bounds are now ineffective and non-uniform, unless the curve is algebraic. The method also works for δ extremely close to 0.

Points-close-to-curve problems often have $m \asymp M$, $n \asymp N$, where N is larger than M. There is an iteration analogous to that of van der Corput, with $T = MN$. Step A, interchanging m and n, replaces $f(x)$ by its inverse function, and replaces M by T/M. Step B, differencing, in its simplest form, introduces a new integer variable $h \leq H$, replaces $f(x)$ by $f(x+h) - f(x) \simeq hf'(x)$, and T by hT/M. In general Step B has r integers $h_1, \ldots, h_r$. We select the integer points (m, n) for which the r points $(m + h_i, n_i)$ are also integer points close to the curve for some $n_1, \ldots, n_r$, and use the $(r + 1) \times (r + 1)$ determinant to construct an r-th differenced function $f[x; h_1, \ldots, h_r]$. A related method (see Filaseta and Trifonov [10]) is to write down the $(r + 1) \times (r + 1)$ determinant of $r + 1$ consecutive integer points, and use it to define major and minor arcs analogously. The major arcs correspond to polynomials in $Q[x]$ of degree at most $r - 1$.

More complicated spacing problems give points close to a curve $y = f(x)$ where one or both of x and y is a rational number of small height, not necessarily an integer (Huxley [15], [18], [20]). The determinant method can be applied, and the major arcs correspond to rational functions in the latter two papers. Konyagin [25] has an interesting alternative approach in which the approximation on a major arc is found by Dirichlet's pigeon-hole principle, not constructed explicitly. No Swinnerton-Dyer refinement has been accomplished in these problems.

3 The Bombieri-Iwaniec-Mozzochi Method

This method was introduced by Bombieri and Iwaniec [1] for the van der Corput sum $\sum e(F(m))$, and adapted by Iwaniec and Mozzochi [23] for the lattice point sum $\sum \rho(g(m))$ by way of the double exponential sum $\sum \sum e(hg(m))$ over h and m, and by Heath-Brown and Huxley [13] for the double sum

$$\sum \sum e(F(m + h) - F(m - h)), \tag{3.1}$$

which is related to the short interval mean square of a van der Corput sum. The method is most easily motivated for the lattice point case, with a domain D bounded by a smooth curve $y = g(x)$. The size parameters are H, the cut-off in the Fourier series for $\rho(t)$, M, the range for x, and T, with the dimensions of area, with $y \ll T/M$. They correspond to the T and M in the discussion above of the van der Corput exponential sum. The curve is divided up into arcs, the Farey arcs, by the vertices of a Voronoi polygon, labelled by the rational gradient a/q of the polygon side. The discrepancy contribution of each Farey arc is expressed as an exponential sum as in van der Corput, which is transformed by Poisson summation in both h and m to get new integer variables k and ℓ. If the Farey arcs correspond to intervals in x of length $\asymp N$, then the lengths K and L of the ranges for k and ℓ are proportional to the denominator q. When a/q has small height, the major arc case, then we get an estimate at once. Other Farey arcs are minor arcs. The denominator q has normal order R, where

$$N R^2 \asymp M^3/T.$$

Larger denominators occur for Farey arcs close to major arcs of small height. In this notation the ranges for k and ℓ have sizes

$$K \asymp Nq/R^2, \quad L \asymp Hq/R^2. \tag{3.2}$$

The estimate for major arcs is so good that the major arcs can be taken longer than N. Their contribution becomes larger, but then there are no Farey arcs with q very large. We can also arrange that $q \ll R^2/H$ on major arcs, $q \gg R$ on minor arcs. If the major arcs are so long that there are no minor arcs ($R \asymp H$), then we recover the simplest van der Corput bound for the discrepancy, $O(T^{1/3+\epsilon})$.

The savings come from a mean-to-max argument on the minor arcs, using the large sieve for the exponential sum bilinear form

$$E = \sum\sum a(k)b(j)e\left(\mathbf{x}^{(\mathbf{k})} \cdot \mathbf{y}^{(j)}\right), \tag{3.3}$$

which comes from the fourth power of the transformed exponential sum over k and ℓ, so there is a formal analogy with the Swinnerton-Dyer refinement for integer points close to curves. There are ingenious technical devices to remove smaller order terms, and to make the ranges for k and ℓ the same for all Farey arcs with q in a range $Q \leq q \leq 2Q$, 'but that's not important right now'. The multi-index $\mathbf{k} = (k_1, \ell_1, k_2, \ell_2)$ consists of the summands k and ℓ from two factors. The vectors $\mathbf{x}$ and $\mathbf{y}$ have four entries, from the four leading terms in the exponent, with

$$\mathbf{x}^{(\mathbf{k})} = \left(\ell_1 k_1 - \ell_2 k_2, \ell_1 \sqrt{k_1} - \ell_2 \sqrt{k_2}, \ell_1 - \ell_2, \ell_1/\sqrt{k_1} - \ell_2/\sqrt{k_2}\right).$$

The index j runs through the minor arcs, and the **y** vectors are constructed from the approximation to $g(x)$ on the minor arc. The **x** vectors in Heath-Brown and Huxley [13] are similar but more complicated. For the single exponential sum the **x** vectors are

$$\left(\sum k_i^2, \sum k_i^{3/2}, \sum k_i, \sum k_i^{1/2}\right). \tag{3.4}$$

The **y** vectors are essentially the same in all three cases, with $F'(x)$ replacing $g(x)$. The Large Sieve inequality (Huxley [16, Lemma 8.4.1]) says

$$|E|^2 = O\left(\frac{ABKL^4NR^2V}{Q^3}\right) = O\left(\frac{ABH^4N^2Q^2V}{R^8}\right),$$

where $V \geq 1$ is a free parameter, and A is the sum of $|a(\mathbf{k})a(\mathbf{k'})|$ taken over a 'neighbourhood of the diagonal', all pairs of vectors $\mathbf{x}^{(\mathbf{k})}$, $\mathbf{x}^{(\mathbf{k'})}$ whose difference lies in a box of size determined by the ranges for the components of the **y** vectors. The factor B is analogous, but the size of the box depends on V as well as on the ranges for the **x** vectors.

The coefficients $a(\mathbf{k})$ are bounded, so we estimate A by counting pairs of vectors in a neighbourhood of the diagonal. The trivial solutions have $\mathbf{k'}$ obtained from $\mathbf{k}$ by permuting the entries, and the First Spacing Problem is to show that most solutions of the inequalities are trivial. This is what Heath-Brown calls a paucity problem. Theorem 13.2.4 of Huxley [16] gives

$$A = O\left(KL^3 + K^2L^2\log K + \eta K^3L^2\log K\right), \tag{3.5}$$

where $\eta \asymp R^2/HN$, so $\eta K \asymp Q/H$, a satisfactory result. For the double sum (3.1) Lemma 13.3.1 on the next page of Huxley [16] gives

$$A = O\left(KL^4 + K^2L^2\log K + \eta K^3L^2\log K\right), \tag{3.6}$$

where the first term is bigger than we want. For the single exponential sum we can vary the number of summands $i = 1, \ldots, r$ in (3.4). We indicate this as $A(r)$. For reasonable Q we have

$$A(3) = O(K^3\log K),$$
$$A(4) = O(K^4 + \eta K^5\log^3 K),$$
$$A(5) = O(\eta K^7);$$

see Huxley [16, Lemma 17.1.1] and Watt [35]. Here $\eta \asymp R^2/N^2$, so $\eta K \asymp Q/N$. We conjecture that $A(5) = O(K^5)$ and $A(6) = O(K^6)$ for relevant values of η, implying better bounds for the exponential sum.

4   The Second Spacing Problem

The **y** vectors are indexed by minor arcs of the boundary curve of the domain D. The difference of two **y** vectors is close to the diagonal when the two Farey arcs can be superposed (to the required accuracy) by an affine map

$$(x'y') = \begin{pmatrix} a & b \\ c & d \end{pmatrix} \begin{pmatrix} x \\ y \end{pmatrix} + \begin{pmatrix} e \\ f \end{pmatrix} \tag{4.1}$$

with a, b, c, d, e, f integers, $ad - bc = 1$. If we fix the 'magic matrix' $\begin{pmatrix} a & b \\ c & d \end{pmatrix}$, then such coincidences (which may persist for a block of consecutive Farey arcs) correspond to integer points close to some curve in a dual space, called the 'resonance curve' in Huxley [16]. There is a choice of magnification in defining the resonance curve, given by a matrix $\begin{pmatrix} A & B \\ C & D \end{pmatrix}$ acting on the gradients a/q, with A, B, C, D integers, $AD - BC = 1$. A beautiful result is that the resonance curve is magnified by an affine map

$$(x' \quad y') = (x \quad y) \begin{pmatrix} A & B \\ C & D \end{pmatrix} + (\xi \quad \eta),$$

where ξ and η are very close to integers. Professor Coates approves this appearance of functoriality in analytic number theory. This is quite hard to prove; approximation theory wants to be linear, but the Poisson summation makes everything non-linear.

The estimates in the Second Spacing Problem are more complicated. There are $O(M/N)$ minor arcs, of which $O(MR^2/NQ^2)$ have q in a range $Q \leq q \leq 2Q$. The coincidence of two four-dimensional **y** vectors is measured by four numbers

$$\Delta_1 \asymp \frac{R^4}{HNQ^2V}, \qquad \Delta_2 \asymp \frac{R^2}{HN}, \qquad \Delta_3 \asymp \frac{R^2}{HQ}, \qquad \Delta_4 \asymp \frac{Q}{H},$$

and the probabilistic estimate for the number of consecutive pairs is

$$B = O\left(\Delta_1 \Delta_2 \Delta_3 \Delta_4 \frac{M^2Q^2}{N^2R^2}\right).$$

The parameter Δ_1 bounds the entries of the magic matrix in (4.1) by $c = O(\Delta_1 Q^2)$. Taking V larger makes the resonance curves longer. The parameter Δ_2 measures the distortion in area under the affine map (4.1), and Δ_3 and Δ_4 measure the shift in position.

The Second Spacing Problem is not a paucity problem: for good choices of the parameters N, R and V, the non-trivial (off-diagonal) coincidences dominate. Without using resonance curves, Bombieri and Iwaniec [1] got

$$B = O\left(\Delta_1(\Delta_1 + \Delta_2)\frac{M^2Q^2}{N^2R^2}\right) \tag{4.2}$$

for the special case when $g(x) = C/x$ with C constant. Huxley and Watt [22] generalised (4.2). It would be useful to replace the factor $\Delta_1 + \Delta_2$ by Δ_2.

The resonance curves are constructed by differential equations. They have endpoints where the gradient is zero and minus infinity, and usually a cusp in between. In Huxley [16] we chose the parameters so that the resonance curves had bounded length, and the number of integer points close to a resonance curve was also bounded. For further progress we assume an hypothesis about the number of integer points close to a smooth plane curve.

Hypothesis $H(\kappa, \lambda)$. *When $F(x)$ is a bounded function, three times continuously differentiable on an interval containing the unit interval $[1, 2]$, with $F''(x) \neq 0$, $F^{(3)}(x) \neq 0$, when M and N are real parameters with*

$$M \geq 2, \qquad \sqrt{M} \leq N \leq M^2,$$

and $f(x) = NF(x/M)$, and when δ is real with $0 \leq \delta \leq 1/2$, then R, the number of integer points (m, n) with

$$|n - f(m)| \leq \delta, \qquad M \leq m \leq 2M,$$

satisfies

$$R = O\left(\delta M + (MN)^\kappa (\log MN)^\lambda\right).$$

The implied constant depends on $f(x)$ only by way of the upper and lower bounds for the derivatives of $F(x)$.

Using Hypothesis $H(\kappa, \lambda)$ we get, in the dominant case, with suitable choices of N, R, and V,

$$B = O\left(\frac{M^2 R^4 (\log N)^\lambda}{H^2 N^4 V_0^{2/3} V}\right),$$

which corresponds to a saving by $V_0^{2/3}$ over (3.6). Here

$$V_0 = \left(\frac{H}{R}\right)^{6\kappa/(9\kappa - 1)}. \tag{4.3}$$

If $\alpha = (\log M)/(\log T)$ is near $1/2$, then we take $V = V_0$ (plan A). If $\alpha \leq (67\kappa - 6)/(156\kappa - 14)$, then we choose V so that all magic matrices are upper triangular (plan B). If $\alpha \geq (89\kappa - 8)/(156\kappa - 14)$, then we choose V so that all magic matrices are lower triangular (plan C).

The hypothesis $H(\kappa, \lambda)$ cannot be used near the cusp or the endpoints. There are additional contributions from cusps of resonance curves, and the regions near the endpoints where the gradient of the resonance curve is very large or very small.

The main contributions to the Iwaniec-Mozzochi sum $\sum \sum e(hg(m))$ and to the exponential sum with a difference (3.1) are

$$O\left(\frac{MR^3}{NQ^2}\sqrt{\frac{|E|N}{M}}\log^2 N\right). \qquad (4.4)$$

The main contribution to the single exponential sum $\sum e(F(m))$ is

$$O\left(\frac{MR^3}{NQ^{3/2}}\left(\frac{|E|N}{M}\right)^{1/r}\log N\right),$$

where r is the number of summands in (3.4).

The full force of $H(\kappa, \lambda)$ is known only for the van der Corput values $\kappa = 1/3$, $\lambda = 0$. The results of Huxley [17] give the bound corresponding to $H(3/10, 1/10)$ for shorter ranges of N. By subdividing the major arcs according to whether the continued fraction for the gradient a/q has a large partial quotient early on, we get the bounds which would follow from $H(3/10, 57/140)$ for the exponential sum (Huxley [21]), and the lattice point problem (Huxley [19]). The new exponents in the classical problems are:

$$\text{Circle problem:} \quad \frac{131}{208} = 0.6298\ldots < \frac{46}{73} = 0.6301\ldots$$

$$\text{Divisor problem:} \quad \frac{131}{416} = 0.3149\ldots < \frac{23}{73} = 0.3151\ldots$$

$$\text{Lindelöf problem:} \quad \frac{32}{205} = 0.156098\ldots < \frac{89}{570} = 0.156140\ldots$$

The Bombieri-Iwaniec method is essentially restricted to sums involving a function of only one variable, because simultaneous approximation to several independent partial derivatives is not accurate enough. The only method is the van der Corput iteration in several variables (see Krätzel [26]), arranged, if possible, to end with an appeal to the Bombieri-Iwaniec-Mozzochi method (Krätzel and Nowak [27]). Other applications of the large sieve for exponential sums are found in Fouvry and Iwaniec [11] and Sargos [30].

5   Mean Values of Exponential Sums

We sketch the proof of new results on the short interval mean square of exponential sums and the mean square of the Riemann zeta function. The

Second Spacing Problem is the same as for the Iwaniec-Mozzochi sum $\sum\sum e(hg(m))$, with $g(x) = 2TF'(x/M)/M$, so the same choices of N, R and V are made as in Huxley [19]. The details of the argument sketched in section 3 are as in Huxley [16, Chapter 9]. There is an extra requirement (9.1.7):

$$H^3 \ll N^2 R, \tag{5.1}$$

and the bound in (3.6) for the First Spacing Problem is bigger than (3.5) by a factor

$$1 + O\left(\frac{L^2}{K}\right) = 1 + O\left(\frac{H^2 Q}{N R^2}\right) \tag{5.2}$$

for $Q \leq q \leq 2Q$. This factor is greater than 1 at the top end of the range for H. The factor Q in (5.2) causes the powers of Q in (4.4) to cancel, so that all ranges for Q contribute equally, and there is an extra logarithm factor in the upper bounds of Proposition 1 in the terms in (5.7), (5.9) and (5.11), and the bounds for H in (5.6), (5.8) and (5.10) have a term from (5.1). Otherwise Proposition 1 would have the same conclusions as the bounds for the Iwaniec-Mozzochi sum in Proposition 1 of Huxley [19].

Proposition 1. *Suppose that $H(\kappa, \lambda)$ holds for some κ, λ with $1/4 \leq \kappa \leq 1/3$, $\lambda \geq 0$. Let $F(x)$ be a real function four times continuously differentiable for $1 \leq x \leq 2$, and let $g(x)$, $G(x)$ be bounded functions of bounded variation on $1 \leq x \leq 2$. Let $C_2, \ldots, C_7$ be real numbers ≥ 1. Suppose that*

$$\left|F^{(r)}(x)\right| \leq C_r \tag{5.3}$$

for $r = 2, 3, 4$,

$$\left|F^{(r)}(x)\right| \geq 1/C_r \tag{5.4}$$

for $r = 2, 3$. In some ranges we require extra conditions, either (5.4) for $r = 4$, or

$$\left|F''(x)F^{(4)}(x) - 3F^{(3)}(x)^2\right| \geq 1/C_5 \tag{5.5}$$

Let H, M and T be large parameters, and let S denote the sum

$$S = \sum_{h=H}^{2H-1} g\left(\frac{h}{H}\right) \sum_{m=M+2H-1}^{2M-2H} G\left(\frac{m}{M}\right) e\left(TF\left(\frac{m+h}{M}\right) - TF\left(\frac{m-h}{M}\right)\right).$$

Then we have the following bounds, in which B_1 and B_2 are positive constants constructed from $C_2, \ldots, C_7$, κ, λ and from the functions $g(x)$ and $G(x)$.
(A) In the range

$$C_6^{-1} T^{67\kappa-6}(\log T)^{(45\kappa-4)\lambda} \leq M^{156\kappa-14} \leq C_6 T^{89\kappa-8}(\log T)^{-(45\kappa-4)\lambda},$$

$$H \le B_1 M T^{-\frac{119\kappa-11}{384\kappa-36}} (\log T)^{\frac{(9\kappa-1)\lambda}{128\kappa-12}}, \tag{5.6}$$

we have

$$S \le B_2 H \left(\frac{H}{M}\right)^{\frac{6\kappa-1}{100\kappa-10}} T^{\frac{67\kappa-7}{200\kappa-20}} (\log T)^{\frac{9}{4}+\frac{(39\kappa-4)\lambda}{200\kappa-20}}$$

$$+ B_2 H \left(\frac{H}{M}\right)^{\frac{126\kappa-13}{200\kappa-20}} T^{\frac{207\kappa-21}{400\kappa-40}} (\log T)^{\frac{13}{4}+\frac{(69\kappa-7)\lambda}{400\kappa-40}}, \tag{5.7}$$

subject to the extra conditions that (5.4) holds for $r = 4$ when

$$M^{156\kappa-14} \le C_6 T^{69\kappa-6} (\log T)^{-3(9\kappa-1)\lambda},$$

that (5.5) holds when

$$M^{156\kappa-14} \ge C_6^{-1} T^{87\kappa-8} (\log T)^{3(9\kappa-1)\lambda},$$

that

$$H \ge C_7^{-1} T^4 (\log T)^{3\lambda}/M^9 \quad \text{for} \quad M \le C_6^{-1} T^{7/16} (\log T)^{\lambda/32\kappa},$$

and that

$$H \ge C_7^{-1} M^{11} (\log T)^{3\lambda}/T^6 \quad \text{for} \quad M \ge C_6 T^{9/16} (\log T)^{-\lambda/32\kappa}.$$

(B) *If (5.4) holds for $r = 4$, then in the ranges*

$$C_6^{-1} T^{1/3} \le M \le C_6 T^{7/16} (\log T)^{3\lambda/16},$$

$$H \le \min\left(B_1 M^{\frac{35\kappa-3}{53\kappa-5}} T^{-\frac{26\kappa-2}{159\kappa-15}} (\log T)^{\frac{(9\kappa-1)\lambda}{53\kappa-5}}, \right.$$

$$\left. B_1 M^{\frac{9}{7}} T^{-\frac{3}{7}}, \quad C_7 T^4 M^{-9} (\log T)^{3\lambda} \right), \tag{5.8}$$

we have

$$S \le B_2 H^{\frac{87\kappa-9}{80\kappa-8}} M^{\frac{15\kappa-1}{80\kappa-8}} T^{\frac{9\kappa-1}{40\kappa-4}} (\log T)^{\frac{9}{4}+\frac{(9\kappa-1)\lambda}{80\kappa-8}}$$

$$+ B_2 H^{\frac{267\kappa-27}{160\kappa-16}} M^{-\frac{45\kappa-5}{160\kappa-16}} T^{\frac{29\kappa-3}{80\kappa-8}} (\log T)^{\frac{13}{4}+\frac{(9\kappa-1)\lambda}{160\kappa-16}}$$

$$+ B_2 H^{\frac{107\kappa-13}{12(9\kappa-1)}} M^{-\frac{7\kappa-1}{4(9\kappa-1)}} T^{\frac{43\kappa-5}{12(9\kappa-1)}} (\log T)^{\frac{9+\lambda}{4}}$$

$$+ B_2 H^{\frac{331\kappa-39}{24(9\kappa-1)}} M^{-\frac{59\kappa-7}{8(9\kappa-1)}} T^{\frac{131\kappa-15}{24(9\kappa-1)}} (\log T)^{\frac{13+\lambda}{4}}. \tag{5.9}$$

(C) *If* (5.5) *holds, then in the ranges*

$$C_6^{-1} T^{9/16} (\log T)^{-5\lambda/16} \le M \le C_6 T^{2/3},$$

$$H \le \min\left(B_1 M^{\frac{71\kappa-7}{53\kappa-5}} T^{-\frac{80\kappa-8}{159\kappa-15}} (\log T)^{\frac{(9\kappa-1)\lambda}{53\kappa-5}}, \right.$$
$$\left. B_1 M^{\frac{5}{7}} T^{-\frac{1}{7}}, C_6 M^{11} T^{-6} (\log T)^{3\lambda} \right), \tag{5.10}$$

we have

$$S \le B_2 H^{\frac{87\kappa-9}{80\kappa-8}} M^{-\frac{29\kappa-3}{80\kappa-8}} T^{\frac{1}{2}} (\log T)^{\frac{9}{4}+\frac{(9\kappa-1)\lambda}{80\kappa-8}}$$
$$+ B_2 H^{\frac{267\kappa-27}{160\kappa-16}} M^{-\frac{169\kappa-17}{160\kappa-16}} T^{\frac{3}{4}} (\log T)^{\frac{13}{4}+\frac{(9\kappa-1)\lambda}{160\kappa-16}}$$
$$+ B_2 H^{\frac{107\kappa-13}{12(9\kappa-1)}} M^{\frac{23\kappa-1}{12(9\kappa-1)}} T^{\frac{7\kappa-1}{4(9\kappa-1)}} (\log T)^{\frac{9+\lambda}{4}}$$
$$+ B_2 H^{\frac{331\kappa-39}{24(9\kappa-1)}} M^{-\frac{53\kappa-9}{24(9\kappa-1)}} T^{\frac{23\kappa-3}{8(9\kappa-1)}} (\log T)^{\frac{13+\lambda}{4}}. \tag{5.11}$$

After some tidying steps, described in Huxley [16, Chapter 19], the estimation of the short interval mean square of an exponential sum can be reduced to the estimation of the sums in Proposition 1. A 'mathematics made difficult' interpretation of the A step in van der Corput's method is as a mean-to-max from the value at a fixed T to the mean square over a short interval of T, which is then expressed in terms of sums like those in Proposition 1.

Proposition 2. *Suppose that $H(\kappa, \lambda)$ holds with some κ in $1/4 \le \kappa \le 1/3$ and some $\lambda \ge 0$. Let the functions $F(x)$ and $G(x)$ satisfy the conditions of Proposition 1, with (5.3) and (5.4) also holding for $r = 1$. Let $S(t)$ be the exponential sum*

$$S(t) = \sum_{M}^{2M-1} G\left(\frac{m}{M}\right) e\left(tF\left(\frac{m}{M}\right)\right),$$

and let Δ be any real number with

$$\Delta \gg T^{\frac{207\kappa-21}{652\kappa-66}} (\log T)^{\frac{130(10\kappa-1)+(69\kappa-7)\lambda}{652\kappa-66}}.$$

Then for

$$T^{545\kappa-57} (\log T)^{\theta} \le M^{1304\kappa-32} \le T^{759\kappa-75} (\log T)^{-\theta} \tag{5.12}$$

where

$$\theta = 26(34\kappa - 1) + (399\kappa - 41)\lambda,$$

we have the mean square bound

$$\int_{T-\Delta}^{T+\Delta} |S(t)|^2 dt \ll \Delta M.$$

Proof of Proposition 2. From the discussion in Huxley [16, Chapter 19] we must sum Proposition 1 over values of H, with H doubling at each step, and obtain a total bound $O(M)$. The main contribution to the upper bound from (4.4) can be written as

$$O\left(\frac{M}{V_0^{1/6}} \left(\left(\frac{H^2}{NR} \right)^{\frac{1}{2}} (\log T)^{\frac{9+\lambda}{4}} + \left(\frac{H^2}{NR} \right)^{\frac{3}{4}} (\log T)^{\frac{13+\lambda}{4}} \right) \right),$$

where V_0 is given by (4.3). If H is so large that this expression is $O(M)$, then $H^2 \gg NR$, and the even-numbered terms in (5.7), (5.9) and (5.11) dominate. When we substitute for V_0, then the structural condition (5.1) becomes

$$R^{19\kappa-3} \ll N^{19\kappa-3} (\log T)^{(39\kappa+3\lambda)(9\kappa-1)},$$

a consequence of the structural condition $R \le N$. In case (A) we take the largest value of H for which the sum is $O(M)$, and then $\Delta \asymp M/H$ in (5.11). In cases (B) and (C) we take the same upper bound for H, and (5.12) is the condition for the terms in (5.8) or (5.10) to be $O(M)$. The values of H in cases (B) and (C) satisfy (5.7) and (5.8) respectively. $\square$

Following the arguments of Heath-Brown and Huxley [13], we obtain the asymptotic formula for the mean square of the zeta function. At the maximum value of H, the sums in Proposition 1 are $O(M \log M)$, so the exponents are related to those in Proposition 2.

Proposition 3. *Suppose that $H(\kappa, \lambda)$ holds with some κ in $1/4 \le \kappa \le 1/3$ and some $\lambda \ge 0$. Then for T large,*

$$\int_0^T \left| \zeta\left(\frac{1}{2} + it \right) \right|^2 dt = T \log \frac{T}{2\pi} + (2\gamma - 1)T + E(T),$$

with

$$E(T) \ll T^{\frac{207\kappa-21}{652\kappa-66}} (\log T)^{\frac{2204\kappa-212+(69\kappa-7)\lambda}{652\kappa-66}}.$$

With the van der Corput values $\kappa = 1/3$, $\lambda = 0$, Propositions 1, 2 and 3 become the results given in Huxley [16, section 19.1] (with the logarithm factor corrected; the extra logarithm from the ranges of Q was overlooked in Huxley [16]). As in Huxley [21], [19], using the bounds by Swinnerton-Dyer's method from Huxley [18], and a further subdivision if the continued fraction for a/q has some partial quotient $\gg \log M$, gives the result that would follow from $H(3/10, 57/140)$.

Theorem. *The results of Propositions 1, 2 and 3 with* $\kappa = 3/10$, $\lambda = 57/140$ *are true unconditionally.*

We note that with $\kappa = 3/10$ the main exponent in Propositions 2 and 3, $(207\kappa - 21)/(652\kappa - 66)$, is

$$\frac{137}{432} = 0.31713 \cdots < 0.31718 \cdots = \frac{72}{227},$$

the latter being the exponent in Huxley [16] from $\kappa = 1/3$.

References

[1] E. Bombieri and H. Iwaniec, *On the order of* $\zeta(1/2 + it)$, Ann. Scuola Norm. Sup. Pisa Cl. Sci. (4) **13** (1986), 449–472.

[2] E. Bombieri and J. Pila, *The number of integral points on arcs and ovals*, Duke Math. J. **59** (1989), 337–357.

[3] M. Branton and P. Sargos, *Points entiers au voisinage d'une courbe plane à très faible courbure*, Bull. Sci. Math. **118** (1994), 15–28.

[4] D. A. Burgess, *On character sums and L-series. II*, Proc. London Math. Soc. **13** (1963), 524–536.

[5] J. G. van der Corput, *Over Roosterpunten in het Platte Vlak*, Noordhof, Groningen, 1919.

[6] _____, *Über Gitterpunkte in der Ebene*, Math. Ann. **81** (1920), 1–20.

[7] _____, *Zahlentheoretische Abschätzungen*, Math. Ann. **84** (1921), 53–79.

[8] _____, *Verschärfung der Abschätzung beim Teilerproblem*, Math. Ann. **87** (1922), 39–65.

[9] _____, *Neue zahlentheoretische Abschätzungen*, Math. Ann. **89** (1923), 215–254.

[10] M. Filaseta and O. Trifonov, *The distribution of fractional parts with applications to gap results in number theory*, Proc. London Math. Soc. **73** (1996), 241–278.

[11] E. Fouvry and H. Iwaniec, *Exponential sums for monomials*, J. Number Theory **33** (1989), 311–333.

[12] S. W. Graham and G. Kolesnik, *Van der Corput's method for exponential sums*, London Math. Soc. Lecture Notes 126, Cambridge University Press, 1991.

[13] D.R. Heath-Brown and M. N. Huxley, *Exponential sums with a difference*, Proc. London Math. Soc. **61** (1990), 227–250.

[14] E. Hlawka, *Integrale auf konvexen Körpern*, Monatsh. Math. **54** (1950), 1–36, 81–99.

[15] M. N. Huxley, *The rational points close to a curve*, Ann. Scuola Norm. Sup. Pisa Cl. Sci. (4) **21** (1994), 357–375.

[16] ————, *Area, lattice points and exponential sums*, London Math. Soc. Monographs 13, Oxford University Press, 1996.

[17] ————, *The integer points close to a curve III*, Number Theory in Progress, de Gruyter, Berlin, 1999, pp. 911–940.

[18] ————, *The rational points close to a curve II*, Acta Arith. **93** (2000), 201–219.

[19] ————, *Exponential sums and lattice points III*, Proc. London Math. Society, to appear.

[20] ————, *The rational points close to a curve III*, to appear.

[21] ————, *Exponential sums and the Riemann zeta function V*, to appear.

[22] M. N. Huxley and N. Watt, *Exponential sums and the Riemann zeta function*, Proc. London Math. Soc. **57** (1988), 1–24.

[23] H. Iwaniec and C. J. Mozzochi, *On the divisor and circle problems*, J. Number Theory **29** (1988), 60–93.

[24] D. G. Kendall, *On the number of lattice points inside a random oval*, Quart. J. Math. (Oxford) **19** (1948), 1–26.

[25] S. Konyagin, *Estimates of the least prime factor of a binomial coefficient*, Mathematika **45** (1999), 41–55.

[26] E. Krätzel, *Lattice points*, Deutscher Verlag Wiss., Berlin, 1988.

[27] E. Krätzel and W. G. Nowak, *Lattice points in large convex bodies II*, Acta Arith **62** (1992), 285–295.

[28] E. Landau, *Die Bedeutung der Pfeiffer'schen Methode für die analytische Zahlentheorie*, Sitzungsberichte Akad. Wiss. Wien, math.-naturwiss. Kl. **121** (1912), 2195–2322.

[29] _____, *Die Bedeutungslosigkeit der Pfeiffer'schen Methode für die analytische Zahlentheorie*, Monatsh. Math. Phys. **34** (1926), 1–36.

[30] P. Sargos, *Points entiers au voisinage d'une courbe, sommes trigonométriques courtes et paires d'exposants*, Proc. London Math. Soc. **70** (1995), 285–312.

[31] W. Sierpiński, *Sur un problème du calcul des fonctions asymptotiques*, Prace Mat.-Fiz. **17** (1906), 77–118.

[32] H. P. F. Swinnerton-Dyer, *The number of lattice points on a convex curve*, J. Number Theory **6** (1974), 128–135.

[33] G. Voronoi, *Sur un problème du calcul des fonctions asymptotiques*, J. Reine Angew. Math. **126** (1903), 241–282.

[34] _____, *Sur une fonction transcendente et ses applications à la sommation de quelques séries*, Ann. École Norm. Sup. **21** (1904), 207–267, 459–533.

[35] N. Watt, *A problem on semicubical powers*, Acta Arith. **52** (1989), 119–140.

Recent Developments in Automorphic Forms and Applications

Wen-Ching Winnie Li[1]

1 Introduction

In the past decade there has been tremendous progress in the area of automorphic forms. The crown jewel was Wiles' spectacular proof of Fermat's Last Theorem in 1994; this was a by-product of his proof of the Taniyama-Shimura conjecture for semi-stable elliptic curves defined over $\mathbb{Q}$. The remaining case of the Taniyama-Shimura conjecture was completely settled through the joint effort of Breuil, Conrad, Diamond, and Taylor. In another direction, the proof of the Local Langlands conjecture for $GL(n)$ in all characteristics is now complete, with the finite characteristic case proved by Laumon, Rapoport and Stuhler and the characteristic zero case proved by Harris and Taylor and by Henniart. Finally, the proof of the Global Langlands conjecture for $GL(n)$ over function fields is well underway. In addition to these developments, substantial progress has been made on many related subjects.

The purpose of this review article is two-fold. First, we describe the local and global Langlands conjectures on correspondences between representations of $GL(n)$ and representations of Galois/Weil groups, and briefly summarize what is currently known about them. Second, we explain recent results giving explicit examples of matching automorphic and algebraic-geometric L-functions, with applications to conjectures in number theory, and results giving applications of automorphic forms in constructing good combinatorial objects.

The paper is organized as follows. In Sections 2 and 3 we introduce the basic local and global invariants, namely the L- and ε-factors attached to a representation of a Galois group, resp. a representation of $GL(n)$. They are used in Sections 4 and 5 to describe the local and global Langlands conjecture for $GL(n)$. We shall view the Taniyama-Shimura conjecture as part of

[1]The research of the author was supported in part by a grant from the National Science Foundation no. DMS-9970651. This work was done when the author was visiting the Institute for Advanced Study at Princeton, NJ, supported by the Ellentuck Fund, to which she expresses sincere thanks.

the global Langlands conjecture for GL(2) over $\mathbb{Q}$, and discuss its conjectural generalization to abelian varieties over $\mathbb{Q}$. In Section 6 we give explicit examples of the Langlands global correspondence over function fields; this is a joint work of the author with C.-L. Chai. We derive certain interesting consequences, among them the establishment of the Kloosterman sum conjecture over function fields and new examples of automorphic forms for GL(2) for which the Sato-Tate conjecture holds. These are discussed in Sections 7 and 8. Applications of automorphic forms to combinatorial constructions are presented in the last section. We review Ramanujan graphs constructed by Lubotzky, Phillips and Sarnak based on quaternion algebras and extend these to constructing Ramanujan 3-hypergraphs based on division algebras of degree 9 over a function field.

2   Invariants Attached to Representations of Galois Groups

One of the major goals of algebraic number theory is to study algebraic extensions of a given field, either a global field—a number field or a function field of one variable over a finite field—or a local field—the completion of a global field at a place. Typical examples of global fields are the field of rational numbers $\mathbb{Q}$ and the field of rational functions $\mathbb{F}_q(t)$; examples of local fields are the field of p-adic numbers $\mathbb{Q}_p$ and the field of Laurent series $\mathbb{F}_q((t))$. Let K be a global or local field. Denote by $\bar{K}$ the separable closure of K and by G_K the Galois group of $\bar{K}$ over K. To understand algebraic extensions of K amounts to understanding G_K, which is a profinite group. According to Tannaka duality, to know a profinite group is equivalent to knowing all of its finite-dimensional representations, not just as a set, but also with the tensor product. Therefore we study (linear) representations of G_K. To each representation we shall attach certain invariants, called L- and ε-factors, such that the family of invariants attached to the representation itself as well as its tensor product with certain representations will characterize the representation up to equivalence.

Assume that K is a global field. We describe representations of the Galois group G_K arising from geometry. Let X be a smooth variety defined over K. The action of G_K on X induces an action of G_K on the étale cohomology group $H^i_{et}(X_{/\bar{K}}, \overline{\mathbb{Q}_\ell})$. Let V be an irreducible subquotient of dimension n of this action. Hence we get a continuous irreducible ℓ-adic representation

$$\rho \colon G_K \longrightarrow \text{Aut}(V).$$

It is known that ρ is unramified at all except finitely many places of K. Thus at a nonarchimedean place v where ρ is unramified, for each chosen decomposition group at v, the restriction of ρ to the inertia subgroup is trivial, and consequently ρ restricted to the Frobenius conjugacy class is single-valued. To take into account the different choices in G_K of a decomposition group at v, which are conjugate to each other, we obtain a conjugacy class $\rho(\mathrm{Frob}_v)$. Nonetheless the eigenvalues $\rho_{1,v}, \ldots, \rho_{n,v}$ of $\rho(\mathrm{Frob}_v)$ are well-defined. Put

$$L_v(s,\rho) = \prod_{1 \le i \le n} \frac{1}{1 - \rho_{i,v} Nv^{-s}},$$

which is a local invariant of ρ restricted to a decomposition group at v. Here Nv, the norm of v, is the cardinality of the residue field of K_v, the completion of K at v. At the places where the representation ρ ramifies, as well as the archimedean places if K is a number field, local invariants $L_v(s,\rho)$ can be defined in a more complicated way. Put together, one defines the L-function attached to ρ, which is a global invariant attached to ρ:

$$L(s,\rho) := \prod_v L_v(s,\rho) \approx \prod_{v \text{ good}} \prod_{1 \le i \le n} \frac{1}{1 - \rho_{i,v} Nv^{-s}},$$

Here $\approx$ means that we ignore finitely many local L-factors.

An example of what we described above is the case of an elliptic curve E defined over K. Given a prime ℓ, not equal to the characteristic of K if $\mathrm{char}(K) > 0$, the points on E with order dividing ℓ^n form an abelian group $E[\ell^n]$, which is isomorphic to $(\mathbb{Z}/\ell^n\mathbb{Z}) \times (\mathbb{Z}/\ell^n\mathbb{Z})$. The inverse limit of $E[\ell^n]$ as n approaches infinity is called a Tate module $T_\ell(E)$ of E, which is isomorphic to $\mathbb{Z}_\ell \times \mathbb{Z}_\ell$. In other words, it is a $\mathbb{Z}_\ell$-module of rank two. The Galois group G_K permutes the points in $E[\ell^n]$, and hence one gets a two-dimensional ℓ-adic representation of G_K. The associated L-function is the Hasse-Weil L-function of E:

$$L(s,E) \approx \prod_{v \text{ good}} \frac{1}{1 - a_v Nv^{-s} + Nv^{1-2s}},$$

where $1 + Nv - a_v$ is the number of rational points of the reduction of E at v over the residue field at v, which is independent of ℓ.

When ρ is nontrivial, $L(s,\rho)$ is expected to have a holomorphic continuation to the whole s-plane, bounded in each vertical strip of finite width, and satisfies a functional equation

$$L(s,\rho) = \varepsilon(s,\rho) L(1-s,\hat{\rho}),$$

where $\hat{\rho}$ denotes the contragredient of ρ. Here the ε-factor, $\varepsilon(s,\rho)$, is another global invariant attached to ρ. Deligne has shown in [13] that $\varepsilon(s,\rho)$

is a product of local ε-factors, each of which is a local invariant attached
to the restriction of ρ to a decomposition group.

These analytic properties for $L(s, \rho)$ were proved by Grothendieck when
K is a function field. When K is a number field, these are mostly unproved.
The Hasse-Weil L-function attached to an elliptic curve defined over $\mathbb{Q}$ is
one case where they are proved, as a consequence of the Taniyama-Shimura
conjecture explained in Section 5.

When $K = \mathbb{Q}$, more properties of ρ are known:

(a) The restriction of ρ to a decomposition group is "potentially semi-
stable" in the sense of Fontaine.

(b) The eigenvalues $\rho_{1,v}, \ldots, \rho_{n,v}$ are algebraic for almost all v.

A conjecture of Fontaine and Mazur says that a continuous irreducible
finite-dimensional ℓ-adic representation ρ of $G_{\mathbb{Q}}$ arises from geometry if and
only if it is ramified at finitely many places and statement (a) above holds.
Some progress on this conjecture has been made by Skinner, Taylor, and
Wiles.

3 Invariants Attached to Automorphic Representations

For a global field K as described in Section 2, denote by $\mathbb{A}_K$ the ring of ade-
les of K. Let π be an automorphic irreducible representation of $\mathrm{GL}_n(\mathbb{A}_K)$.
Then π is a restricted tensor product $\bigotimes'_v \pi_v$, where π_v is an admissible
irreducible representation of $\mathrm{GL}_n(K_v)$ with K_v being the completion of
K at v, which is unramified for all except finitely many places. The unramified local representations are determined by the action of the Hecke
algebra $\mathcal{H}_v$, which is isomorphic to the algebra of symmetric polynomials in
$\mathbb{C}[z_1, z_1^{-1}, \ldots, z_n, z_n^{-1}]$. Therefore π_v is determined by n complex numbers

$$z_1(\pi_v), \ldots, z_n(\pi_v).$$

Define

$$L_v(s, \pi) := L(s, \pi_v) = \prod_{1 \leq i \leq n} \frac{1}{1 - z_i(\pi_v) N v^{-s}},$$

which is a local invariant attached to π_v. At the places where π ramifies
and at the archimedean places of K if K is a number field, one can also
define local L-factors $L(s, \pi_v)$ in a more complicated way. The product of
local L-factors is a global invariant attached to π,

$$L(s, \pi) := \prod_v L_v(s, \pi_v) \approx \prod_{\pi_v \text{ unramified}} \prod_{1 \leq i \leq n} \frac{1}{1 - z_i(\pi_v) N v^{-s}},$$

called an automorphic L-function of $\mathrm{GL}_n(\mathbb{A}_K)$.

Many familiar functions are examples of global automorphic L-functions. For instance, the Riemann zeta function $\zeta(s)$ and the Dirichlet L-function $L(s, \chi)$ are automorphic L-functions of $\mathrm{GL}_1(\mathbb{A}_\mathbb{Q})$, the Dedekind zeta function $\zeta_K(s)$ is an automorphic L-function for $\mathrm{GL}_1(\mathbb{A}_K)$ with K a number field, and the L-function attached to a newform f of weight k, level N, and character χ,

$$L(s, f) = \prod_{p|N} \frac{1}{1 - a_p p^{-s}} \prod_{p \nmid N} \frac{1}{1 - a_p p^{-s} + \chi(p) p^{k-1-2s}},$$

is an automorphic L-function for $\mathrm{GL}_2(\mathbb{A}_\mathbb{Q})$.

An automorphic L-function $L(s, \pi)$ is known to have meromorphic continuation to the whole s-plane, bounded at infinity in each vertical strip of finite width if K is a number field, while $L(s, \pi)$ is a rational function in q^{-s} if K is a function field with q elements in its field of constants, and $L(s, \pi)$ satisfies the functional equation

$$L(s, \pi) = \varepsilon(s, \pi) L(1 - s, \hat{\pi}).$$

Here $\hat{\pi}$ is the contragradient of π. Moreover, if π is a cuspidal representation, then $L(s, \pi)$ is holomorphic everywhere. In the case when K is a function field, $L(s, \pi)$ is in fact a polynomial in q^{-s}. When an L-function has the aforementioned analytic property, it is said to *behave nicely*. Therefore, automorphic L-functions behave nicely and the L-function attached to an ℓ-adic representation ρ of a Galois group G_K arising from geometry behaves nicely if K is a function field; when K is a number field, the analytic behavior of $L(s, \rho)$ is unknown for most ρ.

The $\varepsilon(s, \pi)$ occurring in the functional equation is another global invariant attached to π. It is a product of local ε-factors attached to π_v.

4 The Local Langlands Conjecture

Let K be a non-archimedean local field with residue field F. The residue field of $\bar{K}$ is an algebraic closure $\bar{F}$ of F. Since a Galois automorphism preserves integrality, it naturally induces an automorphism of $\bar{F}$ over F. Such a map from $\mathrm{Gal}(\bar{K}/K)$ to $\mathrm{Gal}(\bar{F}/F)$ is surjective with kernel equal to the inertia subgroup I_K of $\mathrm{Gal}(\bar{K}/K)$. Since $\mathrm{Gal}(\bar{K}/K)$ is compact, we replace it by a dense subgroup W_K, called the Weil group, consisting of the preimages in $\mathrm{Gal}(\bar{K}/K)$ of the dense subgroup generated by the Frobenius automorphism in $\mathrm{Gal}(\bar{F}/F)$. Endowed with the induced topology, the Weil group W_K not only captures the essence of $\mathrm{Gal}(\bar{K}/K)$, it also affords more continuous representations. It turns out to be the right group to work with.

Local Langlands Conjecture. *There is a natural bijection from the set of equivalence classes of degree n irreducible complex continuous representations of W_K to the set of equivalence classes of admissible irreducible supercuspidal representations of $\mathrm{GL}_n(K)$, so that the corresponding representations as well as their twists have the same L- and ε-factors.*

Remark. Henniart proved in [22] that the invariant L- and ε-factors as stated in the conjecture do determine the class of representations of $\mathrm{GL}_n(K)$.

Theorem. *The local Langlands conjecture for $\mathrm{GL}_n(K)$ holds for all local fields K.*

For $n = 1$, this is local class field theory. For $n = 2$, this was proved by Kutzko [28]. For $n = 3$, this is a result of Henniart [21]. Kutzko and Moy [29] proved the conjecture for the case where n is a prime. The general case with K of positive characteristic was proved by Laumon, Rapoport and Stuhler [32] in 1991. In 1998 the remaining case with K of zero characteristic was settled by M. Harris and R. Taylor [20]. Later Henniart [23] gave a much shorter proof.

5   The Global Langlands Conjecture

The situation for the case of a global field K is more complicated, and much less is known. To explain correspondences, we consider three kinds of representations:

(I)   The set G_n of equivalence classes of degree n continuous irreducible complex representations of G_K.

(II) The set $G_{n,\ell}$ of equivalence classes of degree n continuous irreducible ℓ-adic representations of G_K arising from geometry.

(III) The set A_n of equivalence classes of cuspidal automorphic irreducible representations of $\mathrm{GL}_n(\mathbb{A}_K)$.

Note that the representations in G_n are those in $G_{n,\ell}$ whose underlying varieties consist of finitely many points.

Global Langlands Conjecture Over a Number Field K. *There is an injection from G_n into A_n which extends to a map from $G_{n,\ell}$ into A_n such* *that if a representation ρ of G_K corresponds to a representation $\pi = \otimes'\pi_v$ of $\mathrm{GL}_n(\mathbb{A}_K)$, then the corresponding representations as well as their twists have the same attached L- and ε-factors. Further, at each place v of K, the restriction of ρ to a decomposition group at v corresponds to π_v according to the local Langlands conjecture.*

An immediate consequence of the global correspondence is that the global L-functions attached to representations of G_K arising from geometry behave nicely.

When $n = 1$, this is global class field theory. We describe below what is known of the case $n = 2$ and $K = \mathbb{Q}$. A degree two representation ρ of $G_{\mathbb{Q}}$ is called *odd* if the action (via ρ) of complex conjugation has determinant -1. The cuspidal representations of $\mathrm{GL}_2(\mathbb{A}_{\mathbb{Q}})$ arise in two ways: either from classical holomorphic cuspidal newforms of integral weight $k \geq 1$ or from real analytic cuspidal new Maass wave forms. Not much is known about the latter, while the former has been studied extensively. Deligne and Serre [15] established an injection from the set of holomorphic cuspidal (normalized) newforms of weight 1 to the set $G_{2,\mathrm{odd}}$ of odd representations in G_2 satisfying the Langlands conjecture. For forms with higher weight, there is an injection from the set of holomorphic cuspidal newforms of weight $k \geq 2$ to the set $G_{2,\ell,\mathrm{odd}}$ of odd representations in $G_{2,\ell}$ satisfying Langlands' conjecture. This was proved by Eichler-Shimura for $k = 2$ and forms with rational coefficients, and by Deligne [12] in general for $k \geq 2$. For $n = 2$ and K a totally real field of degree r, Rogawski-Tunnell [44] together with a technical result from H. Reimann (see, for example, [42]) extended the result of Deligne-Serre to holomorphic forms of weight $(1,\ldots,1)$, while Brylinski-Labesse [5] and Blasius-Rogawski [2] generalized Deligne's result to forms of weight $(k_1,\ldots,k_r)$, where $k_1,\ldots,k_r$ have the same parity modulo 2.

The progress made on the global Langlands conjecture had a far reaching impact on the advancement of number theory in the past decade. Here we elaborate a little. Recall first the result by Eichler and Shimura.

Theorem. *Let f be a weight 2 holomorphic cuspidal normalized newform with coefficients in $\mathbb{Q}$. Then $L(s,f) = L(s,E)$ for some elliptic curve E defined over $\mathbb{Q}$.*

In fact, E is a quotient A_f of the Jacobian of the modular curve $X_0(N)$ attached to f. Here N is the level of f. As explained in Section 2, to an elliptic curve E there is the associated ℓ-adic representation of $G_{\mathbb{Q}}$ arising from its action on the Tate module $T_\ell(E)$. This is the representation in $G_{2,\ell,\mathrm{odd}}$ to which f corresponds. The conjecture of Taniyama-Shimura says that the relation $L(s, f) = L(s, E)$ actually holds for all elliptic curves defined over $\mathbb{Q}$.

Taniyama-Shimura Conjecture. *Every elliptic curve E defined over $\mathbb{Q}$ is modular, that is, there is a holomorphic cuspidal normalized newform f of weight 2 such that $L(s, E) = L(s, f)$.*

Using the isogeny conjecture proved by Faltings [18], we can restate the above conjecture: every elliptic curve defined over $\mathbb{Q}$ is isogenous to some A_f.

Wiles [46] and Taylor-Wiles [45] proved the Taniyama-Shimura conjecture for semi-stable elliptic curves, which, combined with the work of Ribet [43] and others, establishes Fermat's Last Theorem. The remaining case of the conjecture was completely settled through the joint efforts of Breuil, Conrad, Diamond, and Taylor in 1999 [4].

If f is a holomorphic cuspidal newform of weight 2 for $\Gamma_1(N)$, that is, of level N and some nontrivial character, then so are its conjugates by elements in $G_{\mathbb{Q}}$. In fact, f has g distinct conjugates, where g is the degree over $\mathbb{Q}$ of the field of Fourier coefficients of f. The variety A_f attached to f in this case, being a quotient of the Jacobian of the modular curve $X_1(N)$, is a simple abelian variety over $\mathbb{Q}$ of dimension g. (It is also the variety attached to any conjugate of f.) Furthermore, the ring of endomorphisms of A_f contains an order of a field of degree g over $\mathbb{Q}$. An abelian variety with the properties described above is said to be of GL_2-type. A natural question along the same lines as the Taniyama-Shimura conjecture is to characterize the abelian varieties isogenous to A_f. The conjectural answer is stated below.

Generalized Taniyama-Shimura Conjecture. *Every simple abelian variety defined over $\mathbb{Q}$ which is of GL_2-type is modular, that is, isogenous to A_f for some holomorphic cuspidal newform f of weight 2.*

The global Langlands conjecture over a function field can be stated more precisely as follows.

Global Langlands Conjecture Over a Function Field K. *There is a bijection from the subset $G_{n,\ell}^f$ of $G_{n,\ell}$ consisting of representations whose determinants are of finite order to the subset A_n^f of A_n consisting of representations whose central characters are of finite order such that the corresponding representations as well as their twists have the same L- and ε-factors. Further, if a representation ρ of G_K corresponds to the representation $\pi = \otimes_v' \pi_v$ of $\mathrm{GL}_n(\mathbb{A}_K)$, then for each place v of K, the restriction* *of ρ to a decomposition group at v corresponds to π_v as described by the local Langlands conjecture.*

For $n = 1$ this is global class field theory. Drinfeld [16] proved the conjecture for the case $n = 2$, and Lafforgue's proof of the case of general n is well underway.

The Taniyama -Shimura conjecture over a function field K asserts that the Hasse-Weil L-function attached to an elliptic curve defined over K is an automorphic L-function of $GL_2(\mathbb{A}_K)$. This was proved by Deligne [13] using Grothendieck's result on the analytic behavior of the L-functions attached to the elliptic curve as well as its twists by characters, and the converse theorem for GL_2 proved by Jacquet and Langlands [24]. This is totally different from the approach in the case $K = \mathbb{Q}$.

6 Explicit Examples of Global Langlands Correspondence Over a Function Field

For arithmetic applications one sometimes wants more than the abstract existence of the Langlands global correspondence, namely, specific details for individual representations of interest. In this section we describe results in the function field case, which give completely explicit examples of the global correspondence between n-dimensional ℓ-adic representations of the Galois group and cuspidal representations of GL_n. The correspondence is expressed in terms of the attached L-functions. These results are joint work with C.-L. Chai (cf. [9], [10]).

We first illustrate through a familiar example the prototype of the L-functions which we shall be concerned with. Let K be a function field of one variable with field of constants F. Denote by F_v the residue field of the completion K_v of K at the place v. Its cardinality is Nv, the norm of v.

Let E be an elliptic curve defined over K. Supposing, for convenience, that the characteristic of K is not 2 or 3, we may assume that an affine part of E is given by the equation

$$y^2 = g(x),$$

where g is a polynomial over K of degree 3. Then

$$L(s, E) \approx \prod_{v \text{ good}} \frac{1}{1 - a_v Nv^{-s} + Nv^{1-2s}},$$

where

$$a_v = -\sum_{x \in F_v} \chi \circ N_{F_v/F}(g_v(x))$$

with χ the quadratic character of $F^\times$, $N_{F_v/F}$ the norm map from F_v to F, and g_v the polynomial over F_v obtained from the reduction of g mod v. Let η_v denote the quadratic idele class character of the rational function field $F_v(t)$ attached to the quadratic extension $y^2 = g_v(t)$. Then at a place w of $F_v(t)$ where η_v is unramified, its value at a uniformizer ϖ_w is given by

$$\eta_v(\varpi_w) = \chi \circ N_{F_v/F}\left(\prod_\alpha g_v(\alpha)\right).$$

Here α runs through all points of the projective line in an algebraic closure of F_v which constitute the closed point w. While each $g_v(\alpha)$ might lie outside F_v, after taking the product over all α, the resulting value does lie in F_v. Note that the L-function attached to η_v is the reciprocal of the factor at v occurring in $L(s, E)$; in other words,

$$L(s, \eta_v) = 1 - a_v N v^{-s} + N v^{1-2s}.$$

The L-functions we shall consider have the following features:

(1) Each is attached to a (global) rational function over K and either an additive character or a multiplicative character of F.

(2) Each has an Euler product with the local factor at a good place v being the reciprocal of a global L-function attached to an idele class character η_v of a rational function field.

(3) The character η_v in (2) is constructed using the reduction mod v of the global rational function in (1) and the additive/multiplicative character.

The details are as follows.

Let $f(x)$ be a rational function with coefficients in K. Only finitely many places of K occur as zeros or poles of the coefficients of f. These are the 'possibly bad' places, and the remaining places are called 'good' places of f. At a good place v, the coefficients of f are units in the completion of K at v. By passing to the residue field F_v, we get a rational function $f_v := f \mod v$ with coefficients in F_v.

Theorem A. *Let $f(x) \in K(x)$ be a non-constant rational function over K and ψ a nontrivial additive character of F.*

(A1) *At each good place v of f, there exists an idele class character η_{ψ, f_v} of $F_v(t)$ such that at the places w of $F_v(t)$ where η_{ψ, f_v} is unramified, its value at a uniformizer ϖ_w is given by the character sum*

$$\eta_{\psi, f_v}(\varpi_w) = \psi \circ \mathrm{Tr}_{F_v/F}\left(\sum_\alpha f_v(\alpha)\right),$$

where α runs through all points of the projective line in an algebraic closure $\bar{F}$ of F which constitute the closed point w.

(A2) *Suppose that $f(x)$ is not of the form $h_1(x)^p - h_1(x) + h_2$ for any $h_1(x) \in K(x)$ and $h_2 \in K$. Then the L-function $L(s, \eta_{\psi, f_v})$ attached to η_{ψ, f_v} is a polynomial in Nv^{-s} of degree n equal to the degree of the conductor of η_{ψ, f_v} minus 2, and n is the same for almost all v.*

(A3) *With the same assumption as in (A2), there exists a compatible family of ℓ-adic representations of G_K, depending on ψ and f, such that its associated L-function $L(s, \psi, f)$ satisfies*

$$L(s, \psi, f) \approx \prod_{v \ good} L(s, \eta_{\psi, f_v})^{-1}.$$

Further $L(s, \psi, f)$ and its twists by idele class characters of K and by all ℓ-adic representations of G_K behave nicely.

Theorem B. *Let $g(x) \in K(x)$ be a non-constant rational function over K and let χ be a nontrivial character of $F^\times$ of order $d > 1$.*

(B1) *At a good place v of g, there exists an idele class character η_{χ, g_v} of $F_v(t)$ such that at the places w of $F_v(t)$ where η_{χ, g_v} is unramified, its value is given by the character sum*

$$\eta_{\chi, g_v}(\varpi_w) = \chi \circ N_{F_v/F} \left(\prod_\alpha g_v(\alpha) \right),$$

where, as in Theorem A, α runs through all points in the projective line constituting the closed point w.

(B2) *Suppose that $g(x)$ is not of the form $h_1 h_2(x)^d$ for any $h_2(x) \in K(x)$ and any $h_1 \in K$. Then the L-function $L(s, \eta_{\chi, g_v})$ attached to η_{χ, g_v} is a polynomial in Nv^{-s} of degree n equal to the degree of the conductor of η_{χ, g_v} minus 2, and n is the same for almost all v.*

(B3) *Under the same assumption as in (B2), there exists a compatible family of ℓ-adic representations of G_K, depending on χ and g, such that the associated L-function $L(s, \chi, g)$ satisfies*

$$L(s, \chi, g) \approx \prod_{v \ good} L(s, \eta_{\chi, g_v})^{-1}.$$

Further $L(s, \chi, g)$ and its twists by idele class characters of K and by all ℓ-adic representations of G_K behave nicely.

As a consequence of the global Langlands conjecture proved for $n \leq 2$ by Drinfeld [16] and the converse theorem for GL_n proved by Jacquet and Langlands [24] for $n = 2$, by Jacquet, Piatetski-Shapiro and Shalika [25] for $n = 3$, and by Cogdell and Piatetski-Shapiro [11] for $n = 4$, we conclude

Corollary. $L(s, \psi, f)$ (resp. $L(s, \chi, g)$) is an automorphic L-function for $\mathrm{GL}_n(\mathbb{A}_K)$ if the common degree n as described in Theorem A (resp. B) is at most 4.

It would follow from the global Langlands conjecture over function fields that all $L(s, \psi, f)$ and $L(s, \chi, g)$ are automorphic L-functions for $\mathrm{GL}_n(\mathbb{A}_K)$ if the common degree of local factors is n.

More concretely, we consider the following examples.

Example 1. Let $f(x) = x + b/x$ with any nonzero element b in K. We obtain an automorphic L-function $L(s, \psi, f)$ for $\mathrm{GL}_2(\mathbb{A}_K)$.

Example 2. Let $f_1(x) = x^2 + a/x$ and $f_2(x) = x + a/x^2$ for a nonzero element a in K. If char K is not 2, then $L(s, \psi, f_1)$ and $L(s, \psi, f_2)$ are automorphic L-functions for $\mathrm{GL}_3(\mathbb{A}_K)$.

Example 3. $L(s, \psi, f)$ is an automorphic L-function for $\mathrm{GL}_4(\mathbb{A}_K)$ if $f(x) = x^3 + a/x$ and char K is not equal to 3, or $f(x) = x^2 + a/x^2$ and char K is not equal to 2. Here a is any nonzero element in K as before.

Example 4. Suppose char K is odd. Let $g(x) = (x - 1)^2 + 4ax$ for $a \in K^\times$ and χ be a nontrivial character of $F^\times$. At a place v of K which is not a pole of a and where $a \not\equiv 0, 1 \pmod{v}$, there is an idele class character η_{χ, g_v} whose associated L-function is

$$L(s, \eta_{\chi, g_v}) = 1 + \lambda_\chi(F_v; a) N v^{-s} + N v^{1-2s},$$

where

$$\lambda_\chi(F_v; a) = \sum_{\substack{x, y \in \mathbb{F}_v \\ y^2 = g_v(x)}} \chi \circ \mathrm{N}_{F_v/F}(x).$$

One can show

Theorem C. *Given a nonzero a in K, there exists an automorphic form ϕ_a of $\mathrm{GL}_2(\mathbb{A}_K)$ such that*

$$L(s, \phi_a) \approx \prod_{v \text{ good}} \frac{1}{1 + \lambda_\chi(F_v; a) N v^{-s} + N v^{1-2s}}.$$

Example 5. Suppose char K is odd. Denote by F' the quadratic extension of the field of constants F of K. Let ε be the quadratic character of $F^\times$,

ω be a regular character of $F'^{\times}$, and a be a nonzero element in K. At a place v of K which is not a pole of a and where $a \not\equiv 2, -2 \pmod{v}$, there is an idele class character $\eta_{a,\varepsilon,\omega,v}$ whose associated L-function is

$$L(s, \eta_{a,\varepsilon,\omega,v}) = 1 + \lambda_{\varepsilon,\omega}(F_v; a)Nv^{-s} + Nv^{1-2s},$$

with

$$\lambda_{\varepsilon,\omega}(F_v; a) = \sum_{u \in U(F_v)} \varepsilon \circ N_{F_v/F}(\mathrm{Tr}_{F'_v/F_v}(u) + a_v)\omega \circ N_{F'_v/F'}(u).$$

Here a_v denotes $a \pmod{v}$, $U(F_v)$ is the norm 1 subgroup of F'_v over F_v, and $F'_v = F_v \otimes_F F'$, which is a quadratic extension of F_v (resp. $F_v \times F_v$) when $\deg v$ is odd (resp. even).

Similar to the previous example, we have

Theorem D. *Given a nonzero a in K, a quadratic character ε of $F^{\times}$ and a regular character ω of $F'^{\times}$, there exists an automorphic form $\phi_{a,\varepsilon,\omega}$ of $GL_2(\mathbb{A}_K)$ such that*

$$L(s, \phi_{a,\varepsilon,\omega}) \approx \prod_{v \text{ good}} \frac{1}{1 + \lambda_{\varepsilon,\omega}(F_v; a)Nv^{-s} + Nv^{1-2s}}.$$

7 Kloosterman Sum Conjectures

For a prime p denote by ψ_p the additive character of $\mathbb{Z}/p\mathbb{Z}$ given by $\psi_p(x) = \exp(2\pi i x/p)$. Given a finite field k of characteristic p, the composition $\psi_p \circ Tr_{k/(\mathbb{Z}/p\mathbb{Z})}$ is a nontrivial additive character ψ of k. A Kloosterman sum is defined for each $b \in k^{\times}$ as

$$Kl(k; b) = \sum_{x \in k^{\times}} \psi\left(x + \frac{b}{x}\right).$$

Example 1 in Section 6 deserves some elaboration. For a given $b \in K^{\times}$ and $f(x) = x + b/x$, a good place is a place v of K away from the zeros and poles of b. At such a place, $b \bmod v$ is a nonzero element in the residue field F_v, and one finds that

$$L(s, \eta_{\psi,f_v}) = 1 + Kl(F_v; b)Nv^{-s} + Nv^{1-2s},$$

and we know that $L(s, \psi, f)$ is an automorphic L-function for $GL_2(\mathbb{A}_K)$. Stated in another way, we have proven [9]:

Theorem E (Kloosterman Sum Conjecture Over a Function Field).
*Let K be a function field with the field of constants F a finite field. Given
a nonzero element $b \in K$, there exists an automorphic form ϕ_b of $GL_2(\mathbb{A}_K)$
which is an eigenfunction of the Hecke operator T_v with eigenvalue
$-\mathrm{Kl}(F_v; b)$ at each place v of K which is neither a zero nor a pole of b. In
other words,*

$$L(s, \phi_b) \approx \prod_{v \ good} \frac{1}{1 + \mathrm{Kl}(F_v; b)Nv^{-s} + Nv^{1-2s}}.$$

It is interesting to compare this with its counterpart over the field $\mathbb{Q}$;
this is a question raised by Katz in [27].

Kloosterman Sum Question Over $\mathbb{Q}$. *Given a nonzero integer b, does
there exist a Maass wave form f which is an eigenfunction of the Hecke
operator T_p for all $p \nmid b$ with eigenvalue $-\mathrm{Kl}(\mathbb{Z}/p\mathbb{Z}; b)$? In other words,
does there exist a Maass wave form f such that*

$$L(s, f) \approx \prod_{p \nmid b} \frac{1}{1 + \mathrm{Kl}(\mathbb{Z}/p\mathbb{Z}; b)p^{-s} + p^{1-2s}}?$$

Using an idea of Sarnak, A. Booker [3] showed that if there existed a
Maass wave form f with the p-th Fourier coefficient given by $\pm \mathrm{Kl}((\mathbb{Z}/p\mathbb{Z}); 1)$
and of level N equal to a power of 2, then

$$N(\lambda + 3) > 2^{24},$$

where λ is the eigenvalue of the Laplacian at f. He also proved that it is
impossible to show the nonexistence of such a form numerically since there
always exists a Maass wave form for $SL_2(\mathbb{Z})$ whose normalized Fourier
coefficients at the prescribed finitely many places are as close to prescribed
values as we wish provided that the eigenvalue of the Laplacian is allowed
to grow unboundedly.

8   The Sato-Tate Conjecture

If a cuspidal automorphic representation π of $GL_n(\mathbb{A}_K)$ corresponds to a
representation ρ of G_K, then they have the same L-functions. In particular,
the Ramanujan-Petersson conjecture holds for π, that is, when $L(s, \pi)$ is
written as an Euler product

$$L(s, \pi) \approx \prod_v \prod_{1 \le i \le n} \frac{1}{1 - z_i(\pi_v)Nv^{-s}},$$

we have

$$|z_1(\pi_v)| = \cdots = |z_n(\pi_v)|.$$

Hence we may ask how the angles distribute, or equivalently, how the normalized jth coefficients of

$$(1 - z_1(\pi_v)x)\ldots(1 - z_n(\pi_v)x) = 1 + a_{1,v}x + \cdots + a_{n,v}x^n$$

distribute as v varies. When $n = 2$, this is the Sato-Tate conjecture, which originally was stated for elliptic curves.

Sato-Tate Conjecture for Elliptic Curves. *If E is an elliptic curve defined over $\mathbb{Q}$ which does not have complex multiplications, then the family $\{a_p/\sqrt{p}\}$, where p runs through all primes not dividing the conductor N of E, is uniformly distributed with respect to the Sato-Tate measure*

$$\mu_{ST} = \frac{1}{\pi}\sqrt{1 - \frac{x^2}{4}}\,dx \qquad on\ [-2, 2].$$

Here $a_p = 1 + p - N_p$, where N_p is the number of $\mathbb{Z}/p\mathbb{Z}$-rational points on the reduction of E mod p.

Since the Taniyama-Shimura conjecture has been established, we know that these a_p's are in fact eigenvalues of the Hecke operator at p acting on a cuspidal newform of weight 2 and level N. There is an obvious extension of the above conjecture to cuspidal newforms of level N, trivial character, any weight, and not of CM type.

There is good supporting numerical evidence for the conjecture, but it remains unproved for any elliptic curves over $\mathbb{Q}$ or holomorphic cuspidal automorphic forms for $GL_2(\mathbb{A}_{\mathbb{Q}})$.

The situation over a function field K is quite different. Indeed, the Sato-Tate conjecture holds for an elliptic curve defined over K with nonconstant j-invariant, as proved by Yoshida [47] in the 1970s. The Ramanujan-Petersson conjecture for $GL_2(\mathbb{A}_K)$ was proved by Drinfeld [17]. While the status of the global Langlands conjecture over K is unclear at the moment, the work of Lafforgue [31], [30] showed that the Ramanujan-Petersson conjecture always holds for cuspidal automorphic forms for $GL_n(\mathbb{A}_K)$.

In the joint papers [9], [10] we obtain more examples of cuspidal automorphic forms for $GL_2(\mathbb{A}_K)$ for which Sato-Tate conjecture holds.

Theorem F. *If a in Theorem C is not in the field of constants F of K, then the Sato-Tate conjecture holds for ϕ_a in Theorem C.*

Theorem G. *If a in Theorem D is not in the field of constants F of K, then the Sato-Tate conjecture holds for $\phi_{a,\varepsilon,\omega}$ in Theorem D.*

Theorem H. *If b in Theorem D is not in the field of constants F of K, then the Sato-Tate conjecture holds for ϕ_b in Theorem E.*

9   Applications

In this section we present applications of automorphic forms to constructing good combinatorial objects. Let X be a k-regular connected undirected graph on n vertices. Denote by A_X the adjacency matrix of X. Its rows and columns are parametrized by the vertices of X, and the ijth entry of A_X records the number of edges from vertex i to vertex j. The matrix A_X is best viewed as an operator which sends a function f defined on the vertices of X to another function Af whose value at the vertex x is given by

$$A_X f(x) = \sum_{y \text{ is a neighbor of } x} f(y).$$

Since A is a symmetric matrix, its eigenvalues are real. By the maximal modulus principle, one finds them lying between k and $-k$. So we can order them in non-increasing order

$$k = \lambda_1 > \lambda_2 \geq \cdots \geq \lambda_n \geq -k.$$

Here the largest eigenvalue is achieved once since the graph is k-regular and connected; the smallest eigenvalue $-k$ is achieved if and only if the graph X is bipartite. Denote by $\lambda^+(X)$ the largest eigenvalue $< k$ and by $\lambda^-(X)$ the smallest eigenvalue $> -k$. We are interested in the behavior of $\lambda^\pm(X)$ as X varies.

Let $\{X_j\}$ be a family of k-regular graphs with the size of X_j approaching infinity as j tends to infinity. The growth of $\lambda^+(X_j)$ is well-known:

Theorem (Alon-Boppana). $\liminf_{j\to\infty} \lambda^+(X_j) \geq 2\sqrt{k-1}$.

The behavior of $\lambda^-(X_j)$ is not always similar to that of $\lambda^+(X_j)$. For example, if $\{X_j\}$ is taken to be the family of line graphs of k-regular graphs, then the X_j's are $(2k-2)$-regular and all $\lambda^-(X_j) \geq -2$. On the other hand, I proved recently [36] that under an extra hypothesis one obtains a symmetrical result.

Proposition 1. *Assume the extra condition that the length of odd cycles (when they exist) in X_j tends to infinity as j tends to infinity. Then*

$$\limsup_{j\to\infty} \lambda^-(X_j) \leq -2\sqrt{k-1}.$$

This conclusion can also be derived from McKay's result [40] under the assumption that for each fixed length d, the number of cycles in X_j with length d divided by the size of X_j approaches zero as j tends to infinity. The above analysis leads to the following definition.

Definition. A k-regular graph X is Ramanujan if

$$|\lambda^{\pm}(X)| \leq 2\sqrt{k-1}.$$

In other words, a Ramanujan graph has nontrivial eigenvalues that are small in absolute value. Such a graph has a large magnifying constant and large girth. Hence it is a good communication network. The reader is referred to [34] for a comprehensive review of Ramanujan graphs.

The universal cover of a k-regular graph is the k-regular infinite tree. Its spectrum is $[-2\sqrt{k-1}, 2\sqrt{k-1}]$. Hence a Ramanujan graph has its nontrivial eigenvalues stay within the limits of the spectrum of its universal cover, while the nontrivial spectrum of a large regular graph in general has the tendency to "test the limit".

When $k = q+1$ for some prime power q, the infinite k-regular tree is the Bruhat-Tits building attached to $\mathrm{PGL}_2(K)/\mathrm{PGL}_2(\mathcal{O})$, where K is a local field with q elements in its residue field and $\mathcal{O}$ is the ring of integers in K. For example, $K = \mathbb{Q}_p$ and $\mathcal{O} = \mathbb{Z}_p$, or $K = \mathbb{F}_q((t))$ and $\mathcal{O} = \mathbb{F}_q[[t]]$. In this case, vertices are equivalence classes of rank 2 lattices over $\mathcal{O}$; two classes $[L_1]$ and $[L_2]$ are adjacent if and only if one can choose a representative from each class, say, L_1 and L_2, such that L_2 is a sublattice of L_1 with index q.

A systematic way to construct an infinite family of $(q+1)$-regular Ramanujan graphs is to use the theory of automorphic forms on quaternion groups which we describe below.

Fix a prime p. The graphs to be constructed will have valency $k = p+1$. Let H be a definite quaternion algebra defined over $\mathbb{Q}$ unramified at p. Denote by D the multiplicative group of H divided by its center, viewed as an algebraic group. Our graph X has its vertices the double coset space $D(\mathbb{Q})\backslash D(\mathbb{A}_\mathbb{Q})/D(\mathbb{R}) \prod_\ell D(\mathcal{O}_\ell)$, where ℓ runs through all primes. By strong approximation theory, we can choose coset representatives locally at p, that is, X can also be seen as the double coset space $D(\mathbb{Z}[\frac{1}{p}])\backslash D(\mathbb{Q}_p)/D(\mathbb{Z}_p)$, which is the same as $D(\mathbb{Z}[\frac{1}{p}])\backslash \mathrm{PGL}_2(\mathbb{Q}_p)/\mathrm{PGL}_2(\mathbb{Z}_p)$ since H is unramified at p.

The third expression gives the graph structure of X, namely, as a quotient of the infinite $(p+1)$-regular tree $\mathrm{PGL}_2(\mathbb{Q}_p)/\mathrm{PGL}_2(\mathbb{Z}_p)$ by the group $D(\mathbb{Z}[\frac{1}{p}])$ so that X is $(p+1)$-regular. The first expression of X allows us to view functions on vertices of X as automorphic forms on the quaternion group $D(\mathbb{A}_\mathbb{Q})$ left invariant by the global rational points of D, and right invariant by the real points and the product of the standard maximal compact subgroup at each nonarchimedean place ℓ. The space C of such automorphic forms is known to contain the constant functions, and, if X is bipartite, the alternating constant functions taking opposite value on adjacent vertices. The orthogonal complement C' of such functions in C can be viewed as certain cusp forms of weight 2 for $GL_2(\mathbb{A}_\mathbb{Q})$ via Jacquet-Langlands correspondence, as proved in [19]. The restriction of the adjacency matrix A_X to C' is nothing but the Hecke operator T_p, and the nontrivial eigenvalues of A_X are therefore eigenvalues of T_p on certain cusp forms of weight 2, which are known to satisfy the Ramanujan-Petersson conjecture, that is, have absolute value majorized by $2\sqrt{p} = 2\sqrt{k-1}$. This shows that X is a Ramanujan graph.

We can replace $D(\mathbb{Z}[\frac{1}{p}])$ by congruence subgroups to get an infinite family of $(p+1)$-regular Ramanujan graphs. These were first constructed by Lubotzky-Phillips-Sarnak [38], and independently by Margulis [39]. The parallel construction with $\mathbb{Q}$ replaced by a function field K was done by Morgenstern [41]. There the Ramanujan-Petersson conjecture was proved by Drinfeld [17]. For a totally real field K, the Ramanujan-Petersson conjecture holds for certain holomorphic forms for $GL_2(\mathbb{A}_K)$, such as those from the work of [44], [5] and [2], as discussed in Section 5, and from the work of Carayol [6] coupled with the results of Deligne in [14]. Using these, Jordan and Livné [26] extended the construction of Ranamujan graphs to totally real fields.

There are other number-theoretic constructions, which give rise to finitely many Ramanujan graphs for each fixed k. They turn out to be quotients of the Ramanujan graphs constructed over function fields [35]. For example, the Cayley graph on the additive group $\mathbb{F}_{q^2}$ with generator set N_1, the set of elements in $\mathbb{F}_{q^2}$ whose norm to $\mathbb{F}$ is 1, is a $(q+1)$-regular Ramanujan graph, called a norm graph. See [33] for more details. The nontrivial eigenvalues of a norm graph are $-\operatorname{Kl}(\mathbb{F}_q; b)$. Theorem E describes an explicit automorphic form for GL_2 over a function field which is an eigenfunction of Hecke operators at good places with eigenvalues equal to eigenvalues of norm graphs, given by Kloosterman sums.

Another example is the Cayley graph on the cosets $GL_2(\mathbb{F}_q)/M$, where M is an embedded image of $\mathbb{F}_{q^2}^\times$ in $GL_2(\mathbb{F}_q)$, with generator set the M-cosets of a double M coset S, constructed by Terras. It is also a $(q+1)$-regular Ramanujan graph, called a Terras graph [1], [8]. The quantities $-\lambda_\chi(\mathbb{F}_q; a)$ occurring in Section 6 are part of the eigenvalues of Terras graphs. Theorems C and D describe automorphic forms for GL_2 over a function field which are eigenfunctions of Hecke operators at good places with eigenvalues equal to eigenvalues of Terras graphs. Theorems F, G, and H may be regarded as describing the distribution of eigenvalues of Terras graphs and norm graphs, respectively.

We end this section by considering a generalization of Ramanujan graphs to 3-hypergraphs.

Again let K be a nonarchimedean local field with q elements in its residue field. Denote by $\mathcal{O}$ the ring of integers, and by ϖ a uniformizer of $\mathcal{O}$. The structure of the Bruhat-Tits building $\mathcal{B}_{3,K}$ for $\mathrm{PGL}_3(K)/\mathrm{PGL}_3(\mathcal{O})$ is as follows. Its vertices are equivalence classes of lattices of rank 3 over $\mathcal{O}$. There is an edge from vertex $[L_1]$ to vertex $[L_2]$ if these two classes can be represented by lattices L_1 and L_2 respectively such that L_2 is a sublattice of L_1 with index q. Thus lattices L_1, L_2, L_3 represent three vertices $[L_1]$, $[L_2]$, and $[L_3]$ such that there is a loop $[L_1] \to [L_2] \to [L_3] \to [L_1]$ of length 3 if and only if $L_1 \supset L_2 \supset L_3 \supset \varpi L_1$ and each lattice is a sublattice of its predecessor with index q. If this happens, we say that $[L_1]$, $[L_2]$, $[L_3]$ form a face. This makes the Bruhat-Tits building a two-dimensional simplicial complex, which is topologically contractible. Each vertex has $q^2 + q + 1$ out-neighbors and $q^2 + q + 1$ in-neighbors.

There are two Hecke operators T and S acting on functions f defined on vertices of the building as follows:

(a) $(Tf)(x) = \sum_{x \to y} f(y)$,

(b) $(Sf)(x) = \sum_{y \to x} f(y)$

for all vertices x on $\mathcal{B}_{3,K}$. The spectrum of T is known to be the set

$$\Omega_T := \{q(\alpha + \beta + \gamma) \mid \alpha, \beta, \gamma \in \mathbb{C}, |\alpha| = |\beta| = |\gamma| = 1, \alpha\beta\gamma = 1\}.$$

The region Ω_T is invariant by multiplication by a cube root of unity ζ_3, that is, rotation by $2\pi/3$. Its boundary, consisting of the points $q(2e^{i\theta} + e^{-2i\theta})$ for $0 \le \theta \le 2\pi$, is smooth except for three singularities at $3q$, $3\zeta_3 q$, and $3\bar{\zeta}_3 q$, respectively, with a hypocycloid connecting any two of the three singular points. With respect to the usual Hermitian inner product $<,>$ on the space of ℓ^2-functions on vertices of $\mathcal{B}_{3,K}$, S and T are transposes of each other. Hence the spectrum of S, being the complex conjugation of the spectrum of T, is also Ω_T. Further, $ST = TS$. The reader is referred to the article [7] by Cartwright and Młotkowski for details.

A finite (q^2+q+1)-regular oriented 3-hypergraph is a finite 2-dimensional simplicial complex for which the Bruhat-Tits building $\mathcal{B}_{3,K}$ is its universal cover; each 0-dimensional simplex is called a vertex and each 2-dimensional triangular face with oriented 1-dimensional boundary is called an oriented 3-hyperedge. Its 1-dimensional skeleton is a directed graph such that each vertex has $q^2 + q + 1$ out-neighbors and $q^2 + q + 1$ in-neighbors. Let $\{X_j\}$

be a family of such hypergraphs with $|X_j| \to \infty$ as $j \to \infty$. Denote by $T_j := T_{X_j}$ the operator acting on functions on vertices of X_j defined by (a).

Parallel to the Alon-Boppana theorem for graphs is the following unconditional lower bound for hypergraphs.

Theorem I. *On each X_j there is a real valued function f_j, perpendicular to the constant functions and of norm one, such that*

$$\liminf_{j \to \infty} \langle T_j f_j, f_j \rangle \geq 3q.$$

Similar to the conditional upper bound in Proposition 1 for graphs, we have the following conditional result which gives examples of families of hypergraphs whose spectra combined are at least dense in the region Ω_T.

Theorem J. *Let $\Gamma_1 \supset \Gamma_2 \supset \cdots$ be a tower of discrete subgroups of $\mathrm{PGL}_3(K)$ such that*
(1) $X_j := \Gamma_j \backslash \mathrm{PGL}_3(K) / \mathrm{PGL}_3(\mathcal{O})$ is a finite, $(q^2 + q + 1)$-regular 3-hypergraph;
(2) $\lim_{j \to \infty} \Gamma_j = \{identity\}$.
Then, given any $\lambda \in \Omega_T$, there exists a function f_j on X_j with norm 1 such that the norm of $T_j f_j - \lambda f_j$ tends to 0 as j approaches ∞.

Consequently, for every $\lambda \in \Omega_T$, there exist eigenvalues λ_j of T_j such that λ_j converges to λ as $j \to \infty$.

The above discussion leads us to the following definition of a Ramanujan 3-hypergraph, analogous to a Ramanujan graph.

Definition. A finite $(q^2 + q + 1)$-regular oriented 3-hypergraph X is *Ramanujan* if all eigenvalues of T_X other than $q^2 + q + 1, (q^2 + q + 1)\zeta_3, (q^2 + q + 1)\bar\zeta_3$ fall in Ω_T.

In a joint work with Yu [37], we construct, for each fixed q, an infinite family of $(q^2 + q + 1)$-regular oriented Ramanujan 3-hypergraphs.

Theorem K. *Let K be a rational function field $F(t)$ over a finite field F of q elements. Let v be a place of K of degree 1. Let $\mathcal{D}$ be a division algebra of degree 9 over K which is unramified at v but ramified at a place w of degree 1. Denote by D the multiplicative group of $\mathcal{D}$ divided by its center. Then for a congruence subgroup $\mathcal{K}$ of $\prod_{v' \neq v, w} D(\mathcal{O}_{v'})$, the double coset space*

$$X_{\mathcal{K}} = D(K) \backslash D(\mathbb{A}_K) / D(K_w) D(\mathcal{O}_v) \mathcal{K},$$

which can also be represented as local double coset space

$$\Gamma_{\mathcal{K}} \backslash \mathrm{PGL}_3(K_v) / \mathrm{PGL}_3(\mathcal{O}_v),$$

is a Ramanujan $(q^2 + q + 1)$-regular oriented 3-hypergraph. Here $\Gamma_{\mathcal{K}}$ is the intersection of $D(K)$ with $\mathcal{K}$.

The proof is parallel to our description of the explicit construction of Ramanujan graphs. The key ingredients are approximation theory, the correspondence from automorphic representations of division group $D(\mathbb{A}_K)$ to automorphic representations of $PGL_3(\mathbb{A}_K)$ as described in [32], and the Ramanujan conjecture for GL_3 proved by Lafforgue [30].

References

[1] J. Angel, N. Celniker, S. Poulos, A. Terras, C. Trimble, and E. Velasquez, *Special functions on finite upper half planes*, Hypergeometric functions on domains of positivity, Jack polynomials, and applications (Tampa, FL, 1991), Contemp. Math., vol. 138, Amer. Math. Soc., Providence, RI, 1992, pp. 1–26.

[2] D. Blasius and J. Rogawski, *Motives for Hilbert modular forms*, Invent. Math. **114** (1993), 55–87.

[3] A. Booker, *A test for identifying Fourier coefficients of automorphic forms and application to Kloosterman sums*, Experimental Math. **9** (2000), 571–581.

[4] C. Breuil, B. Conrad, F. Diamond, and R. Taylor, *On the modularity of elliptic curves over* $\mathbb{Q}$, J. Amer. Math. Soc. **14** (2001), 843–939.

[5] J.-L. Brylinski and J.-P. Labesse, *Cohomologie d'intersection et fonctions L de certaines variétés de Shimura*, Ann Scient. Éc. Norm. Sup. **17** (1984), 361–412.

[6] H. Carayol, *Sur les représentations ℓ-adiques associées aux formes modulaires de Hilbert*, Ann. Sci. École Norm. Sup. **19** (1986), 409–468.

[7] D. Cartwright and W. Młotkowski, *Harmonic analysis for groups acting on triangle buildings*, J. Austral. Math. Soc. Ser. A **56** (1994), 345–383.

[8] N. Celniker, S. Poulos, A. Terras, C. Trimble, and E. Velasquez, *Is there life on finite upper half planes?*, A tribute to Emil Grosswald: number theory and related analysis, Contemp. Math., vol. 143, Amer. Math. Soc., Providence, RI, 1993, pp. 65–88.

[9] C.-L. Chai and W.-C. W. Li, *Character sums, automorphic forms, equidistribution, and Ramanujan graphs. Part I. The Kloosterman sum conjecture over function fields*, Math Forum, to appear.

[10] _____, *Character sums, automorphic forms, equidistribution, and Ramanujan graphs. Part II. Eigenvalues of Terras graphs*, preprint, 2000.

[11] J. W. Cogdell and I. I. Piatetski-Shapiro, *A converse theorem for GL_4*, Math. Research Letters **3** (1996), 67–76.

[12] P. Deligne, *Formes modulaires et représentations ℓ-adiques*, Séminaire Bourbaki, vol. 1968/69, Lecture Notes in Math., vol. 179, Springer, Berlin, 1971.

[13] _____, *Les constantes des équations fonctionelles des fonctions L*, Lecture Notes in Math., vol. 349, Springer, 1973, pp. 501–595.

[14] _____, *La conjecture de Weil II*, IHES Publ. Math. **52** (1980), 137–252.

[15] P. Deligne and J.-P. Serre, *Formes modulaires de poids 1*, Ann. Sci. École Norm. Sup. **7** (1974), 507–530.

[16] V. G. Drinfel'd, *Proof of the global Langlands conjecture for $GL(2)$ over a function field*, Functional Anal. Appl. **11** (1977), 223–225.

[17] _____, *The proof of Petersson's conjecture for $GL(2)$ over a global field of characteristic p*, Functional Anal. Appl. **22** (1988), 28–43.

[18] G. Faltings, *Endlichkeitssätze für abelsche Varietäten über Zahlkörpern*, Invent. Math. **73** (1983), 349–366.

[19] S. Gelbart and H. Jacquet, *Forms of $GL(2)$ from the analytic point of view*, Automorphic forms, representations and L-functions (Corvallis, Ore., 1977), Part 1, Proc. Sympos. Pure Math., vol. XXXIII, Amer. Math. Soc., 1979, pp. 213–251.

[20] M. Harris and R. Taylor, *On the geometry and cohomology of some simple Shimura varieties*, preliminary version, 1998.

[21] G. Henniart, *La conjecture de Langlands locale pour $GL(3)$*, Mémoire Soc. Math. France, France, 1984.

[22] _____, *Caractérisation de la correspondance de Langlands locals par les facteurs ε de paires*, Invent. Math. **113** (1993), 339–350.

[23] _____, *Une preuve simple des conjectures de Langlands pour GL_n sur un corps p-adique*, Invent. Math. **139** (2000), 439–455.

[24] H. Jacquet and R. P. Langlands, *Automorphic Forms on $GL(2)$*, Springer-Verlag, Berlin-Heidelberg-New York, 1970.

[25] H. Jacquet, I. I Piatetski-Shapiro, and J. Shalika, *Automorphic forms on GL₃, I & II*, Ann. of Math. **109** (1979), 169–258.

[26] B. Jordan and R. Livné, *The Ramanujan property for regular cubical complexes*, Duke Math. J. **105** (2000), 85–103.

[27] N. Katz, *Gauss Sums, Kloosterman Sums, and Monodromy Groups*, Princeton Univ. Press, Princeton, 1988.

[28] P. Kutzko, *The Langlands conjecture for GL₂ of a local field*, Ann. of Math. **112** (1980), 381–412.

[29] P. Kutzko and A. Moy, *On the local Langlands conjecture in prime dimension*, Ann. of Math. **121** (1985), 495–517.

[30] L. Lafforgue, *Chtoucas de Drinfeld et conjecture de Ramanujan-Petersson*, Astérisque **243** (1997).

[31] _____, *La correspondance de Langlands sur les corps de fonctions*, preprint, 2000.

[32] G. Laumon, M. Rapoport, and U. Stuhler, *D-elliptic sheaves and the Langlands correspondence*, Invent. Math. **113** (1993), 217–338.

[33] W.-C. W. Li, *Character sums and abelian Ramanujan graphs*, J. Number Theory **41** (1992), 199–217.

[34] _____, *A survey of Ramanujan graphs*, Arithmetic, geometry and coding theory (Luminy, 1993), de Gruyter, Berlin, 1996, pp. 127–143.

[35] _____, *Eigenvalues of Ramanujan graphs*, Emerging applications of number theory (Minneapolis, MN, 1996), IMA Vol. Math. Appl., vol. 109, Springer, New York, 1999, pp. 387–403.

[36] _____, *On negative eigenvalues of regular graphs*, C. R. Acad. Sci. Paris Sér. I Math. **333** (2001), 907–912.

[37] W.-C. W. Li and R.-K. Yu, *Ramanujan 3-graphs*, preprint, 2000.

[38] A. Lubotzky, R. Phillips, and P. Sarnak, *Ramanujan graphs*, Combinatorica **8** (1988), 261–277.

[39] G. Margulis, *Explicit group theoretic constructions of combinatorial schemes and their application to the design of expanders and concentrators*, J. Prob. Info. Trans. (1988), 39–46.

[40] B. D. McKay, *The expected eigenvalue distribution of a large regular graph*, Linear Algebra Appl. **40** (1981), 203–216.

[41] M. Morgenstern, *Existence and explicit constructions of $q + 1$ regular Ramanujan graphs for every prime power q*, J. Comb. Theory Ser. B **62** (1994), 44–62.

[42] H. Reimann, *The semi-simple zeta function of quaternionic Shimura varieties*, Springer-Verlag, Berlin-Heidelberg-New York, 1997.

[43] K. Ribet, *On modular representations of $Gal(\overline{\mathbb{Q}}/\mathbb{Q})$ arising from modular forms*, Invent. Math. **100** (1990), 431–476.

[44] J. Rogawski and J. Tunnell, *On Artin L-functions associated to Hilbert modular forms of weight one*, Invent. Math. **74** (1983), 1–42.

[45] R. Taylor and A. Wiles, *Ring-theoretic properties of certain Hecke algebras*, Ann. of Math. **141** (1995), 553–572.

[46] A. Wiles, *Modular elliptic curves and Fermat's last theorem*, Ann. of Math. **141** (1995), 443–551.

[47] H. Yoshida, *On an analogue of the Sato-Tate conjecture*, Invent. Math. **19** (1973), 261–277.

Convergence of Corresponding Continued Fractions

Lisa Lorentzen

1 Introduction

There exist several methods to expand a given function $f(z)$ into a continued fraction

$$b_0 + K(a_n/b_n) = b_0 + \frac{a_1}{b_1} + \frac{a_2}{b_2} + \dots = b_0 + \cfrac{a_1}{b_1 + \cfrac{a_2}{b_2 + \dots}}, \qquad (1.1)$$

where the elements a_n and b_n are polynomials or other (mostly entire) functions of z. What we hope is that (1.1) *converges* to f; that is, that its sequence of *approximants*

$$f_n = b_0 + \frac{a_1}{b_1} + \frac{a_2}{b_2} + \dots + \frac{a_n}{b_n}; \qquad n = 0, 1, 2, \dots \qquad (1.2)$$

converges to f in some domain $D \subseteq \mathbb{C}$. We regard continued fraction expansions as alternatives to Laurent series expansions $L =: \mathcal{L}(f)$ of f at $z = 0$ or asymptotic series L of f. What we want is that

- $b_0 + K(a_n/b_n)$ converges in a larger domain than L, uniformly on compact subsets, to f;

- $b_0 + K(a_n/b_n)$ converges faster than L in the convergence disk of L;

- $b_0 + K(a_n/b_n)$ lends itself to easy ways of convergence acceleration;

- $b_0 + K(a_n/b_n)$ lends itself to easy ways of meromorphic continuation;

- the computation of f_n is stable.

And, luckily enough, this is what we get in a number of useful cases.

In this paper we shall mainly concentrate on the problem of when $b_0 + K(a_n/b_n)$ converges to f. There exist a number of theorems giving sufficient conditions for a continued fraction to converge, but very few

149

of these use the additional information that $b_0 + K(a_n/b_n)$ is a continued fraction expansion of some given function.

In Section 2 we describe what it means that $b_0 + K(a_n/b_n)$ is an expansion of f. In Section 3 we give some basic results on the chordal metric and extend some earlier results. In Section 4 we present an alternative definition of convergence which extends the classical concept mentioned above. In Section 5 we give some general results on convergence of $b_0 + K(a_n/b_n)$, and in Section 6 we handle the simple but useful case when $b_0 + K(a_n/b_n)$ is limit periodic.

If the elements a_n and b_n are polynomials, then the approximants $f_n(z)$ are rational functions. Under additional conditions they are *Padé approximants* to f, and for these we can prove stronger results, as shown in Section 7. As an example, in Section 8 we treat *regular C-fractions* $b_0 + K(a_n z/1)$, where all a_n are complex numbers. A particularly beautiful case occurs when $a_n > 0$ for all n. These *Stieltjes fractions* are described in Section 9. Finally, in Section 10 we return to the regular C-fractions for additional results. Throughout this paper, $\widehat{\mathbb{C}} := \mathbb{C} \cup \{\infty\}$ denotes the Riemann sphere, and a *domain* is always an open, connected set in $\mathbb{C}$.

2   Correspondence

We shall here clarify what we mean when we say that (1.1) is a continued fraction expansion of f. Actually, what we really say is that $b_0 + K(a_n/b_n)$ corresponds to a Laurent series at $z = 0$. So let Λ denote the family of formal Laurent series $\sum_{n=n_0}^{\infty} c_n z^n$, $n_0 \in \mathbb{Z}$. In this family, we let $\mathrm{ord}(\sum_{k=n_0}^{\infty} c_k z^k) := n_0$ when $c_{n_0} \neq 0$, with the limit form $\mathrm{ord}(\ell_0) = \infty$ for $\ell_0 \equiv 0$. We let further $\mathcal{M}$ denote the family of functions that are meromorphic at $z = 0$. Then $\mathcal{L}(f) \in \Lambda$ for every $f \in \mathcal{M}$. For simplicity we also define $\mathrm{ord}(\mathcal{L}(f)) := -\infty$ for the function $f(z) \equiv \infty$. We say that a sequence $\{g_n\}$ of functions *corresponds* (at $z = 0$) to some $L \in \Lambda$ if $g_n \in \mathcal{M}$ for sufficiently large n, and

$$\nu_n := \mathrm{ord}(\mathcal{L}(g_n) - L) \to \infty \quad \text{as } n \to \infty. \tag{2.1}$$

ν_n is called the *order of correspondence*, and we write $\{g_n\} \sim L$. Evidently, if $\{g_n\} \sim L$ for some $L \in \Lambda$, then L is unique. We say that $b_0 + K(a_n/b_n)$ corresponds to $L \in \Lambda$ if its approximants satisfy $\{f_n\} \sim L$, and we write $b_0 + K(a_n/b_n) \sim L$.

We say that $b_0 + K(a_n/b_n)$ is a *corresponding continued fraction* if there exists an $L \in \Lambda$ such that $b_0 + K(a_n/b_n) \sim L$. According to Jones and Thron we have:

Theorem 2.1 ([14, Thm. 5.11, p. 151]). *A given sequence $\{g_n\}$ from $\mathcal{M}$ corresponds to some $L \in \Lambda$ if and only if $\lim_{n\to\infty} \mathrm{ord}(\mathcal{L}(g_n) - \mathcal{L}(g_{n+1})) = \infty$.*

Hence, $b_0 + K(a_n/b_n)$ is a corresponding continued fraction if and only if $f_n \in \mathcal{M}$ for sufficiently large n, and

$$\mathrm{ord}(\mathcal{L}(f_n) - \mathcal{L}(f_{n+1})) \to \infty \quad \text{as } n \to \infty. \tag{2.2}$$

If further $f \in \mathcal{M}$ and $b_0 + K(a_n/b_n) \sim L = \mathcal{L}(f)$, we say that $b_0 + K(a_n/b_n)$ is a *continued fraction expansion* of f (at $z = 0$), and we write $b_0 + K(a_n/b_n) \sim f$. We also do so if $D \subset \mathbb{C}$ is a domain with 0 on its boundary, f is meromorphic in D, and L is an asymptotic expansion of f in D when $b_0 + K(a_n/b_n) \sim L$.

One may consider correspondence at points in $\widehat{\mathbb{C}}$ other than the point $z = 0$, but in this paper we shall always mean correspondence at the origin. For more information on correspondence and other basic properties of continued fractions we refer to [14], [18], [25].

3 The Chordal Metric and Normal Families

For continued fractions it is natural to use the chordal metric

$$d(z_1, z_2) = \frac{2|z_1 - z_2|}{\sqrt{1 + |z_1|^2}\sqrt{1 + |z_2|^2}} \quad \text{for } z_1, \ z_2 \in \widehat{\mathbb{C}} \tag{3.1}$$

on the Riemann sphere, where $d(z_1, z_2)$ takes the obvious limit forms if z_1 or z_2 is infinite. The advantage of this metric is that $\{z_n\}$ converges to $z \in \widehat{\mathbb{C}}$ (in the usual sense) if and only if $d(z_n, z) \to 0$, regardless of whether z is finite or infinite. That is, convergence to ∞ is no different from convergence to any other number, when expressed in this metric. This allows us to treat meromorphic functions in the same way as we treat analytic functions. In particular, a family $\mathcal{F}$ of meromorphic functions on a domain D is *normal* if every sequence $\{f_n\}$ from $\mathcal{F}$ has a subsequence which converges spherically uniformly on compact subsets of D. Here *spherically uniform convergence* means uniform convergence with respect to the chordal metric. The following two standard results hold; see, for instance, [27, p. 72–75].

Theorem 3.1. *Let $\{f_n\}$ be a sequence of meromorphic functions on a domain D which converges spherically uniformly on compact subsets of D to f. Then f is either meromorphic in D or identically equal to ∞.*

Theorem 3.2. *Let $\mathcal{F}$ be a family of meromorphic functions on a domain D. If either*

(i) *there exist three distinct points p_1, p_2, $p_3 \in \widehat{\mathbb{C}}$ such that $f(z) \neq p_j$ for all $f \in \mathcal{F}$, $z \in D$ and $j = 1, 2, 3$; or*

(ii) *for every compact subset $E \subset D$ there exists a constant $M > 0$ such that the spherical derivative satisfies*

$$f^{\#}(z) := \frac{2|f'(z)|}{1 + |f(z)|^2} \leq M \quad \text{for all } f \in \mathcal{F} \text{ and } z \in E,$$

then $\mathcal{F}$ is normal in D.

The importance of Theorem 3.2 for our purpose lies in the following result, which generalizes results by Jones and Thron [13] and Baker and Graves-Morris [1].

Theorem 3.3. *Let $\{f_n(z)\}$ be a sequence of meromorphic functions on a domain D containing the origin. If $\{f_n\} \sim L$ and the family $\mathcal{F} := \{f_n : n \geq n_0\}$ is normal in D for some $n_0 \in \mathbb{N}$, then $\{f_n\}$ converges spherically uniformly on compact subsets of D to a function f with $\mathcal{L}(f) = L$ that is meromorphic in D.*

Proof. Since $\mathcal{F}$ is normal, there is a subsequence $\{f_{n_k}\}$ of $\{f_n\}$ which converges spherically uniformly on compact subsets of D to some function f. It suffices to prove that $\mathcal{L}(f) = L$, since this ensures that f is unique and $f \not\equiv \infty$.

Suppose first that $f(0) \neq \infty$. Then there exists a $\delta > 0$ such that f is analytic in the closed disk $E_\delta := \{z : |z| \leq \delta\} \subseteq D$. By the uniformity of the spherical convergence of $\{f_{n_k}\}$ to f on E_δ, it follows that f_{n_k} is analytic in E_δ for sufficiently large k. By Weierstrass' theorem it follows that the derivatives $f_{n_k}^{(m)}$ satisfy $f_{n_k}^{(m)}(0) \to f^{(m)}(0)$ as $k \to \infty$. This can only happen if $\mathcal{L}(f) = L$.

Next, suppose that $f(0) = \infty$. Then $g_{n_k} := 1/f_{n_k}$ converges spherically uniformly on compact subsets of D to $g := 1/f$. In particular, $g(0) = 0$, and thus it follows from the arguments above that $\mathcal{L}(g) = 1/L$; i.e., we have $\mathcal{L}(f) = L$. $\square$

4   General Convergence

We suppose first that the elements a_n and b_n of $b_0 + K(a_n/b_n)$ are complex numbers. We also require that $a_n \neq 0$ for all n, so that the continued fraction is non-terminating. In 1942 Paydon and Wall [24] described $b_0 + K(a_n/b_n)$ by means of the linear fractional transformations

$$s_0(w) := b_0 + w, \qquad s_n(w) := a_n/(b_n + w) \quad \text{for } n = 1, 2, 3, \ldots. \tag{4.1}$$

(Note that s_n is non-singular since $a_n \neq 0$.) They formed the compositions

$$S_n(w) := s_0 \circ s_1 \circ s_2 \circ \cdots \circ s_n(w) = b_0 + \frac{a_1}{b_1} + \frac{a_2}{b_2} + \cdots + \frac{a_n}{b_n + w}, \quad (4.2)$$

which again are (non-singular) linear fractional transformations. Then the approximants of $b_0 + K(a_n/b_n)$ can be written as $f_n = S_n(0)$. In [8] it was shown that for the periodic continued fraction

$$\frac{2}{1} + \frac{1}{1} - \frac{1}{1} + \frac{2}{1} + \frac{1}{1} - \frac{1}{1} + \frac{2}{1} + \frac{1}{1} - \frac{1}{1} + \ldots \quad (4.3)$$

$S_{3n}(w) \to 1/2$ for every $w \neq 0$, $S_{3n+1}(w) \to 1/2$ for every $w \neq \infty$, and $S_{3n+2}(w) \to 1/2$ for every $w \neq -1$. Since $S_{3n}(0) = 0$ for all n, this shows that (4.3) diverges, although it has a very strong convergence behavior. To catch this kind of convergence, a new concept of convergence was introduced. Since we allow convergence to ∞, it was natural to base the concept on the chordal metric (3.1).

Definition 4.1 ([8]). *We say that $b_0 + K(a_n/b_n)$, with elements a_n, b_n from $\mathbb{C}$ and $a_n \neq 0$ for all n, converges generally to f if there exist two sequences $\{u_n\}$ and $\{v_n\}$ of numbers from $\widehat{\mathbb{C}}$ such that*

$$\liminf_{n\to\infty} d(u_n, v_n) > 0 \qquad and \qquad \lim_{n\to\infty} S_n(u_n) = \lim_{n\to\infty} S_n(v_n) = f. \quad (4.4)$$

Looking back at (4.3), any sequences $\{w_n\}$ with $\liminf d(w_{3n+2}, -1) > 0$, $\liminf d(w_{3n+1}, \infty) > 0$ and $\liminf d(w_{3n}, 0) > 0$ can be used for $\{u_n\}$ and $\{v_n\}$ in (4.4) to prove that (4.3) converges generally to $1/2$. This is a typical situation:

Theorem 4.2 ([8]). *Let $b_0 + K(a_n/b_n)$, with elements a_n, b_n from $\mathbb{C}$ and $a_n \neq 0$ for all n, converge generally to f. Then there exists a sequence $\{\zeta_n\}$ of numbers from $\widehat{\mathbb{C}}$ such that $S_n(w_n) \to f$ for every sequence $\{w_n\}$ satisfying $\liminf_{n\to\infty} d(w_n, \zeta_n) > 0$.*

We say that $\{\zeta_n\}$ is an *exceptional sequence* for the generally convergent continued fraction $b_0 + K(a_n/b_n)$. The proof of this theorem is so short that I include it here:

Proof. Without loss of generality we assume that $b_0 + K(a_n/b_n)$ converges to $f \neq \infty$. We can do this, since

$$\frac{1}{b_0} + \frac{a_1}{b_1} + \frac{a_2}{b_2} + \frac{a_3}{b_3} + \ldots = \{b_0 + K(a_n/b_n)\}^{-1} \sim \frac{1}{f}. \quad (4.5)$$

Let $\zeta_n := S_n^{-1}(\infty)$ for all n, let $\{u_n\}$ and $\{v_n\}$ be chosen such that (4.4) holds, and let $\{w_n\} \subset \widehat{\mathbb{C}}$ be such that $\liminf d(w_n, \zeta_n) > 0$. We consider indices n for which $w_n \neq u_n, v_n$. Then u_n, v_n, ζ_n and w_n are distinct points for every sufficiently large n. The invariance of the cross ratio under linear fractional transformations therefore gives that

$$\frac{d(S_n(u_n), S_n(w_n))}{d(S_n(u_n), S_n(v_n))} = \frac{d(u_n, w_n) \cdot d(\zeta_n, v_n)}{d(u_n, v_n) \cdot d(\zeta_n, w_n)}. \tag{4.6}$$

Since the right hand side is bounded and $d(S_n(u_n), S_n(v_n)) \to 0$, we must have $d(S_n(u_n), S_n(w_n)) \to 0$. $\qquad\square$

(Theorem 4.2 can be proved without this trick with the cross ratio. In [17] this was done in a more general setting, not only for convergence, but also for functions $\{S_n\}$ with a limiting structure [11].)

Theorem 4.2 shows that the limit f of a generally convergent continued fraction is unique. Moreover, classical convergence to f implies general convergence to f since $S_n(0) = S_{n+1}(\infty)$. The converse is not true, since (4.3) converges generally to $\frac{1}{2}$. In fact, this new definition captures every continued fraction that ought to converge, but which diverges in the classical sense because its exceptional sequence $\{\zeta_n\}$ has a limit point at 0.

Given $g \in \widehat{\mathbb{C}}$, we say that $g_n := S_n^{-1}(g)$ is a *tail sequence* for $b_0 + K(a_n/b_n)$. If $b_0 + K(a_n/b_n)$ converges generally to f and $\{\zeta_n\}$ is an exceptional sequence for $b_0 + K(a_n/b_n)$, then it follows from Theorem 4.2 that $\lim d(g_n, \zeta_n) = 0$ for every $g \neq f$. That is, every such tail sequence is an exceptional sequence. On the other hand, if $g = f$ and we choose $w_n = g_n$, then $S_n(w_n) = g = f$ for all n, which shows that this tail sequence is a particularly good sequence. In fact, an easy way to accelerate the convergence of $b_0 + K(a_n/b_n)$ is to use approximants $S_n(w_n)$ satisfying $\lim d(w_n, S_n^{-1}(f)) = 0$. This is in particular so for limit periodic continued fractions. In some cases this can even give a meromorphic extension of the limit function; see [15].

Also, the stability of the computation is improved if we choose w_n appropriately. If $f \neq \infty$, we choose w_n such that the denominator of the linear fractional transformation

$$S_n(w_n) = b_0 + \frac{a_1}{b_1 +} \cdots + \frac{a_n}{b_n + w_n} = \frac{A_n w_n + B_n}{C_n w_n + D_n} \tag{4.7}$$

stays away from 0, and if $f \neq 0$, we make sure that its numerator stays away from 0. But how can we choose w_n if f is totally unknown? One idea is to compute

$$S_n(0), \qquad S_n(\infty) = S_{n-1}(0) \quad \text{and} \quad S_n(-b_n) = S_{n-2}(0) \tag{4.8}$$

and use the average of the two values that lie closest to each other on the Riemann sphere. This works if $\{b_n\}$ is bounded away from 0 and ∞.

Otherwise one can replace $S_n(-b_n)$ by $S_n(1)$. Another idea is to make use of *value sets* $\{V_n\}_{n=-1}^{\infty}$; that is, the V_n are proper subsets of $\widehat{\mathbb{C}}$ with non-empty interiors such that $s_n(V_n) \subset V_{n-1}$ for all n. Then there is an exceptional sequence $\{\zeta_n\}$ with $\zeta_n \notin V_n$ for all n; see [18]. Hence, if the chordal diameter of the interior of V_n stays $\geq 2\delta > 0$ for all n, then, for each n, we can choose w_n as the center of a δ–disk in V_n.

Let us return to the situation where $b_0 + K(a_n/b_n)$ is a corresponding continued fraction. It is important that the variable w of S_n not be confused with the variable z of a_n and b_n. For instance, in (4.7) the coefficients of S_n are functions of z, so S_n is really a function of two complex variables. To keep the notation simple, will still write $S_n(w)$. The modifying factors w_n will normally also depend on z. It is natural to choose $\{w_n\}$ such that the functions $S_n(w_n)$ are meromorphic functions of z in the domain D of interest, and to ask for spherically uniform convergence of $S_n(w_n)$ on compact subsets of D.

5   Some General Results

Let f be a function that is analytic at $z = 0$, and let $b_0 + K(a_n/b_n) \sim f$. It is then very natural to assume that if $b_0 + K(a_n/b_n)$ converges in some domain $D \subset \mathbb{C}$, then its limit is $f(z)$ in D, at least if the approximants are analytic in D and the convergence is uniform on compact subsets of D. However, this is not always the case. For instance, the periodic continued fraction

$$\frac{z}{1-z} + \frac{z}{1-z} + \frac{z}{1-z} + \ldots \tag{5.1}$$

corresponds to the entire function $f(z) = z$. However, since

$$S_n(w) = z\,\frac{[1 - (-z)^{n-1}]w + 1 - (-z)^n}{[1 - (-z)^n]w + 1 - (-z)^{n+1}}, \tag{5.2}$$

we find that $S_n(0)$ converges uniformly on compact subsets of $|z| > 1$ to the value $f(z) = -1$. This shows that one has to be a little careful. Correspondence and uniform convergence of analytic approximants is not sufficient to secure convergence to the "right value." However, if we also have $0 \in D$, then life is much easier. From the three theorems in Section 3 we obtain:

Theorem 5.1. *Let $D \subseteq \mathbb{C}$ be a domain with $0 \in D$, and let $\{g_n\}$ be a sequence of functions that are meromorphic in D, such that $\{g_n\} \sim L$ for some $L \in \Lambda$. Then the following statements hold.*

(i) *If $\{g_n\}$ omits three distinct values p_1, p_2, $p_3 \in \widehat{\mathbb{C}}$ for $z \in D$ and n sufficiently large, then $\{g_n\}$ converges spherically uniformly on every closed subset E of D.*

(ii) *$\{g_n\}$ converges spherically uniformly on a closed subset E of D if its spherical derivatives $g_n^{\#}$ are uniformly bounded in E for sufficiently large n.*

(iii) *If $\{g_n\}$ converges spherically uniformly on closed subsets of D, then its limit function g is meromorphic in D and $L = \mathcal{L}(g)$.*

Theorem 5.1 applies to corresponding continued fractions whose classical approximants $f_n = S_n(0)$ are meromorphic in some domain D containing the origin. By turning to approximants $g_n = S_n(w_n)$ of $b_0 + K(a_n/b_n)$, we have even better chances of proving convergence by means of Theorem 5.1. We only have to make sure that if $b_0 + K(a_n/b_n) \sim L$, then also $\{S_n(w_n)\} \sim L$, but that is not difficult:

Theorem 5.2. *Let $b_0 + K(a_n/b_n) \sim L \in \Lambda$ with $a_n, b_n \in \mathcal{M}$ and $a_n \not\equiv 0$ for all n, and let $\zeta_n := S_n^{-1}(\infty)$. Then $S_n(w_n) \sim L$ for every sequence $\{w_n\}$ which satisfies $w_n \in \mathcal{M}$ for sufficiently large n and $\limsup_{n\to\infty} \mathrm{ord}\{\mathcal{L}(1 - w_n/\zeta_n)\} < \infty$.*

Proof. With the notation (4.7) we find that if $C_n D_n \not\equiv 0$, then

$$S_n(w) = \frac{A_n w + B_n}{C_n w + D_n} = \frac{A_n}{C_n} - \frac{A_n D_n - B_n C_n}{C_n^2 w + C_n D_n} = S_n(\infty) - \frac{S_n(0) - S_n(\infty)}{1 - w/\zeta_n},$$

where $S_n(\infty) = S_{n-1}(0)$, and thus $\mathrm{ord}(\mathcal{L}(S_n(\infty)) - L) \to \infty$. That $C_n D_n \not\equiv 0$ for sufficiently large n follows since $f_n = B_n/D_n = A_{n+1}/C_{n+1}$ and $\{f_n\} \sim L$. Since $\mathrm{ord}(\mathcal{L}(S_n(0)) - \mathcal{L}(S_n(\infty))) \to \infty$ by (2.2) and $\mathrm{ord}(A/B) = \mathrm{ord}(A) - \mathrm{ord}(B)$, the result follows. $\qquad\square$

The relationship to Theorem 4.2 is clear: $\{w_n\}$ just has to stay away from the same kind of exceptional sequence. (In [9] we used this to define *general correspondence* in the obvious way, but that is another story.) To prove that the conditions in Theorem 5.1(i) or (ii) hold, we may again use value sets.

6 Limit Periodic Continued Fractions

Here we let again $a_n \neq 0$ and b_n in $b_0 + K(a_n/b_n)$ be complex numbers. If the limit

$$q := \lim_{n\to\infty} q_n, \quad \text{with } q_1 := a_1/b_1, \ q_n := a_n/b_{n-1}b_n \quad (n \geq 2) \qquad (6.1)$$

exists, we say that $b_0 + K(a_n/b_n)$ is limit periodic with period 1. A large class of important functions have limit periodic continued fraction expansions, mostly with period 1 (the case we treat here) or period 2. For limit periodicity with periods ≥ 2 we refer to [7] and [18].

If the limit q belongs to the cut plane

$$Q := \{u \in \mathbb{C} : \ |\arg(1 + 4u)| < \pi\}, \tag{6.2}$$

then $b_0 + K(a_n/b_n)$ converges in the classical sense. Otherwise, if $q = \infty$ or if q is real with $q \leq -1/4$, the situation is more unclear. The continued fraction may converge or diverge, depending on how q_n approaches its limit; see [5], [10], [29]. To keep this survey simple (and not too long), we shall restrict to the case $q \in Q$. Then the following results follow from [25, Satz 2.4, p. 93], [7, Thm. 3.2B], [18, Thm. 28, p. 151] and the equivalence transformation $b_0 + K(a_n/b_n) \approx b_0 + K(q_n/1)$. (If $q_n = \infty$ for some n, this can only happen for finitely many indices, and thus $b_N + K(a_{N+n}/b_{N+n}) \approx b_N(1 + K(q_{N+n}/1))$ for N sufficiently large.)

Theorem 6.1. *Let* $b_0 + K(a_n/b_n)$ *with elements from* $\mathbb{C}$ *and* $a_n \neq 0$ *for all* n *satisfy* (6.1) *with* $q \in Q$, *and let* $x := -\frac{1}{2}(1 - \sqrt{1 + 4q})$, *where* $\mathrm{Re}(\sqrt{1 + 4q} > 0$. *Then* $b_0 + K(a_n/b_n)$ *has the following properties.*

(i) $b_0 + K(a_n/b_n)$ *converges in the classical sense to some* $f \in \widehat{\mathbb{C}}$.

(ii) $\lim_{n \to \infty} S_n^{-1}(f)/b_n = x$.

(iii) $\lim_{n \to \infty} S_n^{-1}(g)/b_n = -1 - x$ *for every* $g \in \widehat{\mathbb{C}}$ *with* $g \neq f$.

In particular, $\{b_n(-1-x)\}$ is an exceptional sequence, and thus $S_n(w_n) \to f$ as long as $\liminf d(w_n, b_n(-1-x)) > 0$. Moreover, the approximants $S_n(b_n x)$ converge faster to f than $S_n(0)$ if $\liminf |b_n x| > 0$:

Theorem 6.2. *Let* $b_0 + K(a_n/b_n)$, f *and* x *be as in Theorem 6.1. Then* $b_0 + K(a_n/b_n)$ *has the following properties.*

(i) If $f \neq \infty$, *then for every* $r > |x|/|1+x|$ *there exists a* $C > 0$ *such that* $|f - S_n(0)| \leq Cr^n$ *for every* n *for which* $S_n(0) \neq \infty$.

(ii) To every $r > |x|/|1+x|$ *there exists a* $C > 0$ *such that* $d(f, S_n(0)) \leq Cr^n$ *for all* $n \geq 1$.

(iii) If $\liminf_{n \to \infty} |b_n x| > 0$ *and* $\liminf_{n \to \infty} d(w_n, b_n(-1-x)) > 0$, *then*

$$\frac{d(f, S_n(w_n))}{d(f, S_n(0))} = \mathcal{O}(d(f^{(n)}, w_n)) \quad \text{as } n \to \infty, \quad f^{(n)} := S_n^{-1}(f).$$

Proof. Part (i) was essentially proved by Thron and Wadeland [31] for the case when $b_n = 1$ for all n and thus $a_n = q_n \to q \in Q$. We shall instead prove (ii), from which (i) follows immediately.

(ii) Let $f \neq \infty$. By using (4.6) with $u_n = f^{(n)}$, $w_n = 0$ and $v_n = \infty$, we get

$$\frac{d(f, S_n(0))}{d(f, S_n(\infty))} = \frac{d(f, f_n)}{d(f, f_{n-1})} = \frac{d(f^{(n)}, 0) \cdot d(\zeta_n, \infty)}{d(f^{(n)}, \infty) \cdot d(\zeta_n, 0)}$$

$$= \left| \frac{f^{(n)}}{\zeta_n} \right| \to \left| \frac{x}{1+x} \right|, \qquad (6.3)$$

which proves the result for this case. If $f = \infty$, then (4.5) converges to $\widehat{f} = 1/f = 0$, and thus its approximants $\widehat{f}_n = 1/f_{n-1}$ satisfy

$$\frac{d(\widehat{f}, \widehat{f}_n)}{d(\widehat{f}, \widehat{f}_{n-1})} \to \left| \frac{x}{1+x} \right|, \qquad (6.4)$$

by (6.3). Since $d(z_1, z_2) = d(1/z_1, 1/z_2)$, the result follows for this case as well.

(iii) Again we first assume that $f \neq \infty$, so that, by (4.6),

$$\frac{d(f, S_n(w_n))}{d(f, S_n(0))} = \frac{d(f^{(n)}, w_n) \cdot d(\zeta_n, 0)}{d(f^{(n)}, 0) \cdot d(\zeta_n, w_n)}$$

$$\sim \frac{d(f^{(n)}, w_n) \cdot d(b_n(-1-x), 0)}{d(b_n x, 0) \cdot d(b_n(-1-x), w_n)} \quad \text{as } n \to \infty,$$

which proves the result for this case. If $f = \infty$, then the result follows by the argument used in the proof of part (ii).   □

This way of improving the rate of convergence was essentially proved by Gill [6] and Thron and Waadeland [30] for $K(a_n/1)$. Indeed, if we can approximate $f^{(n)}$ even better, then we can improve the rate of convergence even further; see [15]. If the elements of $b_0 + K(a_n/b_n)$ are functions of z such that $b_0 + K(a_n/b_n) \sim L$, then we can combine Theorem 6.1 with Theorem 5.1 to obtain:

Corollary 6.3. *Let $D \subseteq \mathbb{C}$ be a domain containing the origin, and let a_n and b_n be functions that are analytic in D such that $a_n(z) \not\equiv 0$ for all n, $b_0 + K(a_n/b_n) \sim L \in \Lambda$ and $\lim_{n \to \infty} a_n(z)/(b_{n-1}(z)b_n(z)) = q(z) \in Q$ (where Q is given by (6.2)), for all $z \in D$. Then $b_0 + K(a_n/b_n)$ converges spherically uniformly to a meromorphic function f on compact subsets of D, and $\mathcal{L}(f) = L$.*

7 Padé Approximants

So far, the best we have done (apart from Corollary 6.3) is Theorem 5.1 which essentially says that if $b_0 + K(a_n/b_n) \sim L$, and if the approximants $\{S_n(0)\}$ converge uniformly on compact subsets of a domain D containing the origin, then $S_n(0) \to f$ in D, with $\mathcal{L}(f) = L$. Moreover, we may replace $\{S_n(0)\}$ by $\{S_n(w_n)\}$ as long as $\{w_n\}$ stays asymptotically away from an exceptional sequence.

In this section we shall see that we can remove the condition $0 \in D$ if the correspondence is good enough and if $L = \mathcal{L}(f)$ for a function which is meromorphic in the whole complex plane $\mathbb{C}$. We shall also assume that the elements a_n and b_n of $b_0 + K(a_n/b_n)$ are polynomials in the complex variable z. Then its classical approximants are rational functions $f_n = S_n(0) = B_n/D_n$. We also require that $\text{ord}(L) \geq 0$ and $D_n(0) = 1$ and that B_n, D_n are the canonical numerators and denominators of $b_0 + K(a_n/b_n)$. Then f_n is a *Padé approximant* to f (or L) if

$$\text{ord}(LD_n - B_n) \geq M_n + N_n + 1, \tag{7.1}$$

where M_n and N_n are the degrees of B_n and D_n respectively. We write $f_n = [M_n/N_n]_f$. We also write $[M_n/N_n; w_n]_f$ for the modified Padé approximant $(B_n + B_{n-1}w_n)/(D_n + D_{n-1}w_n)$ with $[M_n/N_n; \infty]_f := B_{n-1}/D_{n-1}$. Clearly, $[M_n/N_n; w_n]_f$ depends on the *sequence* $\{[M_n/N_n]_f\}$. Examples of continued fractions which have Padé approximants are:

- regular C-fractions $b_0 + K(a_n z/1)$, where $b_0, a_n \in \mathbb{C}$ with $a_n \neq 0$ for all n;

- associated continued fractions

$$b_0 + \frac{a_1 z}{1 + d_1 z} + \frac{a_2 z^2}{1 + d_2 z} + \frac{a_3 z^2}{1 + d_3 z} + \frac{a_4 z^2}{1 + d_4 z} + \dots, \tag{7.2}$$

where $a_n, d_n \in \mathbb{C}$ with $a_n \neq 0$ for all n, or, equivalently, J-fractions $b_0 + K(a_n/(d_n + z))$;

- P-fractions $b_0 + K(1/b_n(z))$, where each $b_n(z)$ is a polynomial in z^{-1} of degree ≥ 1.

The approximants of (5.1) are not Padé approximants, since $M_n = n$ and $N_n = n$, whereas $\text{ord}(LD_n - B_n) = n + 1$. (In fact, (5.1) has 2-point Padé approximants, but that is another story.) If the approximants of $b_0 + K(a_n/b_n)$ are Padé approximants, then the correspondence is best possible. In [16] the following generalization of work by Beardon [2] and Chisholm [4] was proved:

Theorem 7.1 ([16]). *Let f be meromorphic in $\mathbb{C}$ with $f(0) \neq \infty$, let $E \subset \mathbb{C}$ be a compact set containing no poles of f, and let $[M_n/N_n]_f$ be a sequence of Padé approximants for f such that*

$$N_n \leq M_n \leq M_{n+1}, \quad N_n \to \infty, \quad \gamma_n := N_n + M_n - N_{n-1} - M_{n-1} \geq 0 \tag{7.3}$$

for sufficiently large n. If there exists a sequence $\{\beta_n\}$ of numbers from $\widehat{\mathbb{C}}$ such that the poles of $[M_n/N_n; \beta_n z^{\gamma_n}]_f$ have no limit point in E, then $[M_n/N_n; \beta_n z^{\gamma_n}]_f$ converges uniformly to f in E.

Remarks.

(1) The condition $N_n \leq M_n$ in (7.3) can be replaced by the condition $\limsup(N_n - M_n) < \infty$. This follows since $h(z) := z^J f(z)$ has Padé approximants $[M_n + J/N_n]_h = z^J [M_n/N_n]_f$ for $J \in \mathbb{N}$. Moreover, $[M_n + J/N_n; w_n]_h = z^J [M_n/N_n; w_n]_f$.

(2) Since $g(z) := z^J/f(z)$ has Padé approximants $[N_n + J/M_n]_g = z^J/[M_n/N_n]_f$ for $J \in \mathbb{N}$ when $f(0) \neq 0$, it also follows that $[M_n/N_n; \beta_n z^{\gamma_n}]_f$ converges spherically uniformly in E to f if $f(0) \neq 0$, and that

$$M_n \to \infty, \quad \limsup(M_n - N_n) < \infty,$$
$$\gamma_n := M_n + N_n - M_{n-1} - N_{n-1} \geq 0,$$

and the zeros of $[M_n/N_n; \beta_n z^{\gamma_n}]_f$ have no limit point in E.

One way to apply Theorem 7.1 is to let E be a one-point set. If the approximants are finite and their (spherical) derivatives uniformly bounded in E (for sufficiently large n), then the convergence is clear. Value sets can also be helpful. Still, it would have been better if we did not have to worry about the location of the poles! Unfortunately, Wallin [32] has proved that there exists an *entire* function f with a normal Padé table, such that its diagonal Padé approximants $[n/n]_f$ diverge at every $z \in \mathbb{C} \setminus \{0\}$. What happens is that the poles of $[n/n]_f$ are dense in $\widehat{\mathbb{C}}$. One has the feeling that $[n/n; w_n]_f$ still converges to f under mild conditions on w_n, but that has not yet been proved.

It is also a problem that Theorem 7.1 works only for functions that are meromorphic in the *whole* complex plane. There is a similar theorem for functions meromorphic in a disk (see [16]), but the conditions are too strong to allow for continued fraction approximants of the type considered in this paper.

In [23] Nuttall proved that if f is meromorphic in $\mathbb{C}$, then $[n, n]_f$ converges to f in measure in $\mathbb{C}$ as $n \to \infty$. This was generalized by Pommerenke [26] to convergence in logarithmic capacity, and further by Lubinsky [22] to functions meromorphic in some domain D (or even more generally, to functions of the Gonçar–Walsh class). However, to keep the paper to a reasonable length, we shall not go into this here.

8 Regular C-Fractions

As an example, we apply Theorem 7.1 to regular C-fractions $b_0 + K(a_n z/1)$, where b_0 and a_n are complex numbers and all $a_n \neq 0$. Its approximants $f_n = B_n/D_n$ satisfy

$$M_{2n-1} = n, \quad M_{2n} \leq n, \quad N_{2n} = n \quad \text{and} \quad N_{2n+1} \leq n. \tag{8.1}$$

To satisfy (7.3) we shall assume that $b_0 + K(a_n z/1)$ is *normal*; i.e., that (8.1) holds with equalities for sufficiently large n. Then $\gamma_n = 1$, and we get:

Theorem 8.1. *Let the function f be meromorphic in $\mathbb{C}$ and correspond to a normal regular C-fraction $b_0 + K(a_n z/1)$. Let further E be a compact subset of $\mathbb{C}$. Then the following hold.*

(i) *If f is analytic in E and there exists a sequence $\{\beta_n\}$ of numbers from $\widehat{\mathbb{C}}$ such that the poles of $S_n(\beta_n z)$ have no limit point in E, then $S_n(\beta_n z)$ converges uniformly to f in E.*

(ii) *If f omits zero in E and there exists a sequence $\{\beta_n\}$ of numbers from $\widehat{\mathbb{C}}$ such that the zeros of $S_n(\beta_n z)$ have no limit point in E, then $S_n(\beta_n z)$ converges spherically uniformly to f in E.*

(iii) *If $b_0 + K(a_n z/1)$ converges generally in E, and there exists a subsequence $\{S_{n_k}\}$ and a sequence $\{\beta_k\}$ of numbers from $\widehat{\mathbb{C}}$ such that the poles (or zeros) of $S_{n_k}(\beta_k z)$ have no limit point in E, then $b_0 + K(a_n z/1)$ converges generally to f in E.*

Not every meromorphic function f has a normal regular C-fraction expansion. However, Lubinsky [21] proved that for every function f that is analytic at the origin there exists a point u such that if we change the center of our expansions from $z = 0$ to the point $z = u$, then $f \sim b_0 + K(a_n(z-u)/1)$ is normal. Indeed, to every such function f, there is at most a countable number of values for u that has to be avoided.

9 Stieltjes Fractions

Stieljes fractions, or S-fractions, are regular C-fractions with $b_0 > 0$ and $a_n > 0$ for all n. This is a very nice class of continued fractions studied by

Stieltjes [28]. We say that

$$L(z) = \sum_{n=0}^{\infty} (-1)^n c_n z^n \tag{9.1}$$

is a *Stieltjes series* if it corresponds to an S-fraction. All sorts of nice things are true in this situation. For instance, S-fractions are normal, and $L(z)$ is a Stieltjes series if and only if there exists a distribution function ψ on $[0, \infty)$ such that

$$c_n = \int_0^{\infty} t^n \, d\psi(t) \qquad \text{for } n = 0, 1, 2, \ldots \tag{9.2}$$

(ψ is a distribution function on a real interval I if ψ is a real-valued, non-decreasing function taking infinitely many values there.) Hence, correspondence to S–fractions characterizes the sequences $\{c_n\}$ for which the moment problem has a solution on $[0, \infty)$. The convergence properties of S-fractions are also extremely beautiful. With ψ given by (9.2), Stieltjes proved:

Theorem 9.1 ([28]). *Let $\{f_n\}$ be the (classical) approximants of the S-fraction $b_0 + K(a_n z/1)$, and let Ω denote the cut plane $\Omega := \{z \in \mathbb{C} : |\arg z| < \pi\}$. Then the following hold.*

(A) f_n is analytic in $\Omega \cup \{0\}$ for all n.

(B) $\{f_{2n-1}\}$ and $\{f_{2n}\}$ converge uniformly on compact subsets of Ω.

(C) $\{f_n\}$ converges in Ω if and only if $\{f_n(z_0)\}$ converges for some $z_0 \in \Omega$.

(D) $\{f_n\}$ converges in Ω if and only if $\sum b_n = \infty$, where

$$b_1 := \frac{1}{a_1}, \quad b_{2n} := \frac{a_1 a_3 \ldots a_{2n-1}}{a_2 a_4 \ldots a_{2n}}, \quad b_{2n+1} := \frac{a_2 a_4 \ldots a_{2n}}{a_1 a_3 \ldots a_{2n+1}}. \tag{9.3}$$

(E) The moment problem for $\{c_n\}$ on $[0, \infty)$ is determined (i.e., $d\psi$ is essentially unique) if and only if $\{f_n\}$ converges.

(F) If $\{f_n\}$, and thus $b_0 + K(a_n z/1)$, converges, then the convergence is uniform on compact subsets of Ω, and its limit is

$$f(z) = \int_0^{\infty} \frac{d\psi(t)}{1 + zt}. \tag{9.4}$$

Remarks.

(1) Part (A) shows that the poles of f_n are all located on the negative real axis. In fact, Stieltjes [28] proved that the denominators $\{D_n\}$ of $f_n = B_n/D_n$ form an orthogonal polynomial sequence for the distribution function ψ, and thus all the zeros of D_n are simple and interlace the zeros of D_{n-1}.

(2) The limits of $\{f_{2n-1}\}$ and $\{f_{2n}\}$ both have the forms (9.4), but for essentially different distribution functions ψ if $\{f_n\}$ diverges. If $\{f_n\}$ diverges, then the continued fraction also diverges in the general sense, so S-fractions converge generally if and only if they converge in the classical sense. (This is a consequence of the Stern-Stolz theorem; see [18, Thm. 1, p. 94].)

(3) The expression (9.4) allows us to approximate ψ by using approximants of $b_0 + K(a_n z/1)$; see [25, p. 188].

(4) If we expand $(1 + zt)^{-1}$ into a geometric series and formally interchange the summation and integration in (9.4) (without justifying the validity of these operations), we obtain

$$\int_0^\infty \frac{d\psi(t)}{1 + zt} \sim \int_0^\infty \sum_{n=0}^\infty (-zt)^n \, d\psi(t) \sim \sum_{n=0}^\infty (-1)^n \int_0^\infty t^n \, d\psi(t) = L(z)$$

which suggests that $\mathcal{L}(f) = L$. Moreover, since the denominator D_n of the rational function $f_n = B_n/D_n$ only has real and simple zeros $\zeta_{n,k} < 0$, f_n has a partial fraction decomposition which can be written as $\int_0^\infty d\psi_n(t)/(1 + zt)$ with an appropriate step function $\psi_n(t)$ having steps at the points $-1/\zeta_{n,k}$. This suggests that $\psi_n \to \psi$ weakly, and thus that $f_n \to f$. Theorem 9.1 shows that these conclusions, derived here informally, are indeed correct.

A particularly useful result is due to Carleman:

Theorem 9.2 ([3]). *Let (9.1) be a Stieltjes series. If $\sum c_n^{-1/2n} = \infty$, then the corresponding S-fraction converges in Ω.*

To have conditions that are stated in terms of the numbers c_n, instead of the S-fraction itself, can be very useful indeed. Moreover, the condition on $\{c_n\}$ is a rather mild one: for instance, it is satisfied whenever the series (9.1) has a positive radius of convergence.

Theorem 9.3. *Let f be a function that is analytic in $|z| < R$ for some $0 < R \leq \infty$. If f has an S-fraction expansion $c_0 + K(a_n z/1)$, then the following hold.*

(A) f has an analytic extension

$$f^*(z) := \int_0^{1/R} \frac{d\psi(t)}{1+zt} \quad \text{for } z \in \Omega_R := \Omega \cup \{z \in \mathbb{C} : |z| < R\}.$$

(B) $c_0 + K(a_n z/1)$ converges uniformly on compact subsets of Ω_R to f^*.

(C) f is an entire function if and only if $a_n \to 0$.

If the Stieltjes series (9.1) converges only at $z = 0$, it is still possible that its corresponding S-fraction converges in Ω. For instance, this is the case if Carleman's criterion holds; in this case L is an asymptotic expansion of the limit function.

Results similar to Theorems 9.1–9.3 also hold for the associated continued fractions (7.2) with $a_n < 0$ and $d_n \in \mathbb{R}$ or the equivalent real J-fraction $b_0 + K(a_n/(d_n + z))$ with $a_n < 0$ and $d_n \in \mathbb{R}$.

10   More on Regular C-fractions

Inspired by Carleman's theorem for S-fractions, it is tempting to try to find convergence criteria for regular C-fractions based on the coefficients of their corresponding series. This is no easy task, but Lubinsky has come up with some very nice results in this direction. The first result even guarantees that the given entire function f has a normal regular C-fraction expansion. Indeed, it guarantees the stronger result that the Padé table of f is normal, but we restrict the result to C-fractions.

Theorem 10.1 ([19]). *Let $L(z) = \sum_{n=0}^{\infty} c_n z^n$ satisfy $c_n \neq 0$ for all n and $|c_{n-1}c_{n+1}/c_n^2| \leq \rho_0^2$ for all $n \in \mathbb{N}$, where $\rho_0 = 0.4559\ldots$ is the positive root of the equation $2\sum_{n=1}^{\infty} \rho^{n^2} = 1$. Then $L \sim c_0 + K(a_n z/1)$, where $a_n \to 0$, and $c_0 + K(a_n z/1)$ is normal and converges uniformly on compact subsets of $\mathbb{C}$ to an entire function f where $\mathcal{L}(f) = L$.*

Remarks.

(1) The condition $|c_{n-1}c_{n+1}/c_n^2| \leq \rho_0^2$ shows that $|c_{n+1}/c_n| = \mathcal{O}(\rho_0^{2n})$; i.e., it is a very strong condition which not only implies that f is entire, but also that f has order zero.

(2) Lubinsky also proved that $c_0 + K(a_n z/1)$ converges separately in Theorem 10.1. That is, $D_n(z) \to 1$ and $B_n(z) \to f(z)$ as $n \to \infty$, where $f_n = B_n/D_n$.

Theorem 10.2 ([20]). *Let $L(z) = \sum_{n=0}^{\infty} c_n z^n$, with $c_n \neq 0$ for all sufficiently large n, be such that the limit $q := \lim_{n \to \infty} c_{j-1}c_{j+1}/c_j^2$ exists with*

$|q| < 1$. If $L \sim c_0 + K(a_n z/1)$, then $a_n \to 0$ and the regular C-fraction converges uniformly on compact subsets of $\mathbb{C}$ to an entire function f where $\mathcal{L}(f) = L$.

Again the condition implies that $c_{n+1}/c_n \to 0$ at a geometric rate, but this time we have $|c_{n+1}/c_n| = o(\rho^n)$ for every $\rho > |q|$.

References

[1] G.A. Baker, Jr. and P.R. Graves-Morris, *The convergence of Padé approximants*, J. Math. Anal. Appl. **87** (1982), 382 – 394.

[2] A.F. Beardon, *On the convergence of Padé approximants*, J. Math. Anal. Appl. **21** (1968), 344 – 346.

[3] T. Carleman, *Les fonctions quasi analytiques*, Gauthier – Villars, 1926.

[4] J.S.R. Chisholm, *Approximation by sequences of Padé approximants in regions of meromorphy*, J. Math. Phys. **7** (1966), 39 – 44.

[5] J. Gill, *Infinite compositions of Möbius transformations*, Trans. Amer. Math. Soc. **176** (1973), 479 – 487.

[6] ———, *The use of attrative fixed points in accelerating the convergence of limit-periodic continued fractions*, Proc. Amer. Math. Soc. **47** (1975), 119 – 126.

[7] L. Jacobsen, *Convergence of limit k–periodic continued fractions $k(a_n/b_n)$, and of subsequences of their tails*, Proc. London Math. Soc. (3) **51** (1985), 563 – 576.

[8] ———, *General convergence for continued fractions*, Trans. Amer. Math. Soc. **281** (1986), 129 – 146.

[9] ———, *General correspondence for continued fractions*, J. Comp. Appl. Math. **19** (1987), 171 – 177.

[10] L. Jacobsen and D.R. Masson, *On the convergence of limit periodic continued fractions $k(a_n/1)$, where $a_n \to -\frac{1}{4}$. Part III*, Constr. Approx. **6** (1990), 363 – 374.

[11] L. Jacobsen and W.J. Thron, *Limiting structures for sequences of linear fractional transformations*, Proc. Amer. Math. Soc. **99** (1987), 141 – 146.

[12] W.B. Jones and W.J. Thron, *On the convergence of Padé approximants*, SIAM J. Math. Anal. **6** (1975), 9 – 16.

[13] _____ , *Sequences of meromorphic functions corresponding to formal laurent series*, SIAM J. Math. Anal. **10** (1979), 1 – 17.

[14] _____ , *Continued fractions. Analytic theory and applications. Encyclopedia of mathematics and its application, Vol. 11*, Addison-Wesley, Reading, Mass., 1980, now distributed by Cambridge University Press, New York.

[15] L. Lorentzen, *Computation of limit periodic continued fractions. A survey*, Numer. Algorithms **10** (1995), 69 – 110.

[16] _____ , *Ideas from continued fraction theory extended to Padé approximation and generalized iteration*, Proc. International Conference on Rational Approximation, vol. 61, 2000, pp. 185–206.

[17] _____ , *General convergence in quasi-normal families*, Proc. Edinburgh Math. Soc., to appear.

[18] L. Lorentzen and H. Waadeland, *Continued fractions with applications. Studies in computational mathematics, Vol. 3*, North - Holland, 1992.

[19] D.S. Lubinsky, *Padé tables of entire functions of very slow and smooth growth*, Constr. Approx. **1** (1985), 349 – 358.

[20] _____ , *Padé tables of entire functions of very slow and smooth growth, II*, Constr. Approx. **4** (1988), 321 – 339.

[21] _____ , *Power series equivalent to rational functions: A shifting - origin Kronecker type theorem, and normality of Padé tables*, Numer. Math. **54** (1988), 33 – 39.

[22] _____ , *Spurious poles in diagonal rational approximation*, Progress in Approximation Theory, Springer-Verlag, 1992, pp. 191 – 213.

[23] J. Nuttall, *The convergence of Padé approximants of meromorphic functions*, J. Math. Anal. Appl. **31** (1970), 147 – 153.

[24] J.F. Paydon and H.S. Wall, *The continued fraction as a sequence of linear transformations*, Duke Math. J. **9** (1942), 360 – 372.

[25] O. Perron, *Die Lehre von den Kettenbrüchen, Band II*, Teubner, Stuttgart, 1957.

[26] C. Pommerenke, *Padé approximants and convergence in capacity*, J. Math. Anal. Appl. **41** (1973), 775 – 780.

[27] J.L. Schiff, *Normal families*, Springer-Verlag, 1993.

[28] T.J. Stieltjes, *Reserches sur les fractions continues*, Ann. Fac. Sci. Toulouse **8J** (1894), 1 – 47, also in: Oeuvres, Vol. 2, pp. 402-566.

[29] W.J. Thron, *On parabolic convergence regions for continued fractions*, Math. Zeitschr. **69** (1958), 173 – 182.

[30] W.J. Thron and H. Waadeland, *Accelerating convergence of limit periodic continued fractions* $k(a_n/1)$, Numer. Math. **34** (1980), 72 – 90.

[31] _____ , *Truncation error bounds for limit periodic continued fractions*, Math. Comp. **40** (1983), 589 – 597.

[32] H. Wallin, *The convergence of Padé approximants and the size of power series coefficients*, Appl. Anal. 4 (1974), 235 – 251.

On the Analytic Continuation of Various Multiple Zeta-Functions

Kohji Matsumoto

0 Introduction

In this article we give a survey of the history of the problem of analytic continuation of multiple zeta-functions, and we prove some new results in this connection. We begin in Section 1 by describing the work of E. W. Barnes and H. Mellin at the turn of the 20th century. In Sections 2 and 3 we discuss the Euler sum and its multi-variable generalization, which recently have become again the subject of active research. In Section 4 we describe a new method of M. Katsurada which uses the classical Mellin-Barnes integral formula to establish the analytic continuation of the Euler sum. In the final two sections we present new results of the author, obtained by applying the Mellin-Barnes formula to more general multiple zeta-functions.

1 Barnes Multiple Zeta-Functions

The problem of analytic continuation of multiple zeta-functions was first considered by Barnes [7][8] and Mellin [48][49]. Barnes [7] introduced the double zeta-function of the form

$$\zeta_2(s; \alpha, (w_1, w_2)) = \sum_{m_1=0}^{\infty} \sum_{m_2=0}^{\infty} (\alpha + m_1 w_1 + m_2 w_2)^{-s}, \qquad (1.1)$$

where α, w_1, w_2 are complex numbers with positive real parts, and s is the complex variable. The series (1.1) is absolutely convergent in the half-plane $\Re s > 2$. Actually Barnes first defined his function as the contour integral

$$\zeta_2(s; \alpha, (w_1, w_2)) = -\frac{\Gamma(1-s)}{2\pi i} \int_C \frac{e^{-\alpha z}(-z)^{s-1}}{(1 - e^{-w_1 z})(1 - e^{-w_2 z})} dz, \qquad (1.2)$$

where C is the contour which consists of the half-line on the positive real axis from infinity to a small positive constant δ, a circle of radius δ oriented counterclockwise around the origin, and the other half-line on the positive real axis from δ to infinity. It is easy to see that (1.2) coincides with (1.1) when $\Re s > 2$. The expression (1.2) gives the meromorphic continuation of $\zeta_2(s; \alpha, (w_1, w_2))$ to the whole s-plane. Moreover, Barnes [7] studied very carefully how to extend the definition of $\zeta_2(s; \alpha, (w_1, w_2))$ to the case when the real parts of α, w_1, w_2 are not necessarily positive.

Barnes introduced his double zeta-function for the purpose of constructing the theory of double gamma-functions. As for the theory of double gamma-functions, there were several predecessors such as Kinkelin, Hölder, Méray, Pincherle, and Alexeiewsky, but Barnes developed the theory most systematically. Barnes [8] then proceeded to develop a theory of more general multiple gamma-functions, and he introduced the multiple zeta-function defined by

$$\zeta_r(s; \alpha, (w_1, \ldots, w_r)) = \sum_{m_1=0}^{\infty} \cdots \sum_{m_r=0}^{\infty} (\alpha + m_1 w_1 + \cdots + m_r w_r)^{-s}, \quad (1.3)$$

where r is a positive integer and $\alpha, w_1, \ldots, w_r$ are complex numbers. Barnes assumed the following condition to ensure the convergence of the series. Let ℓ be any line in the complex s-plane crossing the origin. Then ℓ divides the plane into two half-planes. Let $H(\ell)$ be one of those half-planes, not including ℓ itself. The assumption of Barnes is that

$$w_j \in H(\ell) \qquad (1 \leq j \leq r). \quad (1.4)$$

Then, excluding the finitely many tuples $(m_1, \ldots, m_r)$ satisfying $m_1 w_1 + \cdots + m_r w_r = -\alpha$ from the sum, we see easily that (1.3) is absolutely convergent when $\Re s > r$. Barnes [8] proved an integral expression similar to (1.2) for $\zeta_r(s; \alpha, (w_1, \ldots, w_r))$, which yields the meromorphic continuation.

On the other hand, Mellin [48][49] studied the meromorphic continuation of the multiple series

$$\sum_{m_1=1}^{\infty} \cdots \sum_{m_k=1}^{\infty} P(m_1, \ldots, m_k)^{-s}, \quad (1.5)$$

where $P(X_1, \ldots, X_k)$ is a polynomial of k indeterminates and of complex coefficients with positive real parts. Mellin's papers include a prototype of the method in the present paper, though he treated the one variable case only. For example, the formula (4.1) appears on p. 21 of [48]. Following Mellin's work, many authors have investigated the series (1.5) and its generalizations; the main contributors include K. Mahler, P. Cassou-Nogués, P. Sargos, B. Lichtin, M. Eie and M. Peter. Most of these authors concentrated on the one variable case, and we do not discuss the details of this work. However, Lichtin's series of papers ([36], [37], [38], [39], and [40]) should be mentioned here. In [36] Lichtin proposed the problem of studying the analytic continuation of the Dirichlet series in several variables

$$\sum_{m_1=1}^{\infty} \cdots \sum_{m_k=1}^{\infty} P_0(m_1, \ldots, m_k)$$
$$\times P_1(m_1, \ldots, m_k)^{-s_1} \cdots P_r(m_1, \ldots, m_k)^{-s_r}, \qquad (1.6)$$

where $P_0, P_1, \ldots, P_r$ are polynomials of k indeterminates, and he carried out this investigation in [37], [38], [39], and [40]. In particular, Lichtin proved that the series (1.6) can be continued meromorphically to the whole space when the associated polynomials are hypoelliptic (and also satisfy some other conditions).

2   The Euler Sum

The two-variable double sum

$$\zeta_2(s_1, s_2) = \sum_{m=1}^{\infty} \sum_{n=1}^{\infty} m^{-s_1}(m+n)^{-s_2} \qquad (2.1)$$

is absolutely convergent if $\Re s_2 > 1$ and $\Re(s_1 + s_2) > 2$. The investigation of this sum goes back to Euler, who was interested in the values of (2.1) when s_1 and s_2 are positive integers. Various properties of the values of (2.1) at positive integers were given in Nielsen's book [54]. Ramanujan also had an interest in such problems, and some of their formulas were later rediscovered; see the comments on pp. 252–253 of Berndt [9]. In recent years, the Euler sum has again become an object of active research; see, for instance, [10] and [17].

As far as the author knows, the first investigation of the analytic continuation of $\zeta_2(s_1, s_2)$ was made by Atkinson [6], in his work on the mean square of the Riemann zeta-function $\zeta(s)$. When $\Re s_1 > 1$ and $\Re s_2 > 1$, one has

$$\zeta(s_1)\zeta(s_2) = \zeta(s_1 + s_2) + \zeta_2(s_1, s_2) + \zeta_2(s_2, s_1). \qquad (2.2)$$

Atkinson's aim was to integrate the left-hand side with respect to t, when $s_1 = \frac{1}{2} + it$ and $s_2 = \frac{1}{2} - it$. Hence he was forced to show the analytic continuation of the right-hand side. Atkinson [6] used the Poisson summation formula to deduce a certain integral representation which enabled him to obtain the analytic continuation.

On the other hand, Matsuoka [47] obtained the analytic continuation of the series

$$\sum_{m=2}^{\infty} m^{-s} \sum_{n<m} n^{-1},$$

which is equal to $\zeta_2(1, s)$. Independently, Apostol and Vu [4] proved that $\zeta_2(s_1, s_2)$ can be continued meromorphically with respect to s_1 for each fixed s_2, and also with respect to s_2 for each fixed s_1. The proofs of Matsuoka and Apostol-Vu are both based on the Euler-Maclaurin summation formula. The main aim of those papers was the investigation of special values of $\zeta_2(s_1, s_2)$ at (not necessarily positive) integer points, and the papers give various formulas for such values.

Note that Apostol and Vu [4] also considered the series

$$T(s_1, s_2) = \sum_{m=1}^{\infty} \sum_{n<m} \frac{1}{m^{s_1} n^{s_2}(m+n)} \tag{2.3}$$

and discussed its analytic continuation.

Let q be a positive integer (≥ 2), $\varphi(q)$ the Euler function, χ a Dirichlet character mod q, and $L(s, \chi)$ the corresponding Dirichlet L-function. Inspired by Atkinson's work [6], Meurman [50] and Motohashi [53] independently considered the sum

$$Q((s_1, s_2); q) = \varphi(q)^{-1} \sum_{\chi \bmod q} L(s_1, \chi) L(s_2, \bar{\chi}).$$

Corresponding to (2.2), the decomposition

$$Q((s_1, s_2); q) = L(s_1 + s_2, \chi_0) + f((s_1, s_2); q) + f((s_2, s_1); q)$$

holds, where χ_0 is the principal character mod q and

$$f((s_1, s_2); q) = \sum_{\substack{1 \leq a \leq q \\ (a,q)=1}} \sum_{m=0}^{\infty} \sum_{n=1}^{\infty} (qm + a)^{-s_1} (q(m+n) + a)^{-s_2}. \tag{2.4}$$

This is a generalization of the Euler sum (2.1). Meurman [50] proved the analytic continuation of (2.4) by generalizing the argument of Atkinson [6]. On the other hand, Motohashi derived a double contour integral representation for (2.4), which yields the analytic continuation. By refining Motohashi's argument, Katsurada and the author [32][33] obtained asymptotic expansions for

$$\sum_{\chi \bmod q} |L(s, \chi)|^2 \quad (s \neq 1) \qquad \text{and} \qquad \sum_{\substack{\chi \bmod q \\ \chi \neq \chi_0}} |L(1, \chi)|^2 \tag{2.5}$$

with respect to q. See also Katsurada [28], where a somewhat different argument using confluent hypergeometric functions is given.

Let $\zeta(s, \alpha)$ be the Hurwitz zeta-function defined by the analytic continuation of the series $\sum_{n=0}^{\infty} (\alpha + n)^{-s}$, where $\alpha > 0$. Katsurada and the author [34] gave an asymptotic expansion of the mean value

$$\int_0^1 |\zeta(s, \alpha) - \alpha^{-s}|^2 d\alpha \tag{2.6}$$

with respect to $\Im s$. The starting point of the argument in [34] is the following generalization of (2.2):

$$\zeta(s_1, \alpha)\zeta(s_2, \alpha) = \zeta(s_1 + s_2, \alpha) + \zeta_2((s_1, s_2); \alpha) + \zeta_2((s_2, s_1); \alpha), \tag{2.7}$$

where

$$\zeta_2((s_1, s_2); \alpha) = \sum_{m=0}^{\infty} \sum_{n=1}^{\infty} (\alpha + m)^{-s_1}(\alpha + m + n)^{-s_2}. \tag{2.8}$$

This is again a generalization of (2.1). In [34], the meromorphic continuation of $\zeta_2((s_1, s_2); \alpha)$ was shown using the formula

$$\begin{aligned}
\zeta_2((s_1, s_2); \alpha) &= \frac{\Gamma(s_1 + s_2 - 1)\Gamma(1 - s_1)}{\Gamma(s_2)}\zeta(s_1 + s_2 - 1) \\
&+ \frac{1}{\Gamma(s_1)\Gamma(s_2)(e^{2\pi i s_1} - 1)(e^{2\pi i s_2} - 1)} \int_C \frac{y^{s_2} - 1}{e^y - 1} \\
&\times \int_C h(x + y; \alpha)x^{s_1 - 1} dx dy, \tag{2.9}
\end{aligned}$$

where

$$h(z; \alpha) = \frac{e^{(1-\alpha)z}}{e^z - 1} - \frac{1}{z}.$$

This formula is an analogue of Motohashi's integral expression for (2.4).

The author [42] considered the more general series

$$\zeta_2((s_1, s_2); \alpha, w) = \sum_{m_1=0}^{\infty} \sum_{m_2=0}^{\infty} (\alpha + m_1)^{-s_1}(\alpha + m_1 + m_2 w)^{-s_2}, \tag{2.10}$$

where $w > 0$, and proved its analytic continuation using a similar method, which also gives the asymptotic expansion of $\zeta_2((s_1, s_2); \alpha, w)$ with respect to w when $w \to +\infty$. In particular, this yields an asymptotic expansion of the Barnes double zeta-function $\zeta_2(s; \alpha, (1, w))$ with respect to w, because this function is just $\zeta_2((0, s); \alpha, w)$. These results and also the asymptotic expansion of the double gamma-function are proved in [42]. Note that some claims in [42] on the uniformity of the error terms are not true; they are corrected in [44] (see also [43]).

3   Multi-variable Euler-Zagier Sums

The r-variable generalization of the Euler sum (2.1), defined by

$$\zeta_r(s_1,\ldots,s_r) \tag{3.1}$$
$$= \sum_{m_1=1}^{\infty} \sum_{m_2=1}^{\infty} \cdots \sum_{m_r=1}^{\infty} m_1^{-s_1}(m_1+m_2)^{-s_2}\cdots(m_1+\cdots+m_r)^{-s_r},$$

is absolutely convergent in the region

$$\mathcal{A}_r = \{(s_1,\ldots,s_r) \in \mathbf{C}^r \mid \Re(s_{r-k+1}+\cdots+s_r) > k \quad (1 \leq k \leq r)\}, \tag{3.2}$$

as will be shown in Theorem 3 below (in Section 6). (The condition of absolute convergence given by Proposition 1 of Zhao [66] is not sufficient.) In connection with knot theory, quantum groups and mathematical physics, the properties of (3.1) have been recently investigated by Zagier [64][65], Goncharov [19] and others, and the series is called the Euler-Zagier sum or the multiple harmonic series. The case $r = 3$ of (3.1) had, in fact, already been studied by Sitaramachandrarao and Subbarao [58]. The Euler-Zagier sum also appears in work of Butzer, Markett and Schmidt ([15], [16], [41]).

There are various interesting relations among values of (3.1) at positive integers. Some of these (for small r) can be found in earlier references, but systematic studies were undertaken by Hoffman [22] (see also [23]), who proved a number of relations, including some previous results and conjectures, and stated the sum conjecture and the duality conjecture. The sum conjecture, originally due to M. Schmidt (see Markett [41]) and also to C. Moen, was proved by Granville [21] and Zagier (unpublished). On the other hand, the duality conjecture has turned out to be an immediate consequence of iterated integral representations of Drinfel'd and Kontsevich (cf. Zagier [65]). Further generalizations were given by Ohno [55] and Hoffman-Ohno [24]. Other families of relations, coming from the theory of knot invariants, were discovered by Le-Murakami [35] and Takamuki [60]. Various relations were also discussed by Borwein et al. [11][12], Flajolet-Salvy [18] and Minh-Petitot [51]. For instance, a conjecture mentioned in Zagier [65] was proved in [12][13]. For the latest developments we refer to [14] and [56]. (The recent developments in this field are enormous, and it is impossible to mention all of them here.)

The papers mentioned above were mainly devoted to the study of the values of $\zeta_r(s_1,\ldots,s_r)$ at positive integers. On the other hand, except for the case $r = 2$ mentioned in the preceding section, the study of the analytic continuation of $\zeta_r(s_1,\ldots,s_r)$ has begun only very recently. First, Arakawa and Kaneko [5] proved that if $s_1,\ldots,s_{r-1}$ are fixed, then (3.1) can be continued meromorphically with respect to s_r to the whole complex plane. The analytic continuation of (3.1) to the whole $\mathbf{C}^r$-space as an r-variable function was established by Zhao [66], and independently by Akiyama, Egami and Tanigawa [1]. Zhao's proof is based on properties of generalized functions in the sense of Gel'fand and Shilov. The method in [1] is more elementary and is based on an application of the Euler-Maclaurin summation formula. Akiyama, Egami and Tanigawa [1] further studied the values of $\zeta_r(s_1, \ldots, s_r)$ at non-positive integers (see also Akiyama and Tanigawa [3]). Note that the statements about the trivial zeros of ζ_2 in Zhao [66] are incorrect. T. Arakawa pointed out that the method of Arakawa and Kaneko [5] can also be refined to give an alternative proof of the analytic continuation of $\zeta_r(s_1, \ldots, s_r)$ as an r-variable function.

Akiyama and Ishikawa [2] considered the multiple L-function

$$L_r((s_1, \ldots, s_r); (\chi_1, \ldots, \chi_r)) \tag{3.3}$$

$$= \sum_{m_1=1}^{\infty} \sum_{m_2=1}^{\infty} \cdots \sum_{m_r=1}^{\infty} \frac{\chi_1(m_1)}{m_1^{s_1}} \frac{\chi_2(m_1 + m_2)}{(m_1 + m_2)^{s_2}} \cdots \frac{\chi_r(m_1 + \cdots + m_r)}{(m_1 + \cdots + m_r)^{s_r}},$$

where $\chi_1, \ldots, \chi_r$ are Dirichlet characters. This series had been introduced earlier by Goncharov [20], but the main goal of Akiyama and Ishikawa [2] was to prove the analytic continuation of (3.3). For this purpose, they first wrote (3.3) as a linear combination of functions of the form

$$\zeta_r((s_1, \ldots, s_r); (\alpha_1, \ldots, \alpha_r)) = \sum_{m_1=1}^{\infty} \sum_{m_2=1}^{\infty} \cdots \sum_{m_r=1}^{\infty} (\alpha_1 + m_1)^{-s_1}$$

$$\times (\alpha_2 + m_1 + m_2)^{-s_2} \cdots (\alpha_r + m_1 + \cdots + m_r)^{-s_r}, \tag{3.4}$$

where $\alpha_1, \ldots, \alpha_r$ are positive, and considered the analytic continuation of the latter functions. They established this continuation by generalizing the argument in Akiyama, Egami and Tanigawa [1]. Ishikawa [26] derived additional properties in the special case $s_1 = \cdots = s_r$ of (3.3) and applied those results to the study of certain multiple character sums (Ishikawa [27]).

4   Katsurada's Idea

In Section 2 we mentioned the work of Katsurada and the author on asymptotic expansions of (2.5) and (2.6). The key ingredient in this work is the treatment of the functions (2.4) and (2.8); in [32] and [34] these functions are expressed by certain double contour integrals.

Katsurada [29][30] reconsidered this problem, and discovered a simple elegant alternative way of proving the expansions of (2.5) and (2.6). The key tool of Katsurada's method is the Mellin-Barnes integral formula

$$\Gamma(s)(1+\lambda)^{-s} = \frac{1}{2\pi i} \int_{(c)} \Gamma(s+z)\Gamma(-z)\lambda^z dz, \qquad (4.1)$$

where s and λ are complex with $\Re s > 0$, $|\arg \lambda| < \pi$, $\lambda \neq 0$, and c is real with $-\Re s < c < 0$. The path of integration is the vertical line from $c - i\infty$ to $c + i\infty$. This formula is known (see, e.g., Whittaker and Watson [62], Section 14.51, p. 289, Corollary), or can easily be proved as follows. First assume $|\lambda| < 1$, and shift the path to the right. The relevant poles of the integrand are located at $z = n$ ($n = 0, 1, 2, \ldots$) with the residue $(-1)^{n+1}\Gamma(s+n)\lambda^n/n!$. Hence the right-hand side of (4.1) is equal to

$$\Gamma(s) \sum_{n=0}^{\infty} \binom{-s}{n} \lambda^n = \Gamma(s)(1+\lambda)^{-s},$$

which is the left-hand side. The extension to $|\lambda| \geq 1$ now follows by analytic continuation.

Katsurada [30] used (4.1) to give a simple proof of the analytic continuation and the asymptotic expansion of the function (2.4). Subsequently, Katsurada [29] (this article was published earlier, but written later than [30]) showed that the same idea can be applied to the function (2.8) to obtain its analytic continuation. In [29], this idea is combined with some properties of hypergeometric functions, and hence the technical details are not so simple. Therefore, to illustrate the essence of Katsurada's idea, we give here a simple proof of the analytic continuation of the Euler sum (2.1) by Katsurada's method.

Assume $\Re s_2 > 1$ and $\Re(s_1 + s_2) > 2$. Putting $s = s_2$ and $\lambda = n/m$ in (4.1), and dividing the both sides by $\Gamma(s_2)m^{s_1+s_2}$, we obtain

$$m^{-s_1}(m+n)^{-s_2} = \frac{1}{2\pi i} \int_{(c)} \frac{\Gamma(s_2+z)\Gamma(-z)}{\Gamma(s_2)} m^{-s_1-s_2-z} n^z dz. \qquad (4.2)$$

We may assume $\max\{-\Re s_2, 1 - \Re(s_1 + s_2)\} < c < -1$. Then we can sum both sides of (4.2) with respect to m and n to obtain

$$\zeta_2(s_1, s_2) = \frac{1}{2\pi i} \int_{(c)} \frac{\Gamma(s_2+z)\Gamma(-z)}{\Gamma(s_2)} \zeta(s_1 + s_2 + z)\zeta(-z)dz. \qquad (4.3)$$

We now shift the path to $\Re z = M - \varepsilon$, where M is a positive integer and ε is a small positive number. This shifting is easily justified using Stirling's formula. The relevant poles of the integrand are at $z = -1, 0, 1, 2, \ldots, M -$

1. Counting the residues of those poles, we get

$$\zeta_2(s_1, s_2) = \frac{1}{s_2 - 1}\zeta(s_1 + s_2 - 1) + \sum_{k=0}^{M-1} \binom{-s_2}{k} \zeta(s_1 + s_2 + k)\zeta(-k)$$

$$+ \frac{1}{2\pi i} \int_{(M-\varepsilon)} \frac{\Gamma(s_2 + z)\Gamma(-z)}{\Gamma(s_2)} \zeta(s_1 + s_2 + z)\zeta(-z)dz. \quad (4.4)$$

The last integral can be continued holomorphically to the region

$$\{(s_1, s_2) \in \mathbf{C}^2 \mid \Re s_2 > -M + \varepsilon, \Re(s_1 + s_2) > 1 - M + \varepsilon\},$$

because in this region the poles of the integrand are not on the path of integration. Hence (4.4) gives the analytic continuation of $\zeta_2(s_1, s_2)$ to this region. Since M is arbitrary, the proof of the continuation to the whole $\mathbf{C}^2$-space is complete. Moreover, from (4.4) we can see that the singularities of $\zeta_2(s_1, s_2)$ are located only on the subsets of $\mathbf{C}^2$ defined by one of the equations

$$s_2 = 1, \quad s_1 + s_2 = 2 - \ell \quad (\ell \in \mathbf{N}_0), \quad (4.5)$$

where $\mathbf{N}_0$ denotes the set of non-negative integers.

Katsurada applied (4.1) to various other types of problems. Here we mention his short note [31], in which he introduced (inspired by [42]) the double zeta-function

$$\sum_{m=0}^{\infty} \sum_{n=0}^{\infty} e^{2\pi i(ms_1 + ns_2)} (\alpha + m)^{-s_1} (\alpha + \beta + m + n)^{-s_2},$$

expressed it as an integral similar to (4.3), and obtained some asymptotic results in the domain of absolute convergence.

5 The Mordell-Tornheim Zeta-Function and the Apostol-Vu Zeta-Function

Let $\Re s_j > 1$ $(j = 1, 2, 3)$ and define

$$\zeta_{MT}(s_1, s_2, s_3) = \sum_{m=1}^{\infty} \sum_{n=1}^{\infty} m^{-s_1} n^{-s_2} (m + n)^{-s_3}. \quad (5.1)$$

This series was first considered by Tornheim [61], and the special case $s_1 = s_2 = s_3$ was studied independently by Mordell [52]. We call (5.1) the Mordell-Tornheim zeta-function. Tornheim himself called it the harmonic double series. Zagier [65] quoted Witten's paper [63] and studied (5.1) calling it the Witten zeta-function.

The analytic continuation of $\zeta_{MT}(s_1, s_2, s_3)$ was established by S. Akiyama and also by S. Egami in 1999. Akiyama's method is based on the Euler-Maclaurin summation formula, while Egami's proof is a modification of the method of Arakawa and Kaneko [5]. Both of these proofs are still unpublished.

Here, by using the method explained in the preceding section, we give a simple proof of the following result:

Theorem 1. *The function $\zeta_{MT}(s_1, s_2, s_3)$ can be meromorphically continued to the whole $\mathbf{C}^3$-space, and all of its singularities are located on the subsets of $\mathbf{C}^3$ defined by one of the equations $s_1 + s_3 = 1 - \ell$, $s_2 + s_3 = 1 - \ell$ ($\ell \in \mathbf{N}_0$) and $s_1 + s_2 + s_3 = 2$.*

Proof. Assume $\Re s_1 > 1$, $\Re s_2 > 0$ and $\Re s_3 > 1$. Then the series (5.1) is absolutely convergent. Putting $s = s_3$ and $\lambda = n/m$ in (4.1), and dividing the both sides by $\Gamma(s_3) m^{s_1+s_3} n^{s_2}$, we obtain

$$m^{-s_1-s_3} n^{-s_2}(1 + \frac{n}{m})^{-s_3} = \frac{1}{2\pi i} \int_{(c)} \frac{\Gamma(s_3 + z)\Gamma(-z)}{\Gamma(s_3)} m^{-s_1-s_3-z} n^{-s_2+z} dz.$$

We may assume $-\Re s_3 < c < \min\{\Re s_2 - 1, 0\}$. Summing with respect to m and n we get

$$\zeta_{MT}(s_1, s_2, s_3) = \frac{1}{2\pi i} \int_{(c)} \frac{\Gamma(s_3 + z)\Gamma(-z)}{\Gamma(s_3)} \zeta(s_1 + s_3 + z)\zeta(s_2 - z) dz. \quad (5.2)$$

Let M be a positive integer greater than $\Re s_2 - 1 + \varepsilon$, and shift the path to $\Re z = M - \varepsilon$. First assume that s_2 is not a positive integer. Then all the relevant poles are simple, and we obtain

$$\zeta_{MT}(s_1, s_2, s_3) = \frac{\Gamma(s_2 + s_3 - 1)\Gamma(1 - s_2)}{\Gamma(s_3)} \zeta(s_1 + s_2 + s_3 - 1)$$

$$+ \sum_{k=0}^{M-1} \binom{-s_3}{k} \zeta(s_1 + s_3 + k)\zeta(s_2 - k)$$

$$+ \frac{1}{2\pi i} \int_{(M-\varepsilon)} \frac{\Gamma(s_3 + z)\Gamma(-z)}{\Gamma(s_3)} \zeta(s_1 + s_3 + z)\zeta(s_2 - z) dz. \quad (5.3)$$

When $s_2 = 1 + h$ ($h \in \mathbf{N}_0$, $h \le M - 1$), the right-hand side of (5.3) contains two singular factors, but they cancel each other. In fact, we obtain

$$\zeta_{MT}(s_1, 1+h, s_3)$$

$$= \binom{-s_3}{h} \left\{ \left(1 + \frac{1}{2} + \cdots + \frac{1}{h} - \psi(s_3 + h)\right) \zeta(s_1 + s_3 + h) \right.$$

$$\left. - \zeta'(s_1 + s_3 + h) \right\}$$

$$+ \sum_{\substack{k=0 \\ k \neq h}}^{M-1} \binom{-s_3}{k} \zeta(s_1 + s_3 + k)\zeta(1 + h - k)$$

$$+ \frac{1}{2\pi i} \int_{(M-\varepsilon)} \frac{\Gamma(s_3 + z)\Gamma(-z)}{\Gamma(s_3)} \zeta(s_1 + s_3 + z)\zeta(1 + h - z)dz,$$

$$(5.4)$$

where $\psi = \Gamma'/\Gamma$ and an empty sum is to be interpreted as zero. The desired assertions of Theorem 1 now follow from (5.3) and (5.4), as in the argument described in the preceding section. $\square$

After the papers of Tornheim [61] and Mordell [52], the values of $\zeta_{MT}(s_1, s_2, s_3)$ at positive integers have been investigated by many authors (Subbarao and Sitaramachandrarao [59], Huard, Williams and Zhang [25], and Zagier [65]). In view of Theorem 1, it is an interesting problem to study the properties of the values of $\zeta_{MT}(s_1, s_2, s_3)$ at non-positive integers.

Next, recall the series (2.3) considered by Apostol and Vu [4] who, inspired by the work of Sitaramachandrarao and Sivaramasarma [57], obtained various formulas on the special values of (2.3).

Here we introduce the following three-variable Apostol-Vu zeta-function:

$$\zeta_{AV}(s_1, s_2, s_3) = \sum_{m=1}^{\infty} \sum_{n<m} m^{-s_1} n^{-s_2} (m+n)^{-s_3} \qquad (\Re s_j > 1). \quad (5.5)$$

Note that there is the following simple relation between ζ_{AV} and ζ_{MT}:

$$\zeta_{MT}(s_1, s_2, s_3) = 2^{-s_3}\zeta(s_1 + s_2 + s_3)$$
$$+ \zeta_{AV}(s_1, s_2, s_3) + \zeta_{AV}(s_2, s_1, s_3). \quad (5.6)$$

Also, there is a simple relation between $\zeta_{AV}(s_1, s_2, 1)$ and $\zeta_2(s_1, s_2)$ (see (17) of Apostol and Vu [4]).

We now prove the analytic continuation of the (three-variable) Apostol-Vu zeta-function $\zeta_{AV}(s_1, s_2, s_3)$. The proof is based on the same principle as that of Theorem 1, but the details are somewhat more complicated.

Theorem 2. *The function $\zeta_{AV}(s_1, s_2, s_3)$ can be continued meromorphically to the whole $\mathbf{C}^3$-space, and all of its singularities are located on the subsets of $\mathbf{C}^3$ defined by one of the equations $s_1 + s_3 = 1 - \ell$, and $s_1 + s_2 + s_3 = 2 - \ell$ ($\ell \in \mathbf{N}_0$).*

Proof. Assume $\Re s_j > 1$ ($j = 1, 2, 3$). Similar to (5.2), we obtain

$$\zeta_{AV}(s_1, s_2, s_3) = \frac{1}{2\pi i} \int_{(c)} \frac{\Gamma(s_3 + z)\Gamma(-z)}{\Gamma(s_3)} \sum_{m=1}^{\infty} \sum_{n < m} m^{-s_1 - s_3 - z} n^{-s_2 + z} dz$$

$$= \frac{1}{2\pi i} \int_{(c)} \frac{\Gamma(s_3 + z)\Gamma(-z)}{\Gamma(s_3)} \zeta_2(s_2 - z, s_1 + s_3 + z) dz, \quad (5.7)$$

where $-\Re s_3 < c < 0$. We now shift the path of integration to $\Re z = M - \varepsilon$. It is not difficult to show using (4.4) that $\zeta_2(s_1, s_2)$ is of polynomial order with respect to $\Im s_1$ and $\Im s_2$. Hence this shifting is possible. From (4.5) we see that the only pole of $\zeta_2(s_2 - z, s_1 + s_3 + z)$ (as a function in z), under the assumption $\Re s_j > 1$ ($j = 1, 2, 3$), is $z = 1 - s_1 - s_3$. This pole is located to the left of the line $\Re z = c$, and hence irrelevant now. Counting the residues of the poles at $z = 0, 1, \ldots, M - 1$, we get

$$\zeta_{AV}(s_1, s_2, s_3) = \sum_{k=0}^{M-1} \binom{-s_3}{k} \zeta_2(s_2 - k, s_1 + s_3 + k)$$

$$+ \frac{1}{2\pi i} \int_{(M-\varepsilon)} \frac{\Gamma(s_3 + z)\Gamma(-z)}{\Gamma(s_3)} \zeta_2(s_2 - z, s_1 + s_3 + z) dz. \quad (5.8)$$

This formula already implies the meromorphic continuation except in the case when $s_1 + s_2 + s_3 = 2 - \ell$ ($\ell \in \mathbf{N}_0$), where $\zeta_2(s_2 - z, s_1 + s_3 + z)$ is singular. To investigate the behaviour of the above integral on this polar set, we substitute the formula (4.4) into the integrand on the right-hand side of (5.8). We obtain

$$\zeta_{AV}(s_1, s_2, s_3) = \sum_{k=0}^{M-1} \binom{-s_3}{k} \zeta_2(s_2 - k, s_1 + s_3 + k)$$

$$+ \zeta(s_1 + s_2 + s_3 - 1) P(s_1, s_3)$$

$$+ \sum_{j=0}^{M-1} \zeta(s_1 + s_2 + s_3 + j)\zeta(-j) Q_j(s_1, s_3) + R(s_1, s_2, s_3), \quad (5.9)$$

where

$$P(s_1, s_3) = \frac{1}{2\pi i} \int_{(M-\varepsilon)} \frac{\Gamma(s_3 + z)\Gamma(-z)}{\Gamma(s_3)} \frac{dz}{s_1 + s_3 + z - 1},$$

$$Q_j(s_1, s_3) = \frac{1}{2\pi i} \int_{(M-\varepsilon)} \frac{\Gamma(s_3 + z)\Gamma(-z)}{\Gamma(s_3)} \binom{-s_1 - s_3 - z}{j} dz,$$

and

$$R(s_1, s_2, s_3) = \frac{1}{(2\pi i)^2} \int_{(M-\varepsilon)} \frac{\Gamma(s_3 + z)\Gamma(-z)}{\Gamma(s_3)}$$

$$\times \int_{(M-\varepsilon)} \frac{\Gamma(s_1 + s_3 + z + z')\Gamma(-z')}{\Gamma(s_1 + s_3 + z)} \zeta(s_1 + s_2 + s_3 + z')\zeta(-z')dz'dz.$$

It is easy to see that

(i) $P(s_1, s_3)$ is holomorphic if $\Re s_3 > -M + \varepsilon$ and $\Re(s_1 + s_3) > 1 - M + \varepsilon$,

and

(ii) $Q_j(s_1, s_3)$ is holomorphic for $0 \le j \le M - 1$ if $\Re s_3 > -M + \varepsilon$.

Also, since the inner integral of $R(s_1, s_2, s_3)$ is holomorphic if $\Re(s_1 + s_3 + z) > -M + \varepsilon$ and $\Re(s_1 + s_2 + s_3) > 1 - M + \varepsilon$ as a function of the four variables (s_1, s_2, s_3, z), we see that

(iii) $R(s_1, s_2, s_3)$ is holomorphic if $\Re s_3 > -M + \varepsilon$, $\Re(s_1 + s_3) > -2M + 2\varepsilon$ and $\Re(s_1 + s_2 + s_3) > 1 - M + \varepsilon$.

From (i), (ii), (iii) and (5.9), we find that $\zeta_{AV}(s_1, s_2, s_3)$ can be continued meromorphically to the region

$$\{(s_1, s_2, s_3) \in \mathbf{C}^3 \mid \Re s_3 > -M + \varepsilon, \Re(s_1 + s_3) > 1 - M + \varepsilon,$$
$$\Re(s_1 + s_2 + s_3) > 1 - M + \varepsilon\}.$$

Since M is arbitrary, we obtain the analytic continuation of $\zeta_{AV}(s_1, s_2, s_3)$ to the whole $\mathbf{C}^3$-space. The information on singularities can be deduced from the representation (5.9). The proof of Theorem 2 is complete.   □

6   Generalized Multiple Zeta-Functions

Let $s_1, \ldots, s_r$ be complex variables. Let $\alpha_1, \ldots, \alpha_r, w_1, \ldots, w_r$ be complex parameters, and define the multiple series

$$\zeta_r((s_1, \ldots, s_r); (\alpha_1, \ldots, \alpha_r), (w_1, \ldots, w_r))$$

$$= \sum_{m_1=0}^{\infty} \cdots \sum_{m_r=0}^{\infty} (\alpha_1 + m_1 w_1)^{-s_1} (\alpha_2 + m_1 w_1 + m_2 w_2)^{-s_2}$$

$$\times \cdots \times (\alpha_r + m_1 w_1 + \cdots + m_r w_r)^{-s_r}. \qquad (6.1)$$

We will explain later (in the proof of Theorem 3) how to choose the branch of the logarithms.

When $s_1 = \cdots = s_{r-1} = 0$, then the above series (6.1) reduces to the Barnes multiple zeta-function (1.3). The Euler-Zagier sum (3.1) and its generalization (3.4) are also special cases of (6.1). The multiple series of the form (6.1) was first introduced in the author's article [45], and the meromorphic continuation in the special case $0 < \alpha_1 < \alpha_2 < \cdots < \alpha_r$ and $w_j = 1$ $(1 \leq j \leq r)$ of (6.1) to the whole $\mathbf{C}^r$-space was proved in [45].

To ensure the convergence of (6.1), we impose condition (1.4) on the w_j's, which was first introduced by Barnes for his multiple series (1.3). However, we do not require any condition on the α_j's. If $\alpha_j \notin H(\ell)$ for some j, then there might exist finitely many tuples $(m_1, \ldots, m_j)$ for which

$$\alpha_j + m_1 w_1 + \cdots + m_j w_j = 0 \qquad (6.2)$$

holds. We adopt the convention that the terms corresponding to such tuples are removed from the sum (6.1). Under this convention, we now prove:

Theorem 3. *If the condition (1.4) holds, then the series (6.1) is absolutely convergent in the region $\mathcal{A}_r$, defined by (3.2), uniformly in any compact subset of $\mathcal{A}_r$.*

Proof. We prove the theorem by induction. When $r = 1$, the assertion is obvious. Assume that the theorem is true for ζ_{r-1}. In what follows, an empty sum is to be interpreted as zero.

Let $\theta \in (-\pi, \pi]$ be the argument of the vector contained in $H(\ell)$ and orthogonal to ℓ. Then the line ℓ consists of the points whose arguments are $\theta \pm \pi/2$ (and the origin), and

$$H(\ell) = \left\{ w \in \mathbf{C} \setminus \{0\} \,\middle|\, \theta - \frac{\pi}{2} < \arg w < \theta + \frac{\pi}{2} \right\}.$$

We can write $w_j = w_j^{(1)} + w_j^{(2)}$, with $\arg w_j^{(1)} = \theta - \pi/2$ or $\theta + \pi/2$ (or $w_j^{(1)} = 0$) and $\arg w_j^{(2)} = \theta$. Similarly we write $\alpha_j = \alpha_j^{(1)} + \alpha_j^{(2)}$ with $\arg \alpha_j^{(1)} = \theta - \pi/2$ or $\theta + \pi/2$ (or $\alpha_j^{(1)} = 0$) and $\arg \alpha_j^{(2)} = \theta$ or $-\theta$ (or $\alpha_j^{(2)} = 0$). If the set

$$\mathcal{E} = \left\{ \alpha_j^{(2)} \,\middle|\, \arg \alpha_j^{(2)} = -\theta \quad \text{or} \quad \alpha_j^{(2)} = 0 \right\}$$

is not empty, we denote by $\bar{\alpha}$ (one of) the element(s) of this set whose absolute value is largest. Let μ be the smallest positive integer such that

$\tilde{\alpha} + m_1 w_1^{(2)} \in H(\ell)$ for any $m_1 \geq \mu$, and split the sum (6.1) into

$$
\zeta_r((s_1, \ldots, s_r); (\alpha_1, \ldots, \alpha_r), (w_1, \ldots, w_r))
$$

$$
= \sum_{m_1=0}^{\mu-1} \sum_{m_2=0}^{\infty} \cdots \sum_{m_r=0}^{\infty} + \sum_{m_1=\mu}^{\infty} \sum_{m_2=0}^{\infty} \cdots \sum_{m_r=0}^{\infty} = T_1 + T_2, \quad (6.3)
$$

say. (If $\mathcal{E} = \emptyset$, we put $\mu = 0$.) For any $m_1 \leq \mu - 1$, we put $\alpha_j'(m_1) = \alpha_j + m_1 w_1$. Then

$$
T_1 = \sum_{m_1=0}^{\mu-1} \alpha_1'(m_1)^{-s_1} \sum_{m_2=0}^{\infty} \cdots \sum_{m_r=0}^{\infty} (\alpha_2'(m_1) + m_2 w_2)^{-s_2}
$$

$$
\times \cdots \times (\alpha_r'(m_1) + m_2 w_2 + \cdots + m_r w_r)^{-s_r}
$$

$$
= \sum_{m_1=0}^{\mu-1} \alpha_1'(m_1)^{-s_1}
$$

$$
\times \zeta_{r-1}((s_2, \ldots, s_r); (\alpha_2'(m_1), \ldots, \alpha_r'(m_1)), (w_2, \ldots, w_r)). \quad (6.4)
$$

To evaluate T_2, we put $\alpha_j'(\mu) = \alpha_j + \mu w_1$ and $m_1' = m_1 - \mu$. Then

$$
T_2 = \sum_{m_1'=0}^{\infty} \sum_{m_2=0}^{\infty} \cdots \sum_{m_r=0}^{\infty} (\alpha_1'(\mu) + m_1' w_1)^{-s_1} (\alpha_2'(\mu) + m_1' w_1 + m_2 w_2)^{-s_2}
$$

$$
\times \cdots \times (\alpha_r'(\mu) + m_1' w_1 + m_2 w_2 + \cdots + m_r w_r)^{-s_r}. \quad (6.5)
$$

Since $\alpha_j'(\mu) = (\alpha_j^{(1)} + \mu w_1^{(1)}) + (\alpha_j^{(2)} + \mu w_1^{(2)})$, the definitions of $\tilde{\alpha}$ and μ imply that $\alpha_j'(\mu) \in H(\ell)$. The right-hand side of (6.4) is absolutely convergent by induction assumption. Hence we only need to show the absolute convergence of (6.5). In other words, our remaining task is to prove the absolute convergence of (6.1) under the additional assumption that $\alpha_j \in H(\ell)$ $(1 \leq j \leq r)$. Then all terms $\alpha_j + m_1 w_1 + \cdots + m_r w_r$ are in $H(\ell)$. Each factor on the right-hand side of (6.1) is to be understood as

$$
(\alpha_j + m_1 w_1 + \cdots + m_j w_j)^{-s_j} = \exp(-s_j \log(\alpha_j + m_1 w_1 + \cdots + m_j w_j)),
$$

where the branch of the logarithm is that defined by the condition

$$
\theta - \frac{\pi}{2} < \arg(\alpha_j + m_1 w_1 + \cdots + m_j w_j) < \theta + \frac{\pi}{2}.
$$

Let $\sigma_j = \Re s_j$, $t_j = \Im s_j$, and define $J_+ = \{j \mid \sigma_j \geq 0\}$ and $J_- = $

$\{j \mid \sigma_j < 0\}$. Since

$$|\alpha_j + m_1 w_1 + \cdots + m_j w_j|$$
$$\geq |\alpha_j^{(2)} + m_1 w_1^{(2)} + \cdots + m_j w_j^{(2)}|$$
$$= |\,|\alpha_j^{(2)}|e^{i\theta} + m_1 |w_1^{(2)}|e^{i\theta} + \cdots + m_j |w_j^{(2)}|e^{i\theta}|$$
$$= |\alpha_j^{(2)}| + m_1 |w_1^{(2)}| + \cdots + m_j |w_j^{(2)}|,$$

we have

$$|\alpha_j + m_1 w_1 + \cdots + m_j w_j|^{-\sigma_j} \leq (|\alpha_j^{(2)}| + m_1 |w_1^{(2)}| + \cdots + m_j |w_j^{(2)}|)^{-\sigma_j}$$

for $j \in J_+$. On the other hand, it is clear that

$$|\alpha_j + m_1 w_1 + \cdots + m_j w_j|^{-\sigma_j} \leq (|\alpha_j| + m_1 |w_1| + \cdots + m_j |w_j|)^{-\sigma_j}$$

for $j \in J_-$. Therefore, setting

$$\alpha_j^* = \begin{cases} |\alpha_j^{(2)}| & \text{if } j \in J_+, \\ |\alpha_j| & \text{if } j \in J_-, \end{cases}$$

and

$$w_j^* = \begin{cases} |w_j^{(2)}| & \text{if } j \in J_+, \\ |w_j| & \text{if } j \in J_{-,,} \end{cases}$$

we find that $\alpha_j^* > 0$, $w_j^* > 0$ for all j and that

$$|(\alpha_j + m_1 w_1 + \cdots + m_j w_j)^{-s_j}|$$
$$= |\alpha_j + m_1 w_1 + \cdots + m_j w_j|^{-\sigma_j} \exp(t_j \arg(\alpha_j + m_1 w_1 + \cdots + m_j w_j))$$
$$\leq (\alpha_j^* + m_1 w_1^* + \cdots + m_j w_j^*)^{-\sigma_j} \exp(2\pi |t_j|).$$

Hence

$$|\zeta_r((s_1, \ldots, s_r); (\alpha_1, \ldots, \alpha_r), (w_1, \ldots, w_r))|$$
$$\leq \exp(2\pi(|t_1| + \cdots + |t_r|))$$
$$\times \sum_{m_1=0}^{\infty} \cdots \sum_{m_r=0}^{\infty} (\alpha_1^* + m_1 w_1^*)^{-\sigma_1} (\alpha_2^* + m_1 w_1^* + m_2 w_2^*)^{-\sigma_2}$$
$$\times \cdots \times (\alpha_r^* + m_1 w_1^* + \cdots + m_r w_r^*)^{-\sigma_r}. \tag{6.6}$$

We claim that for any positive integers $k \leq r$, the series

$$S(k) = \sum_{m_{r-k+1}=0}^{\infty} \sum_{m_{r-k+2}=0}^{\infty} \cdots \sum_{m_r=0}^{\infty}$$

$$(\alpha_{r-k+1}^* + m_1 w_1^* + \cdots + m_{r-k+1} w_{r-k+1}^*)^{-\sigma_{r-k+1}}$$

$$\times (\alpha_{r-k+2}^* + m_1 w_1^* + \cdots + m_{r-k+2} w_{r-k+2}^*)^{-\sigma_{r-k+2}}$$

$$\times \cdots \times (\alpha_r^* + m_1 w_1^* + \cdots + m_r w_r^*)^{-\sigma_r}$$

is convergent in the region $\sigma_r > 1$, $\sigma_{r-1} + \sigma_r > 2$,..., $\sigma_{r-k+1} + \cdots + \sigma_r > k$, and the estimate

$$S(k) \ll (\beta_1(k) + m_1 w_1^* + \cdots + m_{r-k} w_{r-k}^*)$$

$$\times (\beta_2(k) + m_1 w_1^* + \cdots + m_{r-k} w_{r-k}^*)^{c(k)} \qquad (6.7)$$

holds, where $\beta_1(k) > \beta_2(k) > 0$,

$$c(k) = k - 1 - (\sigma_{r-k+1} + \cdots + \sigma_r), \qquad (6.8)$$

and the implied constant depends on σ_j, α_j^* and w_j^* $(r - k + 1 \leq j \leq r)$. Note that $c(k) < -1$.

We prove this claim by induction. For any positive real numbers a, b and $\sigma > 1$, we have

$$\sum_{m=0}^{\infty} (a + bm)^{-\sigma} \leq a^{-\sigma} + \int_0^{\infty} (a + bx)^{-\sigma} dx \ll \left(1 + \frac{a}{b}\right) a^{-\sigma}, \qquad (6.9)$$

where the implied constant depends only on σ. Using (6.9) with $m = m_r$, $\sigma = \sigma_r$, $a = \alpha_r^* + m_1 w_1^* + \cdots + m_{r-1} w_{r-1}^*$ and $b = w_r^*$, we easily obtain the case $k = 1$ of the claim with $\beta_1(1) = \alpha_r^* + w_r^*$ and $\beta_2(1) = \alpha_r^*$. Now assume that the claim is true for $S(k - 1)$. Then we have

$$S(k) \ll \sum_{m_{r-k+1}=0}^{\infty} (\alpha_{r-k+1}^* + m_1 w_1^* + \cdots + m_{r-k+1} w_{r-k+1}^*)^{-\sigma_{r-k+1}}$$

$$\times (\beta_1(k - 1) + m_1 w_1^* + \cdots + m_{r-k+1} w_{r-k+1}^*)$$

$$\times (\beta_2(k - 1) + m_1 w_1^* + \cdots + m_{r-k+1} w_{r-k+1}^*)^{c(k-1)}.$$

If $-\sigma_{r-k+1} \geq 0$, we replace α_{r-k+1}^* and $\beta_1(k - 1)$ by $\max\{\alpha_{r-k+1}^*, \beta_1(k - 1)\}$. If $-\sigma_{r-k+1} < 0$, we replace α_{r-k+1}^* and $\beta_2(k-1)$ by $\min\{\alpha_{r-k+1}^*, \beta_2(k-1)\}$. In either case, we get an estimate of the form

$$S(k) \ll \sum_{m_{r-k+1}=0}^{\infty} (B_1 + m_1 w_1^* + \cdots + m_{r-k+1} w_{r-k+1}^*)^{C_1}$$

$$\times (B_2 + m_1 w_1^* + \cdots + m_{r-k+1} w_{r-k+1}^*)^{C_2}, \qquad (6.10)$$

where $B_1 > B_2 > 0$, $C_1 \geq 0$, $C_2 < 0$, and

$$C_1 + C_2 = -\sigma_{r-k+1} + 1 + c(k-1) = c(k). \tag{6.11}$$

Since

$$(B_1 + m_1 w_1^* + \cdots + m_{r-k+1} w_{r-k+1}^*)^{C_1}$$
$$= (B_2 + m_1 w_1^* + \cdots + m_{r-k+1} w_{r-k+1}^*)^{C_1}$$
$$\times \left(1 + \frac{B_1 - B_2}{B_2 + m_1 w_1^* + \cdots + m_{r-k+1} w_{r-k+1}^*} \right)^{C_1}$$
$$\leq \left(1 + \frac{B_1 - B_2}{B_2} \right)^{C_1} (B_2 + m_1 w_1^* + \cdots + m_{r-k+1} w_{r-k+1}^*)^{C_1},$$

(6.10) and (6.11) imply that

$$S(k) \ll \sum_{m_{r-k+1}=0}^{\infty} (B_2 + m_1 w_1^* + \cdots + m_{r-k+1} w_{r-k+1}^*)^{c(k)}. \tag{6.12}$$

The claim for $S(k)$ now follows by applying (6.9) to the right-hand side of (6.12), with $\beta_1(k) = B_2 + w_{r-k+1}^*$ and $\beta_2(k) = B_2$. Hence by induction we find that the claim is true for $1 \leq k \leq r$, and the claim for $k = r$ implies the absolute convergence of the right-hand side of (6.6). This completes the proof of Theorem 3. □

We now apply the method explained in Sections 4 and 5 to the generalized multiple zeta-function (6.1). In addition to (1.4), we assume

$$\alpha_j \in H(\ell) \ (1 \leq j \leq r) \quad \text{and} \quad \alpha_{j+1} - \alpha_j \in H(\ell) \ (1 \leq j \leq r-1). \tag{6.13}$$

We use (4.1) with $s = s_r$ and

$$\lambda = \frac{\alpha_r - \alpha_{r-1} + m_r w_r}{\alpha_{r-1} + m_1 w_1 + \cdots + m_{r-1} w_{r-1}}.$$

Under the assumption (6.13) both the numerator and the denominator of λ are the elements of $H(\ell)$, and hence $|\arg \lambda| < \pi$. Similar to (4.3), (5.2) or (5.7), we obtain

$$\zeta_r((s_1, \ldots, s_r); (\alpha_1, \ldots, \alpha_r), (w_1, \ldots, w_r))$$
$$= \frac{1}{2\pi i} \int_{(c)} \frac{\Gamma(s_r + z)\Gamma(-z)}{\Gamma(s_r)} \zeta_{r-1}((s_1, \ldots, s_{r-2}, s_{r-1} + s_r + z);$$
$$(\alpha_1, \ldots, \alpha_{r-1}), (w_1, \ldots, w_{r-1})) \zeta \left(-z, \frac{\alpha_r - \alpha_{r-1}}{w_r} \right) w_r^z dz. \tag{6.14}$$

Hence, shifting the path of integration, we can prove:

Theorem 4. *Under the conditions* (1.4) *and* (6.13)*, the multiple zeta-function* (6.1) *can be continued meromorphically to the whole* $\mathbf{C}^r$*-space.*

In the present article we confined ourselves to the above very brief outline of the method. The details of the proof, which is by induction on r, will be given in [46].

Finally we mention the analytic continuation of Mordell multiple zeta-functions. In Section 5 we quoted Mordell's paper [52], in which he studied the special case $s_1 = s_2 = s_3$ of (5.1). In the same paper, Mordell also considered the multiple series

$$\sum_{m_1=1}^{\infty} \cdots \sum_{m_r=1}^{\infty} \frac{1}{m_1 m_2 \cdots m_r (m_1 + m_2 + \cdots + m_r + a)}, \qquad (6.15)$$

where $a > -r$. By using Mordell's result on (6.15), Hoffman [22] evaluated the sum

$$\sum_{m_1=1}^{\infty} \cdots \sum_{m_r=1}^{\infty} \frac{1}{m_1 m_2 \cdots m_r (m_1 + m_2 + \cdots + m_r)^s} \qquad (6.16)$$

when s is a positive integer.

Here we introduce the following multi-variable version of (6.16), which is at the same time a generalization of the Mordell-Tornheim zeta-function (5.1):

$$\zeta_{MOR,r}(s_1, \ldots, s_r, s_{r+1})$$
$$= \sum_{m_1=1}^{\infty} \cdots \sum_{m_r=1}^{\infty} m_1^{-s_1} \cdots m_r^{-s_r} (m_1 + \cdots + m_r)^{-s_{r+1}}. \qquad (6.17)$$

Theorem 5. *The series* (6.17) *can be continued meromorphically to the whole* $\mathbf{C}^{r+1}$*-space.*

This and related results will be discussed in a forthcoming paper.

Acknowledgements. The author expresses his sincere gratitude to Professor M. Kaneko, Professor Y. Ohno and the referee for valuable comments and information on recent results concerning multiple zeta values.

References

[1] S. Akiyama, S. Egami, and Y. Tanigawa, *Analytic continuation of multiple zeta functions and their values at non-positive integers*, Acta Arith. **98** (2001), 107–116.

[2]   S. Akiyama and H. Ishikawa, *On analytic continuation of multiple L-functions and related zeta-functions*, Analytic Number Theory (C. Jia and K. Matsumoto, eds.), Kluwer, to appear.

[3]   S. Akiyama and Y. Tanigawa, *Multiple zeta values at non-positive integers*, Ramanujan J., to appear.

[4]   T. M. Apostol and T. H. Vu, *Dirichlet series related to the Riemann zeta function*, J. Number Theory **19** (1984), 85–102.

[5]   T. Arakawa and M. Kaneko, *Multiple zeta values, poly-Bernoulli numbers, and related zeta functions*, Nagoya Math. J. **153** (1999), 189–209.

[6]   F. V. Atkinson, *The mean-value of the Riemann zeta function*, Acta Math. **81** (1949), 353–376.

[7]   E. W. Barnes, *The theory of the double gamma function*, Philos. Trans. Roy. Soc. (A) **196** (1901), 265–387.

[8]   ———, *On the theory of multiple gamma function*, Trans. Cambridge Phil. Soc. **19** (1904), 374–425.

[9]   B. C. Berndt, *Ramanujan's Notebooks, Part I*, Springer, New York, 1985.

[10] D. Borwein, J. M. Borwein, and R. Girgensohn, *Explicit evaluation of Euler sums*, Proc. Edinburgh Math. Soc. **38** (1994), 277–294.

[11] J. M. Borwein, D. M. Bradley, and D. J. Broadhurst, *Evaluations of k-fold Euler/Zagier sums: a compendium of results for arbitrary k*, Electron. J. Comb. **4** (1997), 21 pp.

[12] J. M. Borwein, D. M. Bradley, D. J. Broadhurst, and P. Lisoněk, *Combinatorial aspects of multiple zeta values*, Electron. J. Comb. **5** (1998), 12 pp.

[13] ———, *Special values of multiple polylogarithms*, Trans. Amer. Math. Soc. **353** (2001), 907–941.

[14] D. Bowman and D. M. Bradley, *Resolution of some open problems concerning multiple zeta evaluations of arbitrary depth*, Compositio Math., to appear.

[15] P. L. Butzer, C. Markett, and M. Schmidt, *Stirling numbers, central factorial numbers, and representations of the Riemann zeta function*, Results in Math. **19** (1991), 257–274.

[16] P. L. Butzer and M. Schmidt, *Central factorial numbers and their role in finite difference calculus and approximation*, Approximation Theory (J. Szabados and K. Tandori, eds.), North-Holland, Amsterdam, 1991, pp. 127–150.

[17] R. E. Crandall and J. P. Buhler, *On the evaluation of Euler sums*, Experimental Math. **3** (1995), 275–285.

[18] P. Flajolet and B. Salvy, *Euler sums and contour integral representations*, Experimental Math. **7** (1998), 15–35.

[19] A. B. Goncharov, *Polylogarithms in arithmetic and geometry*, Proc. Intern. Cong. Math. (Zurich, 1994), Vol. 1, Birkhäuser, Basel, 1995, pp. 374–387.

[20] _____, *Multiple polylogarithms, cyclotomy and modular complexes*, Math. Res. Letters **5** (1998), 497–516.

[21] A. Granville, *A decomposition of Riemann's zeta-function*, Analytic Number Theory (Y. Motohashi, ed.), Cambridge Univ. Press, Cambridge, 1997, pp. 95–101.

[22] M. E. Hoffman, *Multiple harmonic series*, Pacific J. Math. **152** (1992), 275–290.

[23] _____, *The algebra of multiple harmonic series*, J. Algebra **194** (1997), 477–495.

[24] M. E. Hoffman and Y. Ohno, *Relations of multiple zeta values and their algebraic expression*, preprint.

[25] J. G. Huard, K. S. Williams, and Nan-Yue Zhang, *On Tornheim's double series*, Acta Arith. **75** (1996), 105–117.

[26] H. Ishikawa, *On analytic properties of multiple L-functions*, Analytic extension formulas and their applications (S. Saitoh, ed.), Kluwer Acad. Publ., Dordrecht, 2001, pp. 105–122.

[27] _____, *A multiple character sum and a multiple L-function*, Arch. Math., to appear.

[28] M. Katsurada, *Asymptotic expansions of the mean values of Dirichlet L-functions II*, Analytic Number Theory and Related Topics (K. Nagasaka, ed.), World Sci. Publishing, River Edge, NJ, 1993, pp. 61–71.

[29] _____, *An application of Mellin-Barnes' type integrals to the mean squares of Lerch zeta-functions*, Collect. Math. **48** (1997), 137–153.

[30] _____ , *An application of Mellin-Barnes type of integrals to the mean square of L-functions*, Liet. Mat. Rink. (Lithuanian Math. J.) **38** (1998), 98–112.

[31] _____ , *Power series and asymptotic series associated with the Lerch zeta-function*, Proc. Japan Acad. **74A** (1998), 167–170.

[32] M. Katsurada and K. Matsumoto, *Asymptotic expansions of the mean values of Dirichlet L-functions*, Math. Z. **208** (1991), 23–39.

[33] _____ , *The mean values of Dirichlet L-functions at integer points and class numbers of cyclotomic fields*, Nagoya Math. J. **134** (1994), 151–172.

[34] _____ , *Explicit formulas and asymptotic expansions for certain mean square of Hurwitz zeta-functions I*, Math. Scand. **78** (1996), 161–177.

[35] T. Q. T. Le and J. Murakami, *Kontsevich's integral for the Homfly polynomial and relations between values of multiple zeta functions*, Topology and its Appl. **62** (1995), 193–206.

[36] B. Lichtin, *Poles of Dirichlet series and D-modules*, Théorie des Nombres/Number Theory (J.-M. DeKoninck, ed.), De Gruyter, Berlin, 1989, pp. 579–594.

[37] _____ , *The asymptotics of a lattice point problem associated to a finite number of polynomials I*, Duke Math. J. **63** (1991), 139–192.

[38] _____ , *Volumes and lattice points - proof of a conjecture of L. Ehrenpreis*, Singularities, Lille 1991 (J.-P. Brasselet, ed.), Cambridge Univ. Press, Cambridge, 1994, pp. 211–250.

[39] _____ , *The asymptotics of a lattice point problem associated to a finite number of polynomials II*, Duke Math. J. **77** (1995), 699–751.

[40] _____ , *Asymptotics determined by pairs of additive polynomials*, Compositio Math. **107** (1997), 233–267.

[41] C. Markett, *Triple sums and the Riemann zeta function*, J. Number Theory **48** (1994), 113–132.

[42] K. Matsumoto, *Asymptotic series for double zeta, double gamma, and Hecke L-functions*, Math. Proc. Cambridge Phil. Soc. **123** (1998), 385–405.

[43] _____, *Asymptotic expansions of double gamma-functions and related remarks*, Analytic Number Theory (C. Jia and K. Matsumoto, eds.), Kluwer, to appear.

[44] _____, *Corrigendum and addendum to "Asymptotic series for double zeta, double gamma, and Hecke L-functions"*, Math. Proc. Cambridge Phil. Soc., to appear.

[45] _____, *Asymptotic expansions of double zeta-functions of Barnes, of Shintani, and Eisenstein series*, preprint.

[46] _____, *The analytic continuation and the asymptotic behaviour of multiple zeta-functions I*, preprint.

[47] Y. Matsuoka, *On the values of a certain Dirichlet series at rational integers*, Tokyo J. Math. **5** (1982), 399–403.

[48] H. Mellin, *Eine Formel für den Logarithmus transcendenter Funktionen von endlichem Geschlecht*, Acta Soc. Sci. Fenn. **29** (1900), no. 4.

[49] _____, *Die Dirichlet'schen Reihen, die zahlentheoretischen Funktionen und die unendlichen Produkte von endlichem Geschlecht*, Acta Math. **28** (1904), 37–64.

[50] T. Meurman, *A generalization of Atkinson's formula to L-functions*, Acta Arith. **47** (1986), 351–370.

[51] H. N. Minh and M. Petitot, *Lyndon words, polylogarithms and the Riemann ζ function*, Discrete Math. **217** (2000), 273–292.

[52] L. J. Mordell, *On the evaluation of some multiple series*, J. London Math. Soc. **33** (1958), 368–371.

[53] Y. Motohashi, *A note on the mean value of the zeta and L-functions I*, Proc. Japan Acad. Ser. A Math. Sci. **61** (1985), 222–224.

[54] N. Nielsen, *Handbuch der Theorie der Gammafunktion*, Teubner, 1906.

[55] Y. Ohno, *A generalization of the duality and sum formulas on the multiple zeta values*, J. Number Theory **74** (1999), 39–43.

[56] Y. Ohno and D. Zagier, *Multiple zeta values of fixed weight, depth, and height*, Indag. Math., to appear.

[57] R. Sitaramachandrarao and A. Sivaramasarma, *Some identities involving the Riemann zeta function*, Indian J. Pure Appl. Math. **10** (1979), 602–607.

[58] R. Sitaramachandrarao and M. V. Subbarao, *Transformation formulae for multiple series*, Pacific J. Math. **113** (1984), 471–479.

[59] M. V. Subbarao and R. Sitaramachandrarao, *On some infinite series of L. J. Mordell and their analogues*, Pacific J. Math. **119** (1985), 245–255.

[60] T. Takamuki, *The Kontsevich invariant and relations of multiple zeta values*, Kobe J. Math. **16** (1999), 27–43.

[61] L. Tornheim, *Harmonic double series*, Amer. J. Math. **72** (1950), 303–314.

[62] E. T. Whittaker and G. N. Watson, *A Course of Modern Analysis*, 4th ed., Cambridge Univ. Press, 1927.

[63] E. Witten, *On quantum gauge theories in two dimensions*, Commun. Math. Phys. **141** (1991), 153–209.

[64] D. Zagier, *Periods of modular forms, traces of Hecke operators, and multiple zeta values*, Studies on automorphic forms and L-functions, Sūrikaiseki Kenkyūsho Kōkyūroku, vol. 843, RIMS, Kyoto Univ., 1993, pp. 162–170.

[65] _____, *Values of zeta functions and their applications*, First European Congress of Mathematics, Vol. II (Paris, 1992), Progr. Math., vol. 120, Birkhäuser, Basel, 1994, pp. 497–512.

[66] J. Zhao, *Analytic continuation of multiple zeta functions*, Proc. Amer. Math. Soc. **128** (2000), 1275–1283.

Quelques Remarques sur la Théorie d'Iwasawa des Courbes Elliptiques

Bernadette Perrin-Riou

Dans ce texte, nous avons cherché à donner un panorama partial et partiel[1] de la théorie d'Iwasawa des courbes elliptiques en utilisant les deux outils que sont les systèmes d'Euler construits par Kato et la théorie de l'exponentielle (ou du logarithme). Nous n'avons regardé ici que la partie interpolation p-adique des valeurs spéciales de la fonction L en 1 tordue par un caractère de Dirichlet[2]. L'idée importante pour nous que nous voudrions faire passer dans ce texte est que l'application logarithme élargi (ou régulateur[3] d'Iwasawa) qui généralise les constructions de Kummer, Iwasawa, Coates-Wiles et l'homomorphisme de Coleman, contient, par sa construction même d'une part et grâce à la loi de réciprocité de Colmez d'autre part, **tous** les renseignements nécessaires permettant le calcul de ses valeurs spéciales et leur lien avec les invariants arithmétiques. Les systèmes d'Euler-Kato permettant de voir la fonction L p-adique comme le régulateur d'Iwasawa d'un système compatible pour les normes construit de manière modulaire grâce au théorème de Kato, la plupart des formules spéciales sur la fonction L p-adique en un entier k s'en déduisent. Dans le cas des courbes elliptiques ayant bonne réduction en p, ces résultats se trouvent déjà dans [31]. Dans le cas des courbes elliptiques semi-stables en p, on obtient une (nouvelle) interprétation du "zéro trivial".

Soit E une courbe elliptique (modulaire) définie sur $\mathbb{Q}$, de conducteur N. La fonction L de Hasse-Weil associée $L(E,s)$ définie pour $Re(s) > 1$ se prolonge à tout le plan complexe en une fonction holomorphe. Si $L(E,s) = \sum_{n=1}^{\infty} a_n n^{-s}$ et si η est un caractère de Dirichlet prolongé par 0 sur les entiers non premiers à son conducteur, on pose $L(E,\eta,s) = \sum_{n=1}^{\infty} \eta(n) a_n n^{-s}$. Si M est un entier, on note $L_{\{M\}}(E,s)$ la fonction L incomplète : $L_{\{M\}}(E,s) = \sum_{(n,M)=1} a_n n^{-s}$.

---

[1] Pour d'autres panoramas, voir par exemple les articles de Greenberg, [15].

[2] Pour la partie arithmétique, voir [31] et [15], [20], [30], [32] dans le cas ordinaire.

[3] Le terme de régulateur désigne à l'origine un nombre et plus particulièrement un volume, l'application permettant de "mesurer" étant le logarithme : par exemple, le volume du tore, quotient du $\mathbb{R}$-espace vectoriel $E = \mathbb{R}^{r_1(F)+r_2(F)-1}$ par le $\mathbb{Z}$-module d'indice fini engendré par l'image des unités d'un corps de nombres F par l'application logarithme est le régulateur du corps F. Il semble que désormais "régulateur" désigne couramment l'application elle-même.

Soit $H^0(E, \Omega^1_{E/\mathbb{Q}})$ le $\mathbb{Q}$-espace vectoriel des formes différentielles invariantes de E définies sur $\mathbb{Q}$ et $\mathrm{Lie}(E)$ l'algèbre de Lie de E. Rappelons la suite exacte

$$0 \to H^0(E, \Omega^1_{E/\mathbb{Q}}) \to H^1_{\mathrm{dR}}(E) \to \mathrm{Lie}(E) \to 0$$

compatible avec l'accouplement canonique

$$H^1_{\mathrm{dR}}(E) \times H^1_{\mathrm{dR}}(E) \to \mathbb{Q}.$$

Si $\omega \in H^0(E, \Omega^1_{E/\mathbb{Q}})$, on note

$$\Omega^\pm_{E,\omega} = \Omega^\pm_E = \int_{E(\mathbb{C})^\pm} \omega$$

les périodes complexes associées ($E(\mathbb{C})^\pm$ est le $\pm$-espace propre de $E(\mathbb{C})$ relatif à l'action de la conjugaison complexe, une orientation est fixée). Soit $D_{\mathrm{dR}}(E)$ le $\mathbb{Q}$-espace vectoriel $H^1_{\mathrm{dR}}(E)^*$ muni de la filtration $\mathrm{Fil}^0 D_{\mathrm{dR}}(E) = \mathrm{Lie}(E)^* \cong H^0(E, \Omega^1_{E/\mathbb{Q}})$.

Fixons un nombre premier p impair et supposons que E a réduction semi-stable en p. Nous serons amenés à distinguer les cas suivants (non exclusifs)

(1) cas (br) : E a bonne réduction en p

  (a) cas (br-ord) : E a bonne réduction ordinaire

  (b) cas (br-ss) : E a bonne réduction supersingulière

(2) cas (mult) : E a mauvaise réduction semi-stable en p

  (a) cas (mult-dépl) : E a réduction multiplicative déployée

  (b) cas (mult-non dépl) : E a réduction multiplicative non déployée

(3) cas (ord) : E a réduction ordinaire ((br-ord)+(mult)).

Si $T_p(E)$ est le module de Tate p-adique de E, c'est-à-dire la limite projective des E_{p^n}, $T_p(E)$ et $V_p(E) = \mathbb{Q}_p \otimes_{\mathbb{Z}_p} T_p(E)$ sont munis d'une action continue des groupes de Galois absolu $G_\mathbb{Q} = \mathrm{Gal}(\overline{\mathbb{Q}}/\mathbb{Q})$ et $G_{\mathbb{Q}_p} = \mathrm{Gal}(\overline{\mathbb{Q}}_p/\mathbb{Q}_p)$. La représentation p-adique $V_p(E)$ est de poids de Hodge-Tate 0 et 1 : si $\mathbb{C}_p$ est la complétion p-adique de $\overline{\mathbb{Q}}_p$, $\mathbb{C}_p \otimes V_p(E)$ est isomorphe en tant que $G_{\mathbb{Q}_p}$-module à $\mathbb{C}_p \oplus \mathbb{C}_p(1)$ où $\mathbb{C}_p(1)$ est le twist de Tate de $\mathbb{C}_p$. D'autre part, le $\mathbb{Q}_p$-espace vectoriel $D_p(E) = \mathbb{Q}_p \otimes D_{\mathrm{dR}}(E)$ est muni d'une structure de (φ, N)-module filtré[4] sur $\mathbb{Q}_p$ se définissant uniquement à l'aide de la représentation p-adique $V_p(E)$ grâce aux isomorphismes de comparaison. La filtration de $D_p(E)$ est compatible avec celle de $D_{\mathrm{dR}}(E)$: $\mathrm{Fil}^0 D_p(E) = \mathbb{Q}_p \otimes \mathrm{Fil}^0 D_{\mathrm{dR}}(E)$. On a avec les conventions choisies

$$\det(1 - X\varphi | D_p(E)^{N=0}) = (1 - a_p p^{-1} X + \epsilon(p) p^{-1} X^2)$$

avec $\epsilon(p) = 0$ si $p|N$ et 1 sinon.

Dans le cas (ord), les valeurs propres de φ agissant sur $D_p(E)$ sont de valuation 0 et -1, on note $D_p(E)_{[0]}$ (resp. $D_p(E)_{[-1]}$) les espaces propres correspondants et $\pi = \pi_{[0]}$ (resp. $\pi_{[-1]}$) la projection de $D_p(E)$ sur $D_p(E)_{[0]}$ (resp. $D_p(E)_{[-1]}$) parallèlement à $D_p(E)_{[-1]}$ (resp. $D_p(E)_{[0]}$). Dans le cas (br-ss), les valeurs propres de φ sont de valuation $-1/2$. Dans le cas (mult), si e_0 est une base de $D_p(E)_{[0]}$, Ne_0 est une base de $D_p(E)_{[-1]}$ et (e_0, Ne_0) forme une base de $D_p(E)$. Lorsque E a réduction multiplicative déployée, on a $\varphi e_0 = e_0$, lorsque la réduction n'est pas déployée, $\varphi e_0 = -e_0$.

Soit μ_{p^∞} le groupe des racines de l'unité d'ordre une puissance de p (dans $\overline{\mathbb{Q}}$ ou $\overline{\mathbb{Q}}_p$). Le groupe de Galois $G_\infty = \mathrm{Gal}(\mathbb{Q}(\mu_{p^\infty})/\mathbb{Q})$ s'identifie naturellement à $\mathrm{Gal}(\mathbb{Q}_p(\mu_{p^\infty})/\mathbb{Q}_p)$. Le caractère cyclotomique $\chi : G_\infty \to \mathbb{Z}_p^*$ défini par $\sigma\zeta = \zeta^{\chi(\sigma)}$ pour $\zeta \in \mu_{p^\infty}$ est un isomorphisme. Si $\mathbb{Q}_\infty$ est la sous-$\mathbb{Z}_p$-extension de $\mathbb{Q}(\mu_{p^\infty})/\mathbb{Q}$, on note $\Delta = \mathrm{Gal}(\mathbb{Q}(\mu_p)/\mathbb{Q}) \cong \mathrm{Gal}(\mathbb{Q}(\mu_{p^\infty})/\mathbb{Q}_\infty)$ et $\Gamma = \mathrm{Gal}(\mathbb{Q}_p(\mu_{p^\infty})/\mathbb{Q}_p(\mu_p))$ (il est isomorphe à $\mathbb{Z}_p$). Si η est un caractère de Dirichlet de conducteur une puissance de p, nous notons de la même manière le caractère de G_∞ qui s'en déduit.

Notons $\mathcal{H}'$ l'algèbre des fonctions (strictement) analytiques en x dans la boule unité de $\mathbb{C}_p$: $|x| < 1$, ayant un développement en série entière dans $\mathbb{Q}_p[[x]]$ et $\mathcal{H}$ la sous-algèbre des éléments f de $\mathcal{H}'$ vérifiant une condition de "croissance logarithmique au bord" : les $(\log \rho)^r ||f||_\rho$ sont bornés[5], lorsque ρ tend vers 1^- pour un réel $r \geq 0$; on dit alors que f est d'ordre $\leq r$.

On note alors $\mathcal{H}(\Gamma) \subset \mathbb{Q}_p[[\gamma - 1]]$ l'image de $\mathcal{H}$ par l'application qui envoie x sur $\gamma - 1$ pour γ un générateur topologique de Γ et $\mathcal{H}(G_\infty) = \mathbb{Z}_p[\Delta] \otimes_{\mathbb{Z}_p} \mathcal{H}(\Gamma)$. L'algèbre $\mathcal{H}(G_\infty)$ contient l'algèbre d'Iwasawa $\mathbb{Z}_p[[G_\infty]]$ (les éléments d'ordre 0 sont exactement les éléments de $\mathbb{Q}_p \otimes \mathbb{Z}_p[[G_\infty]]$).

Si η est un caractère continu de G_∞ dans $\mathbb{C}_p^*$, $\eta(f)$ a un sens et appar-

---

[4]Un (φ, N)-module filtré sur $\mathbb{Q}_p$ est un $\mathbb{Q}_p$-espace vectoriel D (de dimension finie) muni de deux opérateurs φ et N vérifiant $N\varphi = p\varphi N$ avec φ bijectif, et d'une filtration $\mathrm{Fil}^\cdot D$ décroissante exhaustive et séparée ([14]).

[5]Si $f = \sum_{n=0}^\infty a_n x^n \in \mathcal{H}'$ et $0 < \rho < 1$, $||f||_\rho = \sup_n |a_n| \rho^n = \sup_{|x| \leq \rho} |f(x)|$.

tient à $\mathbb{C}_p$:

$$\text{si } f = \sum_{\delta \in \Delta} \sum_{n=0}^{\infty} a_{n,\delta}(\gamma - 1)^n \delta,$$

$$\eta(f) = \sum_{\delta \in \Delta} \sum_{n=0}^{\infty} a_{n,\delta}(\eta(\gamma) - 1)^n \eta(\delta).$$

Un élément de $\mathcal{H}(G_\infty)$ peut donc être vu comme une fonction sur les caractères de G_∞ : on notera indifféremment $\eta(f)$ ou $f(\eta)$. D'autre part, l'action de G_∞ sur les racines de l'unité d'ordre une puissance de p se prolonge en une action de $\mathcal{H}(G_\infty)$: si $f \in \mathcal{H}(G_\infty)$ et $\zeta \in \mu_{p^\infty}$, on note le résultat $f \cdot \zeta \in \mathbb{Q}_p(\zeta) \subset \mathbb{C}_p$. De même, l'action de G_∞ sur $\mathcal{H}$ induite par $\tau(1+x) = (1+x)^{\chi(\tau)}$ induit une opération de $\mathcal{H}(G_\infty)$ sur $\mathcal{H}$ qui est notée $(f, g) \mapsto f \cdot g$.

Fixons un système compatible (ζ_n) de racines de l'unité d'ordre p^n, par exemple les $\exp(2i\pi/p^n)$ dans $\mathbb{C}$. Si η est un caractère de Dirichlet de conducteur p^n, on note $G(\eta)$ la somme de Gauss :

$$G(\eta) = \sum_{\substack{a \bmod p^n \\ (a,p)=1}} \eta(a)\zeta_n^a.$$

Si $\epsilon(\eta) = \eta(-1)$ est la signature de η, $\frac{G(\eta)L(E,\eta^{-1},1)}{\Omega_{E,\omega}^{\epsilon(\eta)}}$ appartient au corps $\mathbb{Q}(\eta)$ engendré sur $\mathbb{Q}$ par les valeurs de η. Ces valeurs spéciales pour η de conducteur une puissance de p ont des propriétés p-adiques très intéressantes qui impliquent le théorème suivant d'interpolation (la première version se trouve en 1974 dans [26] dans le cas de bonne réduction ordinaire, voir [41] et [1] dans le cas supersingulier et [27] dans le cas général).

Théorème (Interpolation). Cas (br-ss) : *Il existe un unique élément* $L_p(E) \in \mathcal{H}(G_\infty) \otimes D_p(E)$ *dont les composantes (dans une base de $D_p(E)$) sont d'ordre $\leq 1/2$ tel que pour tout caractère de Dirichlet non trivial η de conducteur p^n et de signature $\epsilon(\eta)$*

$$\eta(L_p(E)) = \frac{\epsilon(\eta)G(\eta)L(E,\eta^{-1},1)}{\Omega_{E,\omega}^{\epsilon(\eta)}}\varphi^n\omega. \tag{1}$$

Cas (ord) : *Il existe un unique élément $L_p^\pi(E) \in \mathbb{Z}_p[[G_\infty]] \otimes D_p(E)_{[0]}$ tel que pour tout caractère non trivial η de conducteur p^n*

$$\eta(L_p^\pi(E)) = \frac{\epsilon(\eta)G(\eta)L(E,\eta^{-1},1)}{\Omega_{E,\omega}^{\epsilon(\eta)}}\pi(\varphi^n\omega).$$

Pour être complet, donnons aussi la valeur sur le caractère trivial **1**. Dans le premier cas, il s'agit de

$$\mathbf{1}(L_p(E)) = \frac{L_{\{p\}}(E,1)}{\Omega_{E,\omega}^+}(1-\varphi)(1-p^{-1}\varphi^{-1})^{-1}\omega, \qquad (2)$$

dans le second cas[6]

$$\mathbf{1}(L_p^\pi(E)) = \frac{L_{\{p\}}(E,1)}{\Omega_{E,\omega}^+}(1-\varphi)(1-p^{-1}\varphi^{-1})^{-1}\pi(\omega).$$

Faisons quelques remarques.

(1) La somme de Gauss $G(\eta^{-1})$ peut s'interpréter comme $G(\eta^{-1}) = e_\eta(\zeta_n)$ avec e_η l'opérateur de projection sur la η-composante :

$$e_\eta = \sum_{\sigma \in \mathrm{Gal}(\mathbb{Q}(\mu_{p^n})/\mathbb{Q})} \eta^{-1}(\sigma)\sigma.$$

En utilisant la formule $p^n = G(\eta)G(\eta^{-1})\eta(-1)$ et le fait que $e_\eta(f \cdot \zeta_n) = \eta(f)e_\eta(\zeta_n) = \eta(f)G(\eta^{-1})$ pour $f \in \mathcal{H}(G_\infty)$, la formule (1) peut s'écrire

$$e_\eta(L_p(E) \cdot \zeta_n) = \frac{L(E,\eta^{-1},1)}{\Omega_{E,\omega}^{\epsilon(\eta)}}(p\varphi)^n\omega \qquad (3)$$

et de même pour la seconde.

(2) Si a est un entier premier à p, posons pour $n > 0$

$$L(E, a \bmod p^n, s) = \begin{cases} \displaystyle\sum_{\substack{m=1 \\ m \equiv a \bmod p^n}}^{\infty} a_m m^{-s} & \text{si } n > 0 \\ L_{\{p\}}(E,s) & \text{si } n = 0 \end{cases}$$

et

$$L^\pm(E, a \bmod p^n, s) = L(E, a \bmod p^n, s) \pm L(E, -a \bmod p^n, s).$$

---

[6]La formule est ainsi la même dans le cas de bonne et mauvaise réduction ordinaire, si α est la racine de $X^2 - a_pX + p\epsilon(p)$ qui est une unité p-adique, le second membre peut s'écrire eul $\frac{L(E,1)}{\Omega_{E,\omega}^+}$ avec

$$\text{eul} = (1-\alpha^{-1})(1-p^{-1}\alpha)^{-1}(1-\alpha p^{-1})(1-\epsilon(p)\alpha^{-1}),$$

ce qui vaut $(1-\alpha^{-1})^2$ dans le cas de bonne réduction et $1-\alpha^{-1}$ dans le cas de mauvaise réduction.

On aimerait exprimer les formules définissant $L_p(E)$ sans caractères. Mais le comportement particulier pour le caractère trivial empêche que la formule "naturelle" qui vient à l'esprit soit vraie. Il faut donc faire plus compliqué. Faisons-le dans le cas (ss) (dans le cas (ord), il faudrait prendre une projection). Considérons l'opérateur continu φ sur $\mathcal{H}$ tel que $\varphi(x) = (1+x)^p - 1$. Résolvons l'équation

$$(1 - \varphi \otimes \varphi)\mathcal{G}_E = L_p(E) \cdot (1 + x) \tag{4}$$

dans $\mathcal{H} \otimes D_p(E)$, ce qui est possible car les valeurs propres de φ ne sont pas une puissance de p. Alors, pour η caractère non trivial de conducteur p^n,

$$e_\eta(\mathcal{G}_E(\zeta_n - 1)) = e_\eta(L_p(E) \cdot \zeta_n)$$

et $(1 - \varphi)\mathcal{G}_E(0) = \mathbf{1}(L_p(E))$. Les relations "simples" s'expriment en termes de $\mathcal{G}_E^\pm = \mathcal{G}_E \pm \mathcal{G}_E^\iota$ (avec $f^\iota(x) = f((1+x)^{-1} - 1)$) et non de $L_p(E)$: pour a premier à p et n entier strictement positif,

$$\mathcal{G}_E^\pm(\zeta_n^a - 1) = \frac{L^\pm(E, a \bmod p^n, 1)}{\Omega_{E,\omega}^\pm}(p\varphi)^n \omega, \tag{5}$$

et

$$(1 - p^{-1}\varphi^{-1})\mathcal{G}_E^\pm(0) = \frac{L^\pm(E, 1)}{\Omega_{E,\omega}^\pm}\omega. \tag{6}$$

Faisons la démonstration rapidement. La relation (4) prise en $x = 0$ implique que $(1 - \varphi)\mathcal{G}_E^+(0)/2 = \mathbf{1}(L_p(E)) = \frac{L_{\{p\}}(E,1)}{\Omega_{E,\omega}^+}$, d'où

$$(1 - p^{-1}\varphi^{-1})\mathcal{G}_E^+(0) = \frac{L_{\{p\}}^+(E, 1)}{\Omega_{E,\omega}^+}\omega = 2\frac{L(E, 1, 1)}{\Omega_{E,\omega}^+}\omega.$$

On déduit alors de l'équation $\psi(\mathcal{G}_E) = (1 \otimes \varphi)\mathcal{G}_E$ que

$$(1 - p^{-1}\varphi^{-1})\mathcal{G}_E(0) = \mathrm{Tr}_{\mathbb{Q}(\mu_p)/\mathbb{Q}}((p\varphi)^{-1}\mathcal{G}_E^+(\zeta_1 - 1))$$
$$= e_1((p\varphi)^{-1}\mathcal{G}_E^+(\zeta_1 - 1)).$$

Ainsi, pour tout caractère η pair de $(\mathbb{Z}/p^n\mathbb{Z})^*$, on a

$$e_\eta((p\varphi)^{-n}\mathcal{G}_E^+(\zeta_n - 1)) = \frac{L^+(E, \eta^{-1}, 1)}{\Omega_{E,\omega}^+}$$

(cela ne dépend pas de n à cause de la relation $\psi(\mathcal{G}_E) = (1 \otimes \varphi)\mathcal{G}_E$). On en déduit la relation (5) par application de la formule $\mathbf{1}_{a \bmod p^n} =$

$\frac{1}{p^{n-1}(p-1)/2} \sum_{\eta} \eta(a)e_{\eta}$ où la somme est prise sur les caractères de Dirichlet pairs modulo p^n et où $1_{a \bmod p^n}$ est la fonction caractéristique de la classe $a \bmod p^n$. La partie $-$ se démontre de la même manière.

(3) Si D est un φ-module de dimension finie munie d'une norme et $f \in \mathcal{H}(G_{\infty}) \otimes D$, on dit que f est d'ordre $\leq r$ (ou pour être plus précis de φ-ordre $\leq r$) si la suite $p^{-nr}||(1 \otimes \varphi)^{-n}f||_{\rho^{1/p^n}}$ est bornée (pour un ρ avec $0 < \rho < 1$). Ici, r peut prendre des valeurs négatives. La condition sur l'ordre de croissance de $L_p(E)$ peut se dire alors de manière plus "intrinsèque" : $L_p(E)$ est d'ordre ≤ 0.

Soit F une extension finie de $\mathbb{Q}$. Par la théorie de Kummer de la courbe elliptique, la suite de G_F-modules

$$0 \to E_{p^n} \to E(\overline{\mathbb{Q}}) \xrightarrow{p^n} E(\overline{\mathbb{Q}}) \to 0$$

est exacte et induit la suite exacte de cohomologies

$$0 \to E(F)/p^n E(F) \to H^1(F, E_{p^n}) \to H^1(F, E(\overline{\mathbb{Q}}))_{p^n} \to 0.$$

Par passage à la limite projective du premier homomorphisme, on obtient des injections $\mathbb{Z}_p \otimes_{\mathbb{Z}} E(F) \to H^1(F, T_p(E))$ et

$$\mathbb{Q}_p \otimes_{\mathbb{Z}} E(F) \to H^1(F, V_p(E)).$$

Plus précisément l'image de $\mathbb{Q}_p \otimes_{\mathbb{Z}} E(F)$ est contenue dans $H^1_f(F, V_p(E))$ (groupe de Selmer-Bloch-Kato) qui est traditionnellement défini comme le noyau de

$$H^1(F, V_p(E)) \to \prod_v \mathbb{Q}_p \otimes \varprojlim_n H^1(F_v, E(\overline{\mathbb{Q}}))_{p^n}$$

et coïncide avec le noyau de

$$H^1(F, V_p(E)) \to \prod_{v \nmid p} H^1(F_v^{nr}, V_p(E)) \times \prod_{v \mid p} H^1(F_v, B_{\mathrm{cris}} \otimes V_p(E))$$

où F_v^{nr} est la plus grande extension non ramifiée de F_v et B_{cris} l'anneau des périodes p-adiques de Fontaine en p ([13]). Autrement dit, si $H^1_f(F_v, V_p(E))$ est $H^1(F_v^{nr}/F_v, V_p(E)^{G_{F_v,nr}})$ pour v ne divisant pas p et le noyau de $H^1(F_v, V_p(E)) \to H^1(F_v, B_{\mathrm{cris}} \otimes V_p(E))$ pour v divisant p, $H^1_f(F, V_p(E))$ est le sous-espace vectoriel de $H^1(F, V_p(E))$ formé des éléments dont l'image par localisation dans $H^1(F_v, V_p(E))$ est dans $H^1_f(F_v, V_p(E))$. Dans tous les cas,

$$H^1_f(F_v, V_p(E)) = \mathbb{Q}_p \otimes \varprojlim_n E(F_v)/p^n E(F_v).$$

Pour $v|p$, l'espace tangent $t_E(F_v)$ de E/F_v à l'origine s'identifie à $F_v \otimes (D_p(E)/\operatorname{Fil}^0 D_p(E))$ et l'application exponentielle de groupe de Lie

$$\exp_{v,E} \colon t_E(F_v) \overset{\cong}{\to} \mathbb{Q}_p \otimes E(F_v)$$

s'identifie à l'application exponentielle de Bloch-Kato

$$\exp_{v,E} \colon t_E(F_v) \overset{\cong}{\to} H^1_f(F, V_p(E)).$$

Nous utiliserons aussi l'application logarithme inverse de l'exponentielle

$$\log_{v,E} \colon H^1_f(F, V_p(E)) \to t_E(F_v).$$

Par la dualité locale de Tate

$$H^1(F_v, V_p(E)) \times H^1(F_v, V_p(E)) \to H^2(F_v, \mathbb{Q}_p(1)) \cong \mathbb{Q}_p,$$

l'application duale de $\exp_{v,E}$ est

$$\exp^*_{v,E} \colon H^1(F_v, V_p(E)) \to F_v \otimes \operatorname{Fil}^0 D_p(E)$$

dont le noyau est exactement $H^1_f(F_v, V_p(E))$ (on utilise ici le fait que l'accouplement de Weil induit un isomorphisme entre $V_p(E)$ et $V_p(E)^*(1)$).

Le fait que le groupe de Selmer puisse s'exprimer uniquement en termes de la représentation p-adique associée à E a été remarqué indépendamment par Greenberg dans le cas ordinaire et par Bloch et Kato, dans les années 1987-88. Le travail de Bloch et Kato dans [5], qui s'appuie sur la théorie de Fontaine a marqué un tournant dans la théorie d'Iwasawa et a permis de sortir du cadre des représentations p-adiques ordinaires en p.

La construction d'éléments de $H^1(\mathbb{Q}(\mu_{p^n}), T_p(E))$ faite par Kato à partir des éléments de Beilinson ([2], eux-même très liés aux unités de Siegel, de Kubert-Lang ou de Robert ([24], [36]) selon le contexte) et le calcul de leur valeur par $\exp^*_E$ fait par Kato permettent d'aborder sous un jour nouveau le théorème d'interpolation en interprétant les valeurs spéciales de la fonction L en 1 comme "régulateurs p-adiques" d'éléments de la cohomologie galoisienne. Donnons une version un peu vague du théorème de Kato (sans introduire la notion pourtant fondamentale de système d'Euler-Kolyvagin). Les références sont [39], [38, §7] et les articles de Kato à venir.[7]

Théorème (Kato). *Il existe des $c'_n \in H^1(\mathbb{Q}(\mu_{p^n}), T_p(E))$ "construits de manière modulaire", compatibles pour les applications de corestriction et une constante c_E non nulle tels que pour tout caractère η de conducteur p^n*

$$\sum_{\sigma \in \operatorname{Gal}(\mathbb{Q}(\mu_{p^n})/\mathbb{Q})} \eta^{-1}(\sigma) \exp^*_{p,E}(\sigma c'_n) = c_E \frac{L_{\{pN\}}(E, \eta^{-1}, 1)}{\Omega^{\epsilon(\eta)}_{E,\omega}} \omega.$$

---

[7] Note ajoutée : cet article existe désormais ([21]) et on peut enlever le N des fonctions $L_{\{pN\}}$.

Notons

$$H^1_\infty(\mathbb{Q}, V_p(E)) = \mathbb{Q}_p \otimes \varprojlim_n H^1(\mathbb{Q}(\mu_{p^n}), T_p(E)).$$

Les $c_n = c(E)^{-1} c'_n$ forment un système projectif $c_\infty \in H^1_\infty(\mathbb{Q}, V_p(E))$ et on a pour tout caractère de conducteur p^n

$$\sum_{\sigma \in \mathrm{Gal}(\mathbb{Q}(\mu_{p^n})/\mathbb{Q})} \eta^{-1}(\sigma) \exp^*_{p,E}(\sigma c_n) = \frac{L_{\{pN\}}(E, \eta^{-1}, 1)}{\Omega^{\epsilon(\eta)}_{E,\omega}} \omega.$$

Soit

$$Z^1_\infty(\mathbb{Q}_p, V_p(E)) = \mathbb{Q}_p \otimes \varprojlim_n H^1(\mathbb{Q}_p(\mu_{p^n}), T_p(E)).$$

L'homomorphisme de Coleman associé à un groupe formel[8] [8] se généralise en une application $\mathcal{L}_E$ de $Z^1_\infty(\mathbb{Q}_p, V_p(E))$ dans un $\mathcal{H}(G_\infty)$-module $\mathcal{D}_\infty(E)$; $\mathcal{L}_E$ joue le rôle de régulateur pour ce module d'Iwasawa, nous l'appellerons régulateur d'Iwasawa, quant à $\mathcal{D}_\infty(E)$, on peut le décrire en termes uniquement du (φ, N)-module $D_p(E)$. Dans le cas de bonne réduction, $\mathcal{D}_\infty(E)$ a une description très simple : il s'agit de $\mathcal{H}(G_\infty) \otimes D_p(E)$ qu'il est commode de voir comme sous-espace de $\mathcal{H} \otimes D_p(E)$ par la transformée de Mellin :

$$\mathcal{H}(G_\infty) \otimes D_p(E) \to \mathcal{H} \otimes D_p(E)$$
$$f \mapsto f \cdot (1 + x).$$

Ici $\tau \cdot (1 + x) = (1 + x)^{\chi(\tau)}$ pour $\tau \in G_\infty$ et cette action se prolonge naturellement par linéarité et continuité à $\mathcal{H}(G_\infty)$. L'image de $f \mapsto f \cdot (1 + x)$ est égale au noyau de l'opérateur ψ défini[9] par

$$\varphi \circ \psi(g)(x) = p^{-1} \sum_{\zeta \in \mu_p} g(\zeta(1 + x) - 1)$$

avec φ l'opérateur de Frobenius : $\varphi(g)(x) = g((1 + x)^p - 1)$.

Le régulateur d'Iwasawa $\mathcal{L}_E$ est un homomorphisme de G_∞-modules

$$\mathcal{L}_E \colon Z^1_\infty(\mathbb{Q}_p, V_p(E)) \to \mathcal{H}(G_\infty) \otimes D_p(E)$$

dont l'image est formée d'éléments d'ordre ≤ 0. Avec les notations de [35, prop. 5.4.5], [31, §1] au signe près, on a $\mathcal{L}_E = l_0 \Omega^{-1}_{V_p(E),1}$. La condition

---

[8] Il trouve d'ailleurs son origine dans les travaux d'Iwasawa [17] et de Coates et Wiles [7].

[9] Cet opérateur qui est devenu fondamental dans ces constructions se trouve dans les travaux de Dwork ([12], [6]) et dans les constructions de Coleman ([8] sous le nom de $\mathcal{S}$).

sur l'ordre se traduit dans le cas (br-ss) par le fait que les composantes de l'image de $\mathcal{L}_E$ dans une base de $D_p(E)$ sont d'ordre $\leq 1/2$. Dans le cas (br-ord), la composante de $\pi_{[0]}(\mathcal{L}_E(z))$ dans une base de $D_p(E)_{[0]}$ est d'ordre ≤ 0, c'est-à-dire appartient à $\mathbb{Q}_p \otimes \mathbb{Z}_p[[G_\infty]]$, la composante de $\pi_{[-1]}(\mathcal{L}_E(z))$ est d'ordre ≤ 1. Notons

$$\tilde{\mathcal{L}}_E \colon Z^1_\infty(\mathbb{Q}_p, V_p(E)) \to \mathcal{D}_\infty(E) = \mathcal{H}^{\psi=0} \otimes D_p(E)$$

pour éviter les confusions.

Dans le cas (mult), $\mathcal{D}_\infty(E)$ est plus compliqué. Commençons par introduire une algèbre contenant $\mathcal{H}$. Pour cela, considérons l'algèbre $\mathcal{B}$ des séries de Laurent $\sum_{n\in\mathbb{Z}} a_n x^n$, avec $a_n \in \mathbb{Q}_p$, analytiques sur la couronne $\{x \in \mathbb{C}_p, p^{-1/(p-1)} < |x| < 1\}$ vérifiant une condition de croissance comme précédemment, puis l'algèbre $\mathcal{B}[\log x]$ des polynômes en $\log x$ à coefficients dans $\mathcal{B}$. On peut ici voir $\log x$ soit comme une fonction localement analytique vérifiant les propriétés usuelles du logarithme, soit comme un élément formel auquel on impose des règles de calcul : ainsi, $\mathcal{B}[\log x]$ est muni (entre autres) d'opérateurs φ, ψ, d'un opérateur de monodromie N trivial sur $\mathcal{B}$ et tel que $N \log x = 1$, d'une action continue de G_∞, d'une dérivation $D(f) = (1+x)\frac{d}{dx}f$ pour $f \in \mathcal{B}$ et $D \log x = \frac{1+x}{x}$ (remarquons que l'on est obligé d'introduire $\mathcal{B}$ pour définir D et φ, par exemple :

$$\varphi \log x = p \log x + \log \frac{(1+x)^p - 1}{x^p}$$

et un moyen[10] de donner un sens à $\log \frac{(1+x)^p-1}{x^p}$ dans $\mathcal{B}$ est d'écrire

$$\frac{(1+x)^p - 1}{x^p} = 1 + \frac{p}{x} + \ldots + \frac{p}{x^{p-1}}$$

et d'utiliser le développement $\log(1+x) = \sum_{n=1}^\infty (-1)^{n-1} \frac{x^n}{n}$). On définit $\mathcal{B}(G_\infty)$ à partir de $\mathcal{B}$ comme pour $\mathcal{H}(G_\infty)$. On montre que $\mathcal{B}(G_\infty)$ agit naturellement sur $\mathcal{B}[\log x]^{\psi=0}$ et que $\left(\mathcal{B}[\log x]^{\psi=0} \otimes D_p(E)\right)^{N=0}$ est un $\mathcal{B}(G_\infty)$-module de rang[11] 2. Lorsque $N = 0$ sur $D_p(E)$, on a

$$\left(\mathcal{B}[\log x]^{\psi=0} \otimes D_p(E)\right)^{N=0} = \mathcal{B}^{\psi=0} \otimes D_p(E).$$

---

[10]Un autre moyen serait d'écrire

$$\log\left((1+x)^p - 1\right) = \log_p p + \log x + \log(1 + \frac{x}{p} + \ldots + \frac{x^{p-1}}{p}),$$

le troisième terme donne une série convergente sur $|x| < p^{-1/(p-1)}$, ce qui ne peut pas être évaluer sur $\zeta - 1$ pour ζ racine de l'unité non triviale.

[11]Ce qui signifie qu'une fois tensorisé par l'anneau total des fractions $\text{Frac}(\mathcal{B}(G_\infty))$ de $\mathcal{B}(G_\infty)$, il est localement de rang 2 sur $\text{Frac}(\mathcal{B}(G_\infty))$.

On peut décrire à l'intérieur un $\mathcal{H}(G_\infty)$-module $\mathcal{D}_\infty(E)$ de rang 2 : en gros, il s'agit des éléments g de $\left(\mathcal{B}[\log x]^{\psi=0} \otimes D_p(E)\right)^{N=0}$ tels que l'on puisse résoudre dans $\mathcal{B}[\log x] \otimes D_p(E)$ une équation du type $(1 - \varphi \otimes \varphi)G = g + \mathfrak{M}\log(1+x)$ (voir [35, §2.3], ici $\mathcal{D}_\infty(E)$ est le $\mathcal{D}_{\infty,f}$ de [35]). Il se trouve dans la suite exacte

$$0 \to \mathcal{H}^{\psi=0} \otimes \mathbb{Q}_p Ne_0 \to \mathcal{D}_\infty(E) \to \mathcal{H}^{\psi=0} \otimes \mathbb{Q}_p e_0 \to \mathbb{Q}_p \to 0 \qquad (7)$$

où la dernière flèche est l'évaluation en 0 dans le cas déployé et

$$0 \to \mathcal{H}^{\psi=0} \otimes \mathbb{Q}_p Ne_0 \to \mathcal{D}_\infty(E) \to \mathcal{H}^{\psi=0} \otimes \mathbb{Q}_p e_0 \to 0 \qquad (8)$$

dans le cas non déployé. Un élément fe_0 de $\mathcal{H}^{\psi=0} \otimes \mathbb{Q}_p e_0$ (avec $f(0) = 0$ dans le cas déployé) se relève de la manière suivante : on montre qu'il existe un élément F de $(\mathcal{B}[\log x] \otimes D_p(E))^{N=0}$ tel que $(1 - \varphi \otimes \varphi)F \equiv f \bmod \mathcal{B}[\log x] \otimes Ne_0$. Autrement dit, on choisit $F_1 \in \mathcal{H}$ tel que $(1 \mp \varphi)F_1 = f$ avec $\mp = -$ dans le cas déployé (on peut alors supposer que $F_1(0) = 0$) et $+$ dans le cas non déployé ; il existe $F_2 \in \mathcal{B}[\log x]$ tel que $\psi(F_2) = p^{-1}F_2$, $NF_2 = F_1$; un relèvement de fe_0 est alors $(1 - \varphi \otimes \varphi)(F_1e_0 - F_2Ne_0)$. Nous donnons dans l'appendice une manière de calculer un tel relèvement et en particulier ses valeurs sur $\zeta - 1$ avec $\zeta \in \mu_{p^\infty}$.

De manière générale, un élément f de $\mathcal{D}_\infty(E)$ vérifie (outre l'équation fonctionnelle $\psi(f) = 0$) la propriété qu'il existe $F \in (\mathcal{B}[\log x] \otimes D_p(E))^{N=0}$ et $\mathfrak{M} \in \mathbb{Q}_p$ tels que $(1 - \varphi \otimes \varphi)F = f + \mathfrak{M}\log(1+x)Ne_0$ (avec $\mathfrak{M} = 0$ dans le cas non déployé). Le F n'est pas unique dans le cas déployé, un choix possible est d'imposer la nullité de la composante sur e_0 en 0. Nous noterons $\mathcal{S}(f) = \mathcal{S}(f, \mathfrak{M})$ (pour $\mathfrak{M}$ convenable) la solution de $(1 - \varphi \otimes \varphi)F = f + \mathfrak{M}\log(1+x)Ne_0$ dont la composante sur e_0 est nulle en 0 dans le cas déployé.

Remarquons que $(\tau - 1)(1 + x)$ forme une base du noyau de $\mathcal{H}^{\psi=0} \to \mathbb{Q}_p$ pour τ un générateur de G_∞. Notons $\mathcal{T}_\tau$ un relèvement dans $D_\infty(E)$ d'ordre minimal (d'ordre ≤ 1 modulo $\log x\, \mathcal{B} \otimes D_p(E)$). Enfin, $\mathcal{T}_0 = (\tau - 1)^{-1}\mathcal{T}_\tau$ a un sens dans $(\mathcal{B}[\log x]^{\psi=0} \otimes D_\infty(E))^{N=0}$ et a l'avantage de ne pas dépendre de τ. Nous écrirons $\mathcal{T}_0 = (1 + x)e_0 - t_0Ne_0$, $\mathcal{T}_\tau = (\tau - 1)(1 + x)e_0 - t_\tau Ne_0$ avec $t_\tau = (\tau - 1)t_0$ et $\mathcal{T}_{-1} = (1 + x)Ne_0$.

Pour conclure cette description de $D_\infty(E)$, disons que $(\mathcal{T}_{-1}, \mathcal{T}_\tau)$ (resp. $(\mathcal{T}_{-1}, \mathcal{T}_0)$) en forme une base dans le cas déployé (resp. non déployé). Rappelons que lorsque $N = 0$ sur $D_p(E)$, ce sont les $(\mathcal{T}_i = (1 + x)e_i)_{i=0,-1}$ qui forment une base du $\mathcal{H}(G_\infty)$-module $D_\infty(E)$.

Un résultat important ici et démontré dans [35] est qu'il existe un homomorphisme de G_∞-modules naturel $\tilde{\mathcal{L}}_E$ (appelé ici régulateur d'Iwasawa, ailleurs logarithme ou "exponentielle duale")

$$\tilde{\mathcal{L}}_E : Z^1_\infty(\mathbb{Q}_p, V_p(E)) \to \mathcal{D}_\infty(E) \subset \left(\mathcal{B}[\log x]^{\psi=0} \otimes D_p(E)\right)^{N=0}.$$

Avec les notations de [35, prop. 5.4.5], il s'agit de $l_0\Omega_{V_p(E),1}^{-1}$; $\Omega_{V_p(E),1}(g)$ est obtenu en interpolant les valeurs des exponentielles d'éléments construits à partir de G avec $(1 - \varphi \otimes \varphi)G = g - \mathfrak{M}\log(1 + x)$, les exponentielles étant les exponentielles de Bloch-Kato pour la représentation p-adique $V_p(E)$, mais aussi pour ses twists. Le fait que $l_0\Omega_{V_p(E),1}^{-1}$ soit à valeurs dans $\mathcal{D}_\infty(E) \subset \left(\mathcal{B}[\log x]^{\psi=0} \otimes D_p(E)\right)^{N=0}$ se démontre à partir de la loi de réciprocité (Réc) démontrée par Colmez ([11], voir aussi un article annoncé de Kato, Kurihara et Tsuji et [3]) et de calculs sur le déterminant de $\Omega_{V_p(E),1}$ ([35, prop. 5.4.5]). Par composé avec l'application naturelle de localisation

$$H_\infty^1(\mathbb{Q}, V_p(E)) \to Z_\infty^1(\mathbb{Q}_p, V_p(E)),$$

on obtient

$$\tilde{\mathcal{L}}_E : H_\infty^1(\mathbb{Q}, V_p(E)) \to \mathcal{D}_\infty(E).$$

Choisissons un logarithme $\log_p$ sur $\mathbb{C}_p - \{0\}$: il suffit pour cela de fixer $\log_p p$, par exemple $\log_p(p) = 0$. Si η est un caractère non trivial de G_∞ de conducteur p^n, l'application $f \in \mathcal{B}[\log x] \mapsto e_\eta(f(\zeta_n - 1))$ induit une application naturelle d'évaluation

$$\mathrm{ev}_\eta : \mathcal{B}[\log x] \otimes \mathcal{D} \to \mathbb{Q}_p(\eta) \otimes D_p(E).$$

Dans le cas (br), on a

$$\mathrm{ev}_\eta(\tilde{\mathcal{L}}_E(z)) = G(\eta^{-1})\eta(\mathcal{L}_E(z)) = \epsilon(\eta)p^n G(\eta)\eta(\mathcal{L}_E(z)).$$

En utilisant les propriétés de $\mathcal{L}_E$ (voir [35] par exemple) et en procédant comme dans [31], on peut écrire le théorème de Kato sous la forme suivante.

Théorème (Kato). *Pour tout caractère non trivial η de G_∞ de conducteur p^n,*

$$\mathrm{ev}_\eta(\tilde{\mathcal{L}}_E(c_\infty)) = \frac{L_{\{Np\}}(E, \eta^{-1}, 1)}{\Omega_{E,\omega}^{\epsilon(\eta)}}(p\varphi)^n\omega \tag{9}$$

On obtient un moyen détourné de retrouver le théorème d'interpolation et un peu plus (excepté les facteurs locaux aux places divisant N). Regardons d'abord le cas (br) :

Proposition. *Dans le cas* (br), *il existe un élément $L_{p,\{N\}}(E)$ de $\mathcal{H}(G_\infty)\otimes D_p(E)$ d'ordre ≤ 0 tel que, pour tout caractère non trivial η de G_∞ de conducteur p^n,*

$$\eta(L_{p,\{N\}}(E)) = \frac{\epsilon(\eta)G(\eta)L_{\{Np\}}(E, \eta^{-1}, 1)}{\Omega_{E,\omega}^{\epsilon(\eta)}}\varphi^n\omega.$$

Il est unique vérifiant ces conditions dans le cas (ss), *mais pas dans le cas* (br-ord).

On prend[12] $L_{p,\{N\}}(E) = \mathcal{L}_E(c_\infty)$. C'est la fonction L p-adique (incomplète) de $E/\mathbb{Q}$. Si l'on veut obtenir la fonction L p-adique complète, il faut diviser $L_{p,\{N\}}(E)$ par un élément de $\mathbb{Z}_p[[G_\infty]]$ qui n'est pas un diviseur de zéro[13] :

$$L_p(E) = \prod_{\substack{q|N \\ q\neq p}} (1 - a_q\sigma_q^{-1})^{-1} L_{p,\{N\}}(E)$$

où σ_q est l'élément de G_∞ vérifiant $\sigma_q\zeta = \zeta^q$ pour $\zeta \in \mu_{p^\infty}$. On a en particulier dans le cas (br-ord) $\pi_{[0]}(L_p(E)) = L_p^\pi(E)$.

Par définition, la fonction L p-adique est donc le régulateur d'Iwasawa d'un élément "spécial" de $H_\infty^1(\mathbb{Q}, V_p(E))$. Cet élément joue le rôle des unités cyclotomiques.

Traduisons ce que cela signifie dans le cas ordinaire. La racine de $X^2 - a_p X + p\epsilon(p)$ qui est une unité dans $\mathbb{Z}_p$ est notée α et on prend $\beta = p/\alpha$.

Proposition. *Dans le cas* (br-ord), *il existe un unique élément $L_{p,\alpha}$ de $\mathbb{Q}_p \otimes \mathbb{Z}_p[[G_\infty]]$ tel que pour tout caractère η non trivial de conducteur p^n*

$$\eta(L_{p,\alpha}) = \alpha^{-n} \frac{\epsilon(\eta)G(\eta)L_{\{Np\}}(E, \eta^{-1}, 1)}{\Omega_{E,\omega}^{\epsilon(\eta)}}$$

et il existe un élément $L_{p,\beta}$ de $\mathcal{H}(G_\infty)$ d'ordre ≤ 1 vérifiant pour tout caractère non trivial η de conducteur p^n

$$\eta(L_{p,\beta}) = \beta^{-n} \frac{\epsilon(\eta)G(\eta)L_{\{Np\}}(E, \eta^{-1}, 1)}{\Omega_{E,\omega}^{\epsilon(\eta)}}.$$

On passe de la proposition précédente à celle-ci en écrivant $\omega = e_\alpha + e_\beta$ avec $e_\alpha \in D_p(E)_{[0]}$ et $e_\beta \in D_p(E)_{[-1]}$ et en prenant les composantes dans la "base" (e_α, e_β).

---

[12]Il y a une autre définition possible de la fonction L p-adique de E comme élément $L_p^*(E)$ de $\mathcal{H}(G_\infty) \otimes \mathrm{Hom}(D_p(E), \mathbb{Q}_p)$ par

$$L_p^*(E)(n) = L_p(E)^\iota \wedge n$$

pour $n \in D_p(E)$ (ici $\iota\tau = \tau^{-1}$). On a alors

$$\chi^k(L_p^*(E))(n) \stackrel{\text{déf}}{=} \chi^k(L_p^*(E)(n)) = \chi^{-k}(L_p(E)) \wedge n.$$

C'est cette définition qui a été prise dans le cas général de [33] ; les valeurs spéciales correspondant au motif $M(k)$ sont alors liées aux invariants arithmétiques de $M^*(1-k)$ ce qui est plus conforme à ce que l'on fait dans le cas complexe.

[13]Remarquons que la valeur de $1 - a_q q^{-1}\sigma_q^{-1}$ sur $\chi^{-k}\eta$ avec η un caractère d'ordre fini est $1 - a_q q^{k-1}\eta^{-1}(q)$ avec $a_q = \pm 1$ ou 0 et n'est donc jamais nulle sauf peut-être si η est trivial et $k = 1$. Cela ne posera donc pas de problèmes particuliers dans les formules d'évaluation sauf peut-être sur le caractère trivial.

Supposons maintenant que E a réduction multiplicative et exprimons $\tilde{L}_{p,\{N\}}(E) = \tilde{\mathcal{L}}_E(c_\infty)$ sur le système $(\mathcal{T}_0, \mathcal{T}_{-1})$:

$$\tilde{L}_{p,\{N\}}(E) = L_{p,\alpha} \cdot \mathcal{T}_0 + L_{p,\beta} \cdot \mathcal{T}_{-1} \tag{10}$$

avec $L_{p,\alpha}$ et $L_{p,\beta} = L_{p,\beta}^{\mathcal{T}_0}$ appartenant à $\mathcal{H}(G_\infty)$. D'où,

$$\begin{aligned}
\tilde{L}_{p,\{N\}}(E) &= L_{p,\alpha} \cdot ((1+x)e_0 - t_0 N e_0) + L_{p,\beta} \cdot (1+x) N e_0 \\
&= L_{p,\alpha} \cdot (1+x)e_0 + (L_{p,\beta} \cdot (1+x) - L_{p,\alpha} \cdot t_0) N e_0.
\end{aligned}$$

Comme $\tilde{L}_{p,\{N\}}(E) \in D_\infty(E)$, lorsque E a réduction déployée en p, $L_{p,\alpha}$ appartient en fait à $(\tau - 1)\mathcal{H}(G_\infty)$ et même à $\mathbb{Q}_p \otimes (\tau - 1)\mathbb{Z}[[G_\infty]]$, puisque $((\tau - 1)\mathcal{T}_0, \mathcal{T}_1)$ est une base de $\mathcal{D}_\infty(E)$. Donc $\mathbf{1}(L_{p,\alpha})$ est nul. On trouve ainsi une interprétation intéressante du "zéro trivial" traditionnel de la fonction L p-adique dans le cas (mult-dépl). Il provient du fait que par la suite exacte (7), la suite

$$\begin{array}{ccccccccc}
0 & \to & \mathcal{D}_\infty(\mathbb{Q}_p(1)) & \to & \mathcal{D}_\infty(E) & \to & \mathcal{D}_\infty(\mathbb{Q}_p) & \to & 0 \\
 & & \| & & & & \| & & \\
 & & \mathcal{H}^{\psi=0} & & & & \mathcal{H}^{\psi=0} & &
\end{array}$$

n'est pas exacte. Ce zéro est donc d'une autre nature que dans le cas du carré symétrique d'une courbe elliptique ayant bonne réduction où ce qui intervient est le fait que les groupes de Bloch-Kato H_e^1, H_f^1 et H_g^1 sont différents alors qu'ils sont ici égaux.

Question 1. *Que vaut $L_{p,\beta}$ en 1 ?*

Cette valeur ne dépend pas du choix de $\mathcal{T}_0$ puisque $\mathbf{1}(L_{p,\alpha}) = 0$. Je ne vois aucune raison à sa nullité mais ne sais pas le calculer! Pour l'instant, $\tilde{L}_p(E)$ ou plutôt $\tilde{L}_{p,\{N\}}(E)$ n'est pas définie en 0, mais nous allons voir que l'on peut donner un sens à $(1 - p\varphi \otimes \varphi)\tilde{L}_{p,\{N\}}(E)$ en 0. Et cela n'est pas nul. Le "zéro trivial" n'est donc pas un zéro de la fonction L p-adique mais d'une de ses composantes.

Écrivons ω dans la base $(e_0, N e_0)$: quitte à changer e_0 par un multiple, on a $\omega = e_0 - \mathcal{L}(E)N e_0$ avec $\mathcal{L}(E) \in \mathbb{Q}_p^*$. Si q_E est le paramètre de Tate de E, on a $\mathcal{L}(E) = \frac{\log_p q_E}{\operatorname{ord}_p q_E}$.

Proposition. *Dans le cas* (mult)*, il existe un unique élément $L_{p,\alpha}$ de $\mathbb{Q}_p \otimes \mathbb{Z}_p[[G_\infty]]$, nul sur le caractère trivial dans le cas* (mult-dépl) *et tel que pour tout caractère η non trivial de conducteur p^n*

$$\eta(L_{p,\alpha}) = \alpha^{-n} \frac{\epsilon(\eta) G(\eta) L_{\{N\}}(E, \eta^{-1}, 1)}{\Omega_{E,\omega}^{\epsilon(\eta)}} \tag{11}$$

et il existe un élément $L_{p,\beta}$ de $\mathcal{H}(G_\infty)$ d'ordre ≤ 1 vérifiant pour tout caractère η non trivial de conducteur p^n

$$\eta(L_{p,\beta}) = v(\eta)\beta^{-n}\frac{\epsilon(\eta)G(\eta)L_{\{N\}}(E, \eta^{-1}, 1)}{\Omega_{E,\omega}^{\epsilon(\eta)}} \tag{12}$$

avec $v(\eta) = v^{\mathcal{T}_0}(\eta) = -(\mathcal{L}(E) + \epsilon(\eta)G(\eta)\operatorname{ev}_\eta(t_0))$.

Rappelons que l'on peut changer $\mathcal{T}_0$ en $\mathcal{T}_0 + \hat{t} \cdot (1+x)Ne_0$ avec $\hat{t} \in \mathcal{H}(G_\infty)$, ce qui change alors $L_{p,\beta}$ en $L_{p,\beta} - \hat{t}L_{p,\alpha}$. Le terme $v^{\mathcal{T}_0}(\eta)$ est un terme p-adique un peu mystérieux. Il dépend du choix de $\log_p p$. Il dépend surtout de $\mathcal{T}_0$ de même que $L_{p,\beta} = L_{p,\beta}^{\mathcal{T}_0}$.

Remarque. Le fait que les valeurs de $L_p(E)$ sur un caractère non trivial de conducteur p^n appartiennent à $\mathbb{Q}_p(\mu_{p^n}) \otimes \varphi^n \operatorname{Fil}^0 D_p(E)$ (et qui se déduit du fait que l'image de $\exp^*$ est contenue dans $K_n \otimes \operatorname{Fil}^0 D_p(E)$, c'est-à-dire dans une "droite fixe") joue un rôle très important et permet de déduire les formules concernant $L_{p,\beta}^{\mathcal{T}_0}$ de celles concernant $L_{p,\alpha}$. Remarquons que c'est ce genre de propriétés qui permet de donner des résultats locaux sur les normes universelles ([34]) ou sur la croissance du groupe de Tate-Shafarevich ([25]).

Les formules peuvent se réécrire sans caractères comme en (5). Pour cela, on résout l'équation $(1 - \varphi \otimes \varphi)\mathcal{G}_E = \tilde{L}_p(E) + \mathfrak{M}_E \log(1 + x)$. Il est déterminé à un élément de $D(E)^{\varphi=1}$ près, mais on peut le choisir de manière à ce que pour $n > 0$,

$$\mathcal{G}_E^\pm(\zeta_n^a - 1) = \frac{L_{\{p\}}^\pm(E, a \bmod p^n, 1)}{\Omega_{E,\omega}^\pm}(p\varphi)^n \omega \tag{13}$$

avec $\mathcal{G}_E^\pm = \mathcal{G}_E \pm \mathcal{G}_E^\iota$.

Le passage à $n = 0$ est un peu plus délicat car les fonctions $\mathcal{S}(\mathcal{T}_\tau)$ (ou $\mathcal{S}(\mathcal{T}_0)$ dans le cas (mult-non dépl)) et $\mathcal{G}_E$ ne sont pas a priori définies en 0. Par contre, elles vérifient des équation fonctionnelles du type $\psi(G) = 1 \otimes \varphi(G) + m\log(1 + x)$, c'est-à-dire

$$\sum_{\zeta \in \mu_p} G(\zeta(1 + x) - 1) = p(1 \otimes \varphi)G(x) + pm\log(1 + x).$$

Cette équation fonctionnelle implique que G est définie sur $|x| \geq p^{-1/(p-1)}$. On définit alors $[(1 - p\varphi \otimes \varphi)G]$ pour $|x| < p^{-1/(p-1)}$ par

$$[(1 - p\varphi \otimes \varphi)G](x) = -\sum_{\substack{\zeta \in \mu_p \\ \zeta \neq 1}} G(\zeta(1 + x) - 1) + pm\log(1 + x)$$

et

$$[(1 - p\varphi \otimes \varphi)G](0) = -\sum_{\substack{\zeta \in \mu_p \\ \zeta \neq 1}} G(\zeta - 1).$$

Il est facile de voir que si $\tau \in G_\infty$, $[(1-p\varphi\otimes\varphi)\tau(G)](0) = [(1-p\varphi\otimes\varphi)G](0)$.

Remarque. Comme $\psi(\log x) = p^{-1}\log x$, on trouve de même que l'on peut définir

$$[(1 - \varphi)\log](0) = -\log(\prod_{\zeta \in \mu_p, \zeta \neq 1} \zeta - 1) = -\log_p p,$$

ce qui est compatible avec la formule (10).

Faisons le calcul dans le cas (mult-dépl). Rappelons que l'on peut écrire $\mathcal{S}(\mathcal{T}_\tau) = (\tau - 1)T_0 e_0 - S_\tau N e_0$ avec $T_0 \in \mathcal{H}$ vérifiant $(1 - \varphi)T_0 = (1 + x) - 1$, $T_0(0) = 0$ et $(1 - p^{-1}\varphi)S_\tau = t_\tau$ avec $S_\tau \in \mathcal{B}[\log x]$ et $NS_\tau = T_0$. Écrivons aussi $\mathcal{S}(\mathcal{T}_1) = T_1 N e_0$ avec $(1 - p^{-1}\varphi)T_1 = (1 + x) - \log(1 + x)$. De la formule (10), on déduit que

$$\begin{aligned}
\mathcal{G}_E &= L_{p,\beta} \cdot T_1 N e_0 + L_{p,\alpha} \cdot \mathcal{S}(\mathcal{T}_0) + \lambda e_0 \\
&= L_{p,\beta} \cdot T_1 N e_0 + L_{p,\alpha} \cdot T_0 e_0 - (\tau - 1)^{-1} L_{p,\alpha} \cdot S_\tau N e_0 + \lambda e_0 \qquad (14) \\
&= (L_{p,\alpha} \cdot T_0 + \lambda) e_0 + (L_{p,\beta} \cdot T_1 - (\tau - 1)^{-1} L_{p,\alpha} \cdot S_\tau) N e_0
\end{aligned}$$

avec $\lambda \in \mathbb{Q}_p$. Rappelons que $\mathbf{1}(L_{p,\alpha}) = 0$ et posons $L'_{p,\alpha} = \frac{d}{ds}\langle\chi\rangle^s(L_{p,\alpha})|_{s=0}$ avec $\langle\chi\rangle$ le composé de χ avec la projection $\mathbb{Z}_p^* \to 1 + p\mathbb{Z}_p$, d'où

$$\mathbf{1}(L'_{p,\alpha}) = \log\chi(\tau)\mathbf{1}((\tau - 1)^{-1}L_{p,\alpha}).$$

On obtient en prenant la valeur en 0 :

$$\begin{aligned}
[(1 - p\varphi \otimes \varphi)(\mathcal{G}_E)](0) &= \mathbf{1}(L_{p,\beta})\, T_1(0)(1 - p\varphi)N e_0 \\
&\quad - \mathbf{1}(L'_{p,\alpha})\frac{[(1 - p\varphi \otimes \varphi)S_\tau](0)}{\log\chi(\tau)}N e_0 + (1 - p)\lambda e_0 \\
&= -(p - 1)\lambda e_0 + c\mathbf{1}(L'_{p,\alpha})N e_0
\end{aligned}$$

avec $c = -\frac{[(1-p\varphi\otimes\varphi)S_\tau](0)}{\log\chi(\tau)} \in \mathbb{Q}_p$ indépendant de τ, de E et même du choix de S_τ. Nous démontrerons dans l'appendice que $c = 1$. On a d'autre part en utilisant (13) pour $n = 1$

$$(p\varphi)^{-1}[(1 - p\varphi \otimes \varphi)(\mathcal{G}_E)(0)] = -\frac{L_{\{N\}}(E, 1)}{\Omega_{E,\omega}}\omega.$$

On en déduit que

$$(1 - \frac{1}{p})\lambda e_0 - \mathbf{1}(L'_{p,\alpha})N e_0 = \frac{L_{\{N\}}(E, 1)}{\Omega_{E,\omega}}\omega \qquad (15)$$

D'où, en écrivant $\omega = e_0 - \mathcal{L}(E)Ne_0$, la formule

$$1(L'_{p,\alpha}) = \mathcal{L}(E)\frac{L_{\{N\}}(E,1)}{\Omega_{E,\omega}}. \tag{16}$$

La formule (16) est une conjecture de Mazur-Tate-Teitelbaum ([27]) et a été démontrée par Greenberg et Stevens ([16]) (le passage de N à p est immédiat). Elle a été redémontrée par Kato-Kurihara-Tsuji, il y a quelques années. Redisons que nous utilisons la construction de Kato de systèmes d'Euler modulaires et qu'il n'y a donc peut-être rien de nouveau. Nous voulons simplement mettre en valeur la manière dont ces résultats sur les valeurs spéciales peuvent se montrer à partir de l'application régulateur $\mathcal{L}_E$, osons rajouter, de manière simple, c'est-à-dire en faisant des calculs sur des éléments de l'algèbre $\mathcal{B}[\log x]$. Autrement dit, pour $a = (a_n) \in Z^1_\infty(\mathbb{Q}_p, V)$, le régulateur $\mathcal{L}_E(a)$ contient tous les renseignements sur les régulateurs des a_n, et même de leur twist comme nous allons le voir un peu plus loin. Il y a derrière ces calculs l'étude du module $\mathcal{D}_\infty(E)$ et son lien avec $Z^1_\infty(\mathbb{Q}_p, V_p(E))$ qui utilise en particulier la loi de réciprocité. Mais la généralité est finalement grande!

Remarque. Il est à remarquer que les formules s'expriment mieux et sont plus précises en termes de $\mathcal{G}_E$ que de $L_{p,\{N\}}(E)$. Mais il y a d'autre raisons pour préférer $L_p(E)$ puisque c'est sur l'ordre de cette fonction que l'on a des renseignements. Il faut donc peut-être prendre l'habitude de considérer le couple $(L_{p,\{N\}}(E), \mathcal{G}_E)$. Lorsque φ n'a pas de valeurs propres égales à une puissance de p, la première détermine la seconde. Ce qui n'est pas le cas dans le cas (mult-dépl). L'équation (15) est plus précise que (16). On déduit ainsi de (15) que

$$\lambda = \left(1 - \frac{1}{p}\right)^{-1}\frac{L_{\{N\}}(E,1)}{\Omega_{E,\omega}} \ .$$

D'où en remarquant que $N\mathcal{G}_E$ est bien définie en 0 et que le facteur d'Euler en p est $(1 - p^{-s})$,

$$(N\mathcal{G}_E)(0) = \frac{L_{\{N/p\}}(E,1)}{\Omega_{E,\omega}}Ne_0 \tag{17}$$

Le module $\mathcal{D}_\infty(E)$ ne dépend que de la structure de (φ, N)-module de $D_p(E)$ et absolument pas de sa filtration. C'est dans le régulateur d'Iwasawa que cette filtration se manifeste. En particulier, soit f une forme modulaire de poids k pour $\Gamma_0(N)$ avec p divisant exactement N et $a_p(f) = p^{(k-2)/2}$. Soit $D_p(f)$ le (φ, N)-module filtré associé à f en p. Comme plusieurs normalisations sont possibles, précisons que la filtration de $D_p(f)$ a comme poids de Hodge 0 et $-k + 1$. Le (φ, N)-module $D_p(f)[(k - 2)/2]$ admet une base de vecteurs propres (e_0, Ne_0) avec $\varphi e_0 = e_0$. En tant que (φ, N)-module filtré, il est irréductible pour $k \neq 2$. Par contre, en tant que (φ, N)-module, il est réductible, le $\mathcal{H}(G_\infty)$-module $\mathcal{D}_\infty(D_p(f)[(k - 2)/2])$ est isomorphe à celui construit pour une courbe elliptique ayant réduction multiplicative déployée et on a encore une suite exacte de $\mathcal{H}(G_\infty)$-modules

$$0 \to \mathcal{H}^{\psi=0} \to \mathcal{D}_\infty(D_p(f)[(k - 2)/2]) \to \mathcal{H}^{\psi=0} \to \mathbb{Q}_p \to 0.$$

En particulier, tous les calculs faits précédemment sont valables. Mis avec la construction d'un système d'Euler-Kolyvagin à partir des éléments de Beilinson, qui est faite par Kato, on a ainsi démontré la **conjecture de Mazur-Tate-Teitelbaum** [27] pour les formes modulaires de poids k, de conducteur divisible par p (et non par p^2) telles que $a_p = p^{(k-2)/2}$, avec $\mathcal{L}(E)$ la pente de la filtration de $D_p(f)[(k - 2)/2]$ dans la base $e_0, -Ne_0$. Stevens vient de démontrer une formule de ce type avec $\mathcal{L}(E)$ l'invariant défini par Coleman dans [9]. Les deux résultats ensemble montrent l'égalité de ces deux invariants (au moins lorsque $L(f, k/2)$ est non nul!).

Nous allons maintenant nous intéresser aux twists à la Tate de $V_p(E)$. Soit k un entier strictement positif. Soit $H^2_\mathcal{M}(E, k) = (\mathbb{Q} \otimes K_{2k-2}(E))^{(k)}$ le k-ième espace propre pour l'opérateur d'Adams de la K-théorie de Quillen de E. Nous ne rappelons pas la définition des régulateurs complexes par Beilinson. Du point de vue p-adique, on définit l'application régulateur comme le composé Reg des classes de Chern étales p-adiques :

$$H^2_\mathcal{M}(E, k + 1) \to H^1(G_{S,\mathbb{Q}}, V_p(E)(k)) \to H^1(\mathbb{Q}_p, V_p(E)(k))$$

avec $r_k = \log_{E,k}$ le logarithme de Bloch-Kato :

$$r_k : H^1(\mathbb{Q}_p, V_p(E)(k)) = H^1_f(\mathbb{Q}_p, V_p(E)(k)) \overset{\log_{V_p(E)(k)}}{\to} D_p(E)[-k] \cong D_p(E).$$

Ici, S est un ensemble fini de places contenant les places de mauvaise réduction de E et les places divisant p et $G_{S,\mathbb{Q}}$ le groupe de Galois su r $\mathbb{Q}$ de la plus grande extension non ramifiée en dehors de S. Un autre point de vue est de définir directement l'application Reg : $H^2_\mathcal{M}(E, k+1) \to D_p(E)$ (Niziol [29], Nekovář [28], Besser [4]). Nous n'en dirons pas plus ici. Nous notons aussi Reg l'application $H^1(G_{S,\mathbb{Q}}, V_p(E)(k)) \to D_p(E)$. De même, si F est une extension finie de $\mathbb{Q}$, on définit Reg

$$H^2_\mathcal{M}(E/F, k + 1) \to H^1(G_{S,F}, V_p(E)(k))$$
$$\to \prod_{v|p} H^1(F_v, V_p(E)(k)) \to F \otimes_\mathbb{Q} D_p(E).$$

Lorsque k est un entier négatif ou nul, on dispose d'une application

$$r_k : \ \oplus_{v|p} H^1(F_v, V_p(E)(k)) \to H^1_{/f}(F_v, V_p(E)(k))$$

$$\overset{\exp^*_{V_p(E)(1-k)}}{\to} F \otimes_{\mathbb{Q}} D_p(E)[-k] = F \otimes_{\mathbb{Q}} D_p(E)$$

avec $H^1_{/f}(F_v, V_p(E)(k)) = H^1(F_v, V_p(E)(k))/H^1_f(F_v, V_p(E)(k))$ et on note encore Reg le composé : $H^1(G_{S,F}, V_p(E)(k)) \to F \otimes_{\mathbb{Q}} D_p(E)$.

Jannsen ([18], [19]) conjecture que $H^1(G_{S,F}, V_p(E)(k))$ est de dimension $[F : \mathbb{Q}]$, pour $k \neq 0$ et $k \neq \pm 1$ et pour F une extension finie de $\mathbb{Q}$, et plus précisément que si F est une extension abélienne de $\mathbb{Q}$ et η un caractère de $\mathrm{Gal}(F/\mathbb{Q})$, $H^1(G_{S,F}, V_p(E)(k))^{(\eta)}$ (image du projecteur e_η) est de dimension 1 sur $\mathbb{Q}(\eta)$. Il montre l'équivalence avec la nullité de $H^2(G_{S,F}, V_p(E)(k))$ pour ces valeurs de k par application de la formule de caractéristique d'Euler-Poincaré de ces groupes de cohomologie. Lorsque E a bonne réduction en p, il semble raisonnable d'inclure aussi le cas où $k = \pm 1$. Par contre, dans le cas (mult-dépl), on dispose d'une surjection

$$H^2(G_{S,\mathbb{Q}}, V_p(E)(1)) \to H^2(\mathbb{Q}_p, V_p(E)(1)) \cong H^0(\mathbb{Q}_p, V_p(E)^*) = \mathbb{Q}_p.$$

La conjecture est alors que le noyau de cette application est nul. Remarquons à ce propos que, dans ce cas, les groupes $H^1(\mathbb{Q}_p, V_p(E)(1)) = H^1_g(\mathbb{Q}_p, V_p(E)(1))$ et $H^1_f(\mathbb{Q}_p, V_p(E)(1))$ sont différents et diffèrent de

$$(D_p(E)[-1]/ND_p(E)[-1])^{\varphi=1} = (D_p(E)/ND_p(E))^{\varphi=p^{-1}} \cong \mathbb{Q}_p.$$

Revenons à la théorie d'Iwasawa. Elle permet à partir d'éléments compatibles pour la norme des $H^1(G_{S,\mathbb{Q}(\mu_{p^n})}, T_p(E))$ de construire des éléments de $H^1(G_{S,\mathbb{Q}(\mu_{p^n})}, T_p(E)(k))$ pour tout entier k. Les premiers exemples de telles constructions ont été donnés par Soulé (par exemple [40]) et c'est un outil extrêmement puissant. Si c appartient à[14] $H^1_\infty(\mathbb{Q}, V_p(E))$ (resp. à $Z^1_\infty(\mathbb{Q}_p, V_p(E))$), on note $c(k)$ son image dans

$$H^1_\infty(\mathbb{Q}, V_p(E)(k)) = \mathbb{Q}_p \otimes \varprojlim_n H^1(G_{S,\mathbb{Q}(\mu_{p^n})}, T_p(E)(k))$$

(resp. dans $Z^1_\infty(\mathbb{Q}_p, V_p(E)(k))$). Si η est un caractère de conducteur p^n, on note encore ev_η le composé de la projection de $H^1_\infty(\mathbb{Q}, V_p(E)(k))$ sur $H^1(G_{S,\mathbb{Q}(\mu_{p^n})}, T_p(E)(k))$ avec le projecteur e_η (et de même pour les groupes de cohomologie locaux). Ainsi, ev_1 est simplement la projection dans $H^1(G_{S,\mathbb{Q}}, T_p(E)(k))$. On note d'autre part D la dérivation $(1+x)\frac{d}{dx}$ et l'opérateur qu'elle induit sur $\mathcal{D}_\infty(E)$.

---

[14]Rappelons que $\mathbb{Q}_p \otimes \varprojlim_n H^1(G_{S,\mathbb{Q}(\mu_{p^n})}, T_p(E))$ est indépendant de S et égal à $H^1_\infty(\mathbb{Q}, V_p(E))$.

La construction du régulateur d'Iwasawa $\tilde{\mathcal{L}}_E$ utilise de façon essentielle cette opération de twist et ses propriétés impliquent (et même sont) la proposition suivante (ici $\Gamma^*(j)$ est $(j-1)!$ si $j \geq 1$ et $(-1)^j/j!$ si $j \leq 0$) :

Proposition. *Pour tout entier k et tout caractère η de conducteur p^n non trivial, on a*

$$\mathrm{ev}_\eta(D^{-k}(\tilde{\mathcal{L}}_E)(c_\infty)) = \Gamma^*(-k)(p^{1-k}\varphi)^n \operatorname{Reg}(\mathrm{ev}_\eta(c_\infty(k))) \qquad (18)$$

Dans le cas (br), *cela s'écrit aussi*

$$\eta\chi^{-k}(\mathcal{L}_E(c_\infty)) = \Gamma^*(-k)\epsilon(\eta)G(\eta)(p^{-k}\varphi)^n \operatorname{Reg}(\mathrm{ev}_\eta(c_\infty(k))) \qquad (19)$$

On peut alors traduire cela comme un énoncé sur les fonction L p-adiques classiques.

Proposition. *Pour tout entier k, on a*

(1) *dans le cas ordinaire,*

$$\eta\chi^{-k}(L^\pi_{p,\{N\}}(E)) = \Gamma^*(-k)\epsilon(\eta)G(\eta)(p^{-k}\varphi)^n \pi_{[0]}(\operatorname{Reg}(\mathrm{ev}_\eta(c_\infty(k))))$$

(2) *dans le cas supersingulier,*

$$\eta\chi^{-k}(L_{p,\{N\}}(E)) = \Gamma^*(-k)\epsilon(\eta)G(\eta)(p^{-k}\varphi)^n \operatorname{Reg}(\mathrm{ev}_\eta(c_\infty(k))).$$

Mais la formule (18) est plus complète et en termes des fonctions $\tilde{L}_{p,\{N\}}(E)$ ou $L_{p,\{N\}}(E)$, elle s'écrit

$$\mathrm{ev}_\eta(D^{-k}(\tilde{L}_{p,\{N\}}(E))) = \Gamma^*(-k)(p^{1-k}\varphi)^n \operatorname{Reg}(\mathrm{ev}_\eta(c_\infty(k))) \qquad (20)$$

ou

$$\eta\chi^{-k}(L_{p,\{N\}}(E)) = \Gamma^*(-k)\epsilon(\eta)G(\eta)(p^{-k}\varphi)^n \operatorname{Reg}(\mathrm{ev}_\eta(c_\infty(k))). \qquad (21)$$

Des formules analogues sont valables pour le caractère trivial (au moins dans le cas (br)). Par exemple pour $k = 1$, on peut interpréter la valeur de la fonction p-adique en $s = 0$ (avec $L_p(E, s) = \langle\chi\rangle^{1-s}(L_p(E))$) comme le régulateur d'un point C_2 de $K_2(E, \mathbb{Z}_p)$:

$$L_p(E, 0) = (1 - p^{-2}\varphi^{-1})^{-1}(1 - p^{-1}\varphi) \operatorname{Reg} C_2. \qquad (22)$$

Une formule de ce type a été montrée par Coleman et de Shalit [10] dans le cas de multiplication complexe et par Kings [22] dans le même cadre pour tout entier k. L'intérêt supplémentaire qu'il y a dans la formule (22) (outre le fait qu'elle se généralise à tout entier k) est qu'elle tient compte du régulateur complet et non seulement de sa projection sur une droite spéciale.

Restons avec le cas $k = 1$. Par définition même de c_∞, C_2 provient d'un élément de $H^2_{\mathcal{M}}(E, 2)$. Pour obtenir l'équation (22), nous avons ici appliqué deux fois la loi de réciprocité (une fois celle de Kato et une fois celle que j'ai introduite) et twisté une fois vers la gauche et une fois vers la droite. Il devrait donc être possible de s'en passer. Et en effet, il semble possible de montrer directement la formule (22) et ses analogues pour un caractère η. Grâce à la loi de réciprocité de Colmez, on en déduit alors les formules (9) de Kato pour $c_\infty(1)$ au moins dans le cas (br) (article en préparation).

Question 2. *Dans le cas (mult-dépl), $C_2 = \mathrm{ev}_1\left(c_\infty(1)\right)$ appartient-il à $H^1_f(\mathbb{Q}_p, V_p(E)(1))$? dans le cas contraire, que vaut son image dans*

$$H^1_{g/f}(\mathbb{Q}_p, V_p(E)(1)) \cong (D_p(E)/ND_p(E))^{p^{-1}\varphi = p^{-1}} \cong \mathbb{Q}_p e_0?$$

Comme $c_\infty(1)$ est très concret, une réponse est envisagable!

Question 3. *Dans le cas (mult-dépl), $C_0 = \mathrm{ev}_1\left(c_\infty(-1)\right)$ appartient-il à $H^1_f(\mathbb{Q}_p, V_p(E)(-1))$? dans ce cas, que vaut son image dans*

$$H^1_{f/e}(\mathbb{Q}_p, V_p(E)(-1)) \cong D_p(E)^{N=0}/(1 - p\varphi) \cong \mathbb{Q}_p Ne_0?$$

Question 4. *Pourquoi ne semble-t-on envisager la possibilité d'un zéro trivial pour $L_{p,\alpha}$ en χ^{-1} ou χ alors que le facteur d'Euler de la fonction L s'annule ?*

Peut-être parce que ce ne serait pas ici la projection sur $\mathbb{Q}_p e_0$ qui en ferait apparaître un.

L'utilisation des techniques extrêmement puissantes introduites par Kolyvagin dans [23] et relatives à ce qu'il a appelé systèmes d'Euler ont permis à Kato de montrer la divisibilité de la série caractéristique de

$$H^2_{\infty, S}(\mathbb{Q}, V_p(E)) = \mathbb{Q}_p \otimes \varprojlim_n H^2(G_{\mathbb{Q}(\mu_{p^n}), S}, T_p(E))$$

par une fonction liée au système d'Euler. Ce résultat peut se dire de manière simple en utilisant l'idéal arithmétique $I_{\mathrm{arith}}(E)$ introduit dans [31] dans le cas des courbes elliptiques : lorsque $H^1_\infty(\mathbb{Q}, V_p(E))$ est de rang 1, $I_{\mathrm{arith}}(E)$ est simplement le $\mathbb{Z}_p[[G_\infty]]$-module, image dans $\mathcal{D}_\infty(E)$ de $H^1_\infty(\mathbb{Q}, T_p(E))$ multiplié par la série caractéristique de $H^2_{\infty, p}(\mathbb{Q}, T_p(E))$ défini comme le noyau

$$H^2_{\infty, S}(\mathbb{Q}, T_p(E)) \to \varprojlim_n \prod_{v \in S} H^2(\mathbb{Q}(\mu_{p^n})_v, T_p(E))$$

autrement dit, en utilisant le langage des déterminants, l'image par le régulateur d'Iwasawa $\mathcal{L}_E$ de $\det(H^2_{\infty,p}(\mathbb{Q}, T_p(E)))^{-1} \cdot H^1_\infty(\mathbb{Q}, T_p(E))$ dans $\mathcal{D}_\infty(E)$. Par un théorème de Rohrlich [37] sur les valeurs de la fonction L tordue par des caractères, $\tilde{L}_p(E)$ est non nul. Le théorème de Kato peut alors s'énoncer ainsi (bien qu'il ne le fasse pas explicitement)

Théorème (Kato). *Le $\mathbb{Z}_p[[G_\infty]]$-module $H^1_\infty(\mathbb{Q}, V_p(E))$ est de rang 1 et le $\mathbb{Z}_p[[G_\infty]]$-module engendré par $\tilde{L}_p(E)$ est contenu dans $\mathbb{Q}_p \otimes \mathcal{I}_{arith}(E)$.*

Ainsi, la "conjecture principale" de [31] qui prédit que $\mathcal{I}_{\mathrm{arith}}(E) = (\tilde{L}_p(E))$ est à moitié vérifiée.

Nous nous limitons par prudence et manque de temps au cas (br) dans la proposition suivante :

Proposition. (1) *(Kato) Si $L(E, \eta^{-1}, 1)$ est non nul pour un caractère η de conducteur p^n, les η-composantes de $E(\mathbb{Q}(\mu_{p^n}))$ et du groupe de Shafarevich-Tate $\mathrm{III}(E/\mathbb{Q}(\mu_{p^n}))$ sont finis.*

(2) *Soit $k \neq 0$ tel que $\chi^{-k}\eta(L_p(E)) \neq 0$. Alors $H^1(G_{S,\mathbb{Q}(\mu_{p^n})}, V_p(E)(k))^{(\eta)}$ est de rang 1, engendré par $\mathrm{ev}_\eta(c_\infty(k))$ et la conjecture de Jannsen est vraie.*

Sous l'hypothèse faite, les facteurs d'Euler pour les places divisant N ne s'annulent pas. Il serait peut-être quand même raisonnable de vérifier numériquement que $\chi^{-k}(\tilde{L}_p(E))$ n'est pas nul!

Donnons rapidement la démonstration de 2). Comme $\chi^{-k}\eta(L_p(E)) \neq 0$, il en est de même de $\chi^{-k}\eta(\mathcal{I}_{\mathrm{arith}}(E))$. Cela implique en particulier que la série caractéristique de $H^2_{\infty,\{p\}}(\mathbb{Q}, V_p(E)(k))$ ne s'annule pas en η^{-1} et par des arguments classiques que $H^2(G_{S,\mathbb{Q}(\mu_{p^n})}, V_p(E)(k))^{(\eta)}$ est fini et que $H^1(G_{S,\mathbb{Q}(\mu_{p^n})}, V_p(E)(k))^{(\eta)}$ est de rang 1 (formule de caractéristique de Tate-Poitou). On peut en fait calculer les valeurs de $\chi^{-k}\eta(\mathcal{I}_{\mathrm{arith}}(E))$ pour $I_{\mathrm{arith}}(E)$ un générateur de $\mathcal{I}_{\mathrm{arith}}(E)$. Le calcul a été fait dans [33] dans le cas de bonne réduction et pour η caractère trivial (voir aussi [31] dans le cas des courbes elliptiques), et n'est pas plus difficile pour un caractère non trivial.

Finissons par une question. Posons $C_k = \mathrm{ev}_1(c_\infty(k-1))$ et supposons $\mathrm{Reg}\, C_k$ non nul, ce qui est équivalent à ce que $\chi^{-k}(L_p(E))$ soit non nul. En particulier, C_k est non nul et est une base de $H^1(G_{S,\mathbb{Q}}, V_p(E)(k-1))$ pour $k \neq 1$.

Notons ici D_α et D_β les espaces propres de $D_p(E)$ pour les valeurs propres α^{-1} et β^{-1} de φ et supposons que $\mathrm{Fil}^0 D(E)$ n'est pas stable par φ. Lorsque $\mathrm{Reg}\, C_k$ est non nul, pour le repérer dans le plan $D_p(E)$, on peut utiliser le birapport $\ell_k(E)$ des quatre droites D_α, D_β, $\mathrm{Fil}^0 D_p(E)$ et

$\mathbb{Q}_p \operatorname{Reg} C_k = \operatorname{Reg}(H^1(G_{S,\mathbb{Q}}, V_p(E)(k-1)))$. Ainsi, si on écrit $\omega = e_\alpha + e_\beta$ avec $e_\alpha \in D_\alpha$, $e_\beta \in D_\beta$,

$$\operatorname{Reg} C_k \in \mathbb{Q}_p(e_\alpha + \ell_k(E)e_\beta).$$

Lorsque $k = 1$, les deux droites $\operatorname{Fil}^0 D_p(E)$ et $\operatorname{Reg} C_1$ coïncident et $\ell_1(E) = 1$. D'autre part, dans le cas supersingulier, il est facile de voir que $\ell_k(E)$ est de norme 1 dans l'extension quadratique $\mathbb{Q}_p(\alpha)/\mathbb{Q}_p$. On peut aussi écrire

$$\mathbb{Q}_p \operatorname{Reg} C_k = \mathbb{Q}_p(\omega - \lambda_k(E)p\varphi\omega)$$

et $\lambda_k(E) \in \mathbb{Q}_p \cup \infty$ ne dépend pas du choix de $\omega \in \operatorname{Fil}^0 D_p(E)$. Il est facile de voir qu'en fait $\ell_k(E) = (1 + \lambda_k(E)\alpha)/(1 + \lambda_k(E)\beta)$ si $\lambda_k(E) \neq \infty$ et -1 si $\lambda_k(E) = \infty$. On a $\lambda_1(E) = 0$. On peut donner des définitions analogues pour un caractère η de conducteur une puissance $p^{n(\eta)}$ de p et obtenir ainsi des invariants $\ell_{k,\eta}(E)$ et $\lambda_{k,\eta}(E)$. Par exemple, $\ell_{1,\eta}(E) = 1$.

Question 5. *Que peut-on dire de la répartition des $\lambda_k(E)$ dans $\mathbb{P}^1(\mathbb{Q}_p)$? Peut-il exister une formule simple pour $\lambda_{k,\eta}(E)$ pour $k \neq 1$ fixé ?*

Appendice

Dans cet appendice, nous expliquons comment peuvent se calculer des relèvements d'éléments de $\mathcal{D}_\infty(\mathbb{Q}_p)$ dans $\mathcal{D}_\infty(E)$ et en particulier nous calculons la constante c qui intervient dans la démonstration de la formule de Mazur-Tate-Teitelbaum.

Plaçons-nous dans le cas (mult-dépl). Soit $f \in \mathcal{H}^{\psi=0}$ tel que $f(0) = 0$. Soit $F_1 \in \mathcal{H}$ tel que $(1 - \varphi)F_1 = f$ et $F_1(0) = 0$. Nous allons montrer que si $f \in \mathbb{Q}_p \otimes \mathbb{Z}_p[[x]]$, la suite $p^n\psi^n(F_1 \log x)$ converge dans $\mathcal{B}[\log x]$ vers un élément F_2 tel que $\psi(F_2) = p^{-1}F_2$ et $NF_2 = F_1$; $(1 - \varphi \otimes \varphi)(F_1 e_0 - F_2 N e_0)$ est alors un élément de $\mathcal{D}_\infty(E)$.

Posons donc $U_n = p^n\psi^n(F_1 \log x)$. On a

$$
\begin{aligned}
U_{n+1} - U_n &= p^{n+1}\psi^{n+1}\left((1 - p^{-1}\varphi)(F_1 \log x)\right) \\
&= p^{n+1}\psi^{n+1}\left((1 - \varphi)(F_1) \log x + \varphi(F_1)(1 - p^{-1}\varphi)(\log x)\right) \\
&= p^{n+1}\psi^{n+1}\left(f \log x\right) + p^{n+1}\psi^n\left(F_1\psi((1 - p^{-1}\varphi)\log x)\right) \\
&= p^{n+1}\psi^{n+1}\left(f \log x\right)
\end{aligned}
$$

car

$$\psi((1 - p^{-1}\varphi)\log x) = 0$$

$$
\begin{aligned}
&= p^{n+1}\psi^{n+1}\left(f(1 - p^{-1}\varphi)\log x\right) + p^n\psi^{n+1}\left(f\varphi \log x\right) \\
&= p^{n+1}\psi^{n+1}\left(f(1 - p^{-1}\varphi)\log x\right)
\end{aligned}
$$

car $\psi(f) = 0$ et $\psi(f\varphi g) = \psi(f)g$. On remarque alors que $(1 - p^{-1}\varphi)\log x$ est un élément de $\mathbb{Q}_p \otimes \mathbb{Z}_p[[1/x]]$ convergeant pour $|x| > p^{-1/(p-1)}$. Comme $f \in \mathbb{Q}_p \otimes \mathbb{Z}_p[[x]]$, $f(1 - p^{-1}\varphi)\log x$ est la somme d'un tel élément et d'un élément de $\mathbb{Q}_p \otimes \mathbb{Z}_p[[x]]$ et si $p^{-1/(p-1)} < \rho < 1$, les $\|\psi^n(f(1-p^{-1}\varphi)\log x)\|_\rho$ sont bornés par rapport à n. On en déduit la convergence de la suite U_n (comme $V_n = U_n - F_1 \log x \in \mathcal{B}$, cela signifie simplement la convergence de V_n sur les couronnes $\rho_1 \leq |x| \leq \rho_2$ pour $p^{-1/(p-1)} < \rho_1 \leq \rho_2 < 1$). Si F_2 est sa limite, il est clair que $\psi(F_2) = p^{-1}F_2$ et que le coefficient de $\log x$ est F_1, c'est-à-dire que $Nf_2 = F_1$.

Pour $\zeta - 1$ avec $\zeta \in \mu_{p^\infty}$, les valeurs de F_1 ne sont pas difficiles à écrire en fonction de f :

$$F_1(x) = \sum_{m=0}^{n-1} f(\varphi^m(x)) + F_1(\varphi^n(x))$$

et donc pour $\zeta \in \mu_{p^n}$ avec $n \geq 1$

$$F_1(\zeta - 1) = \sum_{m=0}^{n-1} f(\zeta^{p^m} - 1).$$

Passons à F_2. On a alors en fonction de F_1

$$F_2(\zeta_n - 1) = \lim_{s \to \infty} \mathrm{Tr}_{s/n}\left(F_1(\zeta_s - 1)\log(\zeta_s - 1)\right)$$

$$= \lim_{s \to \infty} \sum_{\substack{\zeta \in \mu_{p^s} \\ \zeta \mapsto \zeta_n}} F_1(\zeta - 1)\log(\zeta - 1)$$

où $\zeta \mapsto \zeta_n$ signifie qu'il existe un entier k tel que $\zeta^{p^k} = \zeta_n$ et $\mathrm{Tr}_{m/n}$ désigne la trace de $\mathbb{Q}(\mu_{p^m})$ à $\mathbb{Q}(\mu_{p^n})$, d'où

$$F_2(\zeta_n - 1) = \lim_{s \to \infty} \sum_{m=0}^{n-1} \sum_{\substack{\zeta \in \mu_{p^s} \\ \zeta \mapsto \zeta_n}} f(\zeta^{p^m} - 1)\log(\zeta - 1).$$

Prenons maintenant $f = (\tau - 1) \cdot (1 + x)$, pour $S_\tau = F_2$, calculons

$$c_\tau = -[(1 - p\varphi \otimes \varphi)S_\tau](0) = \sum_{\substack{\zeta \in \mu_p \\ \zeta \neq 1}} S_\tau(\zeta - 1)$$

$$= \lim_{n \to \infty} \mathrm{Tr}_{n/0}(F_1(\zeta_n - 1)\log(\zeta_n - 1)).$$

Posons $G_n = \mathrm{Gal}(\mathbb{Q}_p(\mu_{p^n})/\mathbb{Q}_p)$ et calculons donc

$$X_n = \sum_{\sigma \in G_n} \sum_{m=0}^{n-1} (\tau - 1) \cdot \zeta_{n-m}^\sigma \log(\zeta_n^\sigma - 1).$$

On remarque d'abord que

$$\sum_{\sigma \in G_n} (\tau - 1) \cdot \zeta_{n-m}^\sigma \log(\zeta_n^\sigma - 1) = \sum_{\sigma \in G_m} (\tau - 1) \cdot \zeta_{n-m}^\sigma \log(\zeta_{n-m}^\sigma - 1)$$

car $p\psi(\log x) = \log x$. D'où,

$$X_n = \sum_{m=0}^{n-1} \sum_{\sigma \in G_{n-m}} \zeta_{n-m}^\sigma \log \frac{\zeta_{n-m}^{\tau^{-1}\sigma} - 1}{\zeta_{n-m}^\sigma - 1}.$$

En posant $a_{\tau - 1} = (\tau^{-1} - 1) \log x = \log \frac{(1+x)^{\chi(\tau)^{-1}} - 1}{x} \in \mathcal{H}^{\psi=0}$ et en remarquant que les ζ_{n-m}^σ parcourent toutes les racines p^n-ièmes de l'unité sauf 1, on obtient que

$$X_n = \sum_{\zeta \in \mu_{p^n} - \{1\}} \zeta a_{\tau - 1}(\zeta - 1).$$

Ecrivons $a_{\tau - 1} \equiv \sum_{\substack{0 < k < p^n \\ (k,p)=1}} a^{k,n}(1+x)^k \mod (1+x)^{p^n} - 1$ et remarquons que

$$\sum_{\zeta \in \mu_{p^n}} \zeta^j = \begin{cases} 0 & \text{si } j \not\equiv 0 \bmod p^n \\ p^n & \text{si } j \equiv 0 \bmod p^n \end{cases} \equiv 0 \bmod p^n.$$

Donc

$$X_n \equiv - \sum_{\substack{0 < k < p^n \\ (k,p)=1}} a^{k,n} \bmod p^n$$

$$\equiv -a_{\tau - 1}(0) \bmod p^n$$

et la limite c_τ de X_n lorsque $n \to \infty$ est $-a_{\tau - 1}(0) = \log \chi(\tau)$. Ce qui démontre l'égalité désirée

$$c = -\frac{[(1 - p\varphi \otimes \varphi)S_\tau](0)}{\log \chi(\tau)} = 1.$$

On peut de la même manière écrire une formule pour $\mathrm{ev}_\eta(t_0) = (\eta(\tau) - 1)^{-1}\mathrm{ev}_\eta(S_\tau)$ pour η caractère de conducteur p^n avec ce choix de S_τ :

$$(\eta(\tau) - 1)\,\mathrm{ev}_\eta(t_0) = \sum_{\sigma \in G_n} \eta^{-1}(\sigma) \lim_{s \to \infty} \mathrm{Tr}_{s/n}\left(\sum_{j=0}^{s-1}(\tau - 1)\zeta_s^{j\sigma}\log(\zeta_s^\sigma - 1)\right).$$

On peut la transformer, mais je n'ai pas trouvé de manière de bien la présenter! Il est à remarquer cependant qu'interviennent des traces d'éléments du type $\zeta_s \log(\zeta_s - 1)$. Je ne sais pas si de telles expressions interviennent ailleurs.

Références

[1] Y. Amice and J. Vélu, *Distributions p-adiques associées aux séries de Hecke*, Astérisque (1975), no. 24–25, 119–131.

[2] A. A. Beilinson, *Higher regulators of modular curves*, Applications of algebraic K-theory to algebraic geometry and number theory, Part I, II, (Boulder, Colo., 1983), Amer. Math. Soc., Providence, R.I., 1986, pp. 1–34.

[3] D. Benois, *On Iwasawa theory of crystalline representations*, Duke Math. J. **104** (2000), 211–267.

[4] A. Besser, *Syntomic regulators and p-adic integration. II. K_2 of curves*, Proceedings of the Conference on p-adic Aspects of the Theory of Automorphic Representations (Jerusalem, 1998), vol. 120, 2000, pp. 335–359.

[5] S. Bloch and K. Kato, *L-functions and Tamagawa numbers of motives*, The Grothendieck Festschrift, vol. I, Birkhäuser Boston, Boston, MA, 1990, pp. 333–400.

[6] G. Christol, *Systèmes différentiels linéaires p-adiques, structure de Frobenius faible*, Bull. Soc. Math. France **109** (1981), 83–122.

[7] J. Coates and A. Wiles, *On p-adic L-functions and elliptic units*, J. Austral. Math. Soc. Ser. A **26** (1978), 1–25.

[8] R. F. Coleman, *Division values in local fields*, Invent. Math. **53** (1979), 91–116.

[9] ———, *A p-adic Shimura isomorphism and p-adic periods of modular forms*, p-adic monodromy and the Birch and Swinnerton-Dyer conjecture, (Boston, MA, 1991), Amer. Math. Soc., Providence, RI, 1994, pp. 21–51.

[10] R. F. Coleman and E. de Shalit, *p-adic regulators on curves and special values of p-adic L-functions*, Invent. Math. **93** (1988), 239–266.

[11] P. Colmez, *Théorie d'Iwasawa des représentations de de Rham d'un corps local*, Ann. of Math. **148** (1998), 485–571.

[12] B. Dwork, *p-adic cycles*, Inst. Hautes Études Sci. Publ. Math. **37** (1969), 27–115.

[13] J.-M. Fontaine, *Le corps des périodes p-adiques*, Astérisque **223** (1994), 59–111.

[14] ———, *Représentations p-adiques semi-stables*, Astérisque **223** (1994), 113–184.

[15] R. Greenberg, *Iwasawa theory for elliptic curves*, Arithmetic theory of elliptic curves (Cetraro, 1997), Springer, Berlin, 1999, pp. 51–144.

[16] R. Greenberg and G. Stevens, *On the conjecture of Mazur, Tate, and Teitelbaum*, p-adic monodromy and the Birch and Swinnerton-Dyer conjecture (Boston, MA, 1991), Amer. Math. Soc., Providence, RI, 1994, pp. 183–211.

[17] K. Iwasawa, *Explicit formulas for the norm residue symbol*, J. Math. Soc. Japan **20** (1968), 151–164.

[18] U. Jannsen, *On the Galois cohomology of l-adic representations attached to varieties over local or global fields*, Séminaire de Théorie des Nombres, Paris 1986–87, Birkhäuser Boston, Boston, MA, 1988, pp. 165–182.

[19] ———, *On the l-adic cohomology of varieties over number fields and its Galois cohomology*, Galois groups over **Q** (Berkeley, CA, 1987), Springer, New York, 1989, pp. 315–360.

[20] J. W. Jones, *Iwasawa L-functions for multiplicative abelian varieties*, Duke Math. J. **59** (1989), 399–420.

[21] K. Kato, *p-adic Hodge theory and values of zeta functions of modular forms*, prépublication, 2001.

[22] G. Kings, *The Tamagawa number conjecture for CM elliptic curves*, Invent. Math. **143** (2001), 571–627.

[23] V. A. Kolyvagin, *Euler systems*, The Grothendieck Festschrift, vol. II, Birkhäuser Boston, Boston, MA, 1990, pp. 435–483.

[24] D. S. Kubert and S. Lang, *Modular units*, Springer-Verlag, New York, 1981.

[25] M. Kurihara, *On the Tate-Shafarevich groups over the cyclotomic fields of an elliptic curve with supersingular reduction I*, prépublication, 2000.

[26] B. Mazur and P. Swinnerton-Dyer, *Arithmetic of Weil curves*, Invent. Math. **25** (1974), 1–61.

[27] B. Mazur, J. Tate, and J. Teitelbaum, *On p-adic analogues of the conjectures of Birch and Swinnerton-Dyer*, Invent. Math. **84** (1986), 1–48.

[28] J. Nekovář, *p-adic Abel-Jacobi maps and p-adic heights*, The arithmetic and geometry of algebraic cycles (Banff, AB, 1998), Amer. Math. Soc., Providence, RI, 2000, pp. 367–379.

[29] W. Niziol, *On the image of p-adic regulators*, Invent. Math. **127** (1997), 375–400.

[30] B. Perrin-Riou, *Théorie d'Iwasawa et hauteurs p-adiques*, Invent. Math. **109** (1992), 137–185.

[31] _____, *Fonctions L p-adiques d'une courbe elliptique et points rationnels*, Ann. Inst. Fourier (Grenoble) **43** (1993), 945–995.

[32] _____, *Théorie d'Iwasawa et hauteurs p-adiques (cas des variétés abéliennes)*, Séminaire de Théorie des Nombres, Paris, 1990–91, Birkhäuser Boston, Boston, MA, 1993, pp. 203–220.

[33] _____, *p-adic L-functions and p-adic representations*, Amer. Math. Soc., Providence, RI, 2000, Translated from the 1995 French original by Leila Schneps and revised by the author.

[34] _____, *Représentations p-adiques et normes universelles. I. Le cas cristallin*, J. Amer. Math. Soc. **13** (2000), 533–551.

[35] _____, *Théorie d'Iwasawa des représentations p-adiques semi-stables*, Mém. Soc. Math. Fr. (N.S.) **84** (2001).

[36] G. Robert, *Unités elliptiques*, Bull. Soc. Math. France, Mém. No. 36, vol. 101, Société Mathématique de France, Paris, 1973.

[37] D. E. Rohrlich, *On L-functions of elliptic curves and cyclotomic towers*, Invent. Math. **75** (1984), 409–423.

[38] K. Rubin, *Euler systems and modular elliptic curves*, Galois representations in arithmetic algebraic geometry (Durham, 1996), Cambridge Univ. Press, Cambridge, 1998, pp. 351–367.

[39] A. J. Scholl, *An introduction to Kato's Euler systems*, Galois representations in arithmetic algebraic geometry (Durham, 1996), Cambridge Univ. Press, Cambridge, 1998, pp. 379–460.

[40] C. Soulé, *p-adic K-theory of elliptic curves*, Duke Math. J. **54** (1987), 249–269.

[41] M. M. Vishik, *Non-archimedian measures connected with Dirichlet series*, Math. USSR Sbornik **28** (1976), 216–228.

Computing Rational Points on Curves

Bjorn Poonen[1]

1 Introduction

The solution of diophantine equations (such as $x^{13} + y^{13} = z^{13}$) over the integers often reduces to the problem of determining the *rational* number solutions to a single polynomial equation in two variables. Such an equation describes a curve, and the problem of finding rational number solutions can be interpreted geometrically as finding the rational points on the curve, i.e., the points on the curve with rational coordinates.

Despite centuries of effort, we still do not know if there is a general algorithm that takes the equation of a curve, and outputs a list of its rational points, in the cases where the list is finite.[2] On the other hand, qualitative results such as Faltings' Theorem [26] on the finiteness of the number of rational points on curves of genus at least 2, the wide variety of conjectural effective approaches, and the practical success of recent efforts in determining the rational points on individual curves, have led many to believe that such an algorithm exists.

The reader expecting a thorough introduction or a comprehensive survey of known results may be disappointed by this article. We have selected only a few of the many aspects of the subject that we could have discussed. On the other hand, as we go along, we provide pointers to the literature for the reader wishing to delve more deeply in some particular direction. Finally, most of what we say for the field $\mathbf{Q}$ of rational numbers can be generalized easily to arbitrary number fields.

2 Hilbert's 10th Problem and Undecidability

First,[3] let us consider the problem of what can be computed *in theory*, and let us broaden the perspective to millennial proportions by considering

---

[1]This research was supported by NSF grant DMS-9801104, a Sloan Fellowship, and a Packard Fellowship.

[2]In fact, we do not even know if there is an algorithm that can always decide whether the list *is* finite.

[3]This section is not prerequisite for the rest of the article.

not only curves, but also higher dimensional varieties. Also, before asking whether we can compute all rational points, let's ask first whether we can determine whether a variety *has* a rational point. This leads to "Hilbert's 10th Problem over **Q**":[4]

Is there an algorithm for deciding whether a system of polynomial equations with integer coefficients

$$f_1(x_1, \ldots, x_n) = 0$$
$$f_2(x_1, \ldots, x_n) = 0$$
$$\vdots$$
$$f_m(x_1, \ldots, x_n) = 0$$

has a solution with $x_1, x_2, \ldots, x_n \in \mathbf{Q}$?

The system of equations defines a variety[5] over **Q**, so equivalently we may ask, does there exist an algorithm for deciding whether a variety over **Q** has a rational point?

By "algorithm" we mean Turing machine: see [36] for a definition. The machine is to be fed (for instance) a finite stream of characters containing the TEX code for a system of polynomial equations over **Q**, and is supposed to output yes or no in a finite amount of time, according to whether there is a rational solution or not. There is no insistence that the running time of the algorithm be bounded by a fixed polynomial in the length of the input stream; in Hilbert's 10th Problem, we are happy as long as the algorithm terminates after some unspecified number of steps on each input.

The answer to Hilbert's 10th Problem over **Q** is not known. This can be stated in logical terms as follows: we do not know whether there exists an algorithm for deciding the truth of all sentences such as

$$(\exists x)(\exists y)((2 * x * x + y = 0) \wedge (x + y + 3 = 0))$$

involving only rational numbers, the symbols $+, *, =, \exists$, logical relations $\wedge$ ("and"), $\vee$ ("or"), and variables x, y, ... bound by existential quantifiers. One can try asking for more, namely, for an algorithm to decide the entire *first order theory* of $(\mathbf{Q}, 0, 1, +, *)$; this would mean an algorithm for deciding the truth of sentences such as the one above, but in which in addition the symbols $\forall$ ("for all") and $\neg$ ("not") are allowed to appear. For this

---

[4]The analogous problem with **Q** replaced by **Z** was Problem 10 in the list of 23 problems that Hilbert presented to the mathematical community in 1900. This question over **Z** was settled in the negative [49] around 1970.

[5]In this article, varieties will not be assumed to be irreducible or reduced unless so specified.

more general problem, it is known that there is no algorithm that solves it [67].

To put the situation in perspective, we list the answers for the analogous questions about the existence of algorithms for deciding Hilbert's 10th Problem (existence of solutions to a polynomial system) or for deciding the first order theory, over other commutative rings.[6] Here YES means that there is an algorithm, NO means that no algorithm exists (i.e., Hilbert's 10th Problem of the first order theory is undecidable), and ? means that it is not known whether an algorithm exists.

Ring	Hilbert's 10th Problem	First order theory
$\mathbf{C}$	YES	YES
$\mathbf{R}$	YES	YES
$\mathbf{F}_p$	YES	YES
$\mathbf{Q}_p$	YES	YES
$\mathbf{Q}$	?	NO
$\mathbf{F}_p(t)$	NO	NO
$\mathbf{Z}$	NO	NO

The rings are listed approximately in order of increasing "arithmetic complexity." There is no formal definition of arithmetic complexity, but roughly we can measure the complexity of fields k by the "size" of the absolute Galois group, i.e., the Galois group of the algebraic closure $\overline{k}$ over k. And nonfields can be thought of as more complex than their fields of fractions, for instance because there is "extra structure" coming from the nontriviality of the divisibility relation. Whether or not $\mathbf{F}_p(t)$ is more complex than $\mathbf{Q}$ is debatable, but it is the extra structure coming from the p-th power map on the former that enabled the proof of undecidability of

---

[6]A technical point: in the cases where the commutative ring R is uncountable ($\mathbf{C}$, $\mathbf{R}$, $\mathbf{Q}_p$), we must be careful with our statement of the problem, because for instance, a classical Turing machine cannot examine the entirety of an infinite precision real number in a finite number of steps, and hence cannot even decide equality of two real numbers if fed the strings of their decimal digits on two infinite input tapes. To circumvent the problem, in the uncountable cases we restrict attention to decidability questions in which the constants appearing in the input polynomial system or first order sentence are *integers*. We still, however, require the machine to decide the existence of solutions or truth of the sentence with the variables ranging over all of R. It makes sense to ask this, since the output is to be simply yes or no.

Hilbert's 10th Problem for it.

For the complex numbers **C**, the fact that the first order theory (and hence also Hilbert's 10th Problem) is decidable is a consequence of classical elimination theory. The first order theory has elimination of quantifiers: this means that a first order sentence involving $n \geq 1$ quantifiers $(\exists, \forall)$ can be transformed into a sentence with $n - 1$ quantifiers, in an algorithmic way, such that the latter sentence is true if and only if the former is. Algebraically, this corresponds to the elimination of a single variable from a system of equations, and geometrically it amounts to the fact that the projection from $\mathbf{C}^n$ to $\mathbf{C}^{n-1}$ of a Boolean combination of algebraic subsets of $\mathbf{C}^n$ can be written as a Boolean combination of algebraic subsets of $\mathbf{C}^{n-1}$. See [34, Exercise II.3.19] and the references [13, Exposé 7] and [50, Chapter 2, §6] listed there for a generalization due to Chevalley, which shows that for the same reasons, the first order theory of any algebraically closed field is decidable.

The analogous statement about the decidability of the first order theory of the real numbers **R** was proved by Tarski [76] using the theory of Sturm sequences. Again there is an elimination of quantifiers, provided that one augments the language by adding a symbol for $\leq$. The proof generalizes to real closed fields: see [37].

For the finite field $\mathbf{F}_p$ of p elements, the decidability results are obvious, since a Turing machine can simply loop over all possible values of the variables.

Tarski conjectured that the only fields with a decidable first order theory were the algebraically closed, real closed, and finite fields. This turned out to be false: Ax and Kochen [1] proved decidability for the field $\mathbf{Q}_p$ of p-adic numbers, which is the completion of **Q** with respect to the p-adic absolute value. (See [39] for the definition and basic properties of $\mathbf{Q}_p$.) They [2] gave several other examples of decidable fields. Macintyre [46] showed that there is an elimination of quantifiers for $\mathbf{Q}_p$ analogous to that for **R**.

It is not known whether Hilbert's 10th Problem over the field **Q** of rational numbers is decidable or not; see [52] for a survey. On the other hand, Robinson [67] proved the undecidability of the first order theory of **Q**, using the Hasse principle for quadratic forms. For a statement and proof of the latter, see [69, Chapter IV, §3, Theorem 8].

For the field $\mathbf{F}_p(t)$ of rational functions with coefficients in $\mathbf{F}_p$, Pheidas [60] proved the undecidability of Hilbert's 10th Problem, at least for $p \neq 2$. The $p = 2$ case was settled shortly thereafter by Videla [78]. The simpler problem of proving undecidability of the first order theory was done earlier, by Ershov [24] and Penzin [59] for $p \neq 2$ and $p = 2$, respectively.

Undecidability of Hilbert's 10th Problem itself (over the ring **Z** of integers) was proved by Matiyasevich [49]. The undecidability of the first order theory followed earlier from the fundamental work of Gödel [31]. Hilbert's 10th Problem for the ring of integers $\mathbf{Z}_K$ of a number field K is expected to be undecidable, but has been proved so only for certain K. For an up-to-date account of results in this direction, see [72]. For a survey of Hilbert's 10th Problem over commutative rings in general, see [61].

3 Rational Points on Varieties of Arbitrary Dimension

As discussed in the previous section, we do not know if there is an algorithm to decide in general whether a variety X over $\mathbf{Q}$ has a rational point. In order to pinpoint what is known and what is not, let us subdivide the problem according to the dimension of X. As usual, $X(\mathbf{Q})$ will denote the set of rational points of X, i.e., the set of rational solutions to the system of polynomials defining X.

dim X	$\exists$ algorithm to decide if $X(\mathbf{Q}) \neq \emptyset$?
0	YES
1	not known, but probably YES
≥ 2	?

If dim $X = 0$, then elimination theory lets us reduce to the case where X is a 0-dimensional subset of the affine line, and hence the problem becomes that of deciding whether a polynomial $f \in \mathbf{Q}[x]$ has a rational root. The latter can be done effectively, even in polynomial time [43].

The dim $X = 1$ case is the main subject of this article. Details follow in later sections.

For varieties of higher dimension, very little has been proved about $X(\mathbf{Q})$. On the other hand, below is a sample of some qualitative conjectures/questions that have been thrown around. All of these are known for dim $X \leq 1$, and for certain varieties of higher dimension. No counterexamples are known.

3.1 Bombieri, Lang (Independently)

Define the *special set* $S \subset X$ as the Zariski closure of the union of all positive dimensional images of morphisms of abelian varieties to X. Is it true that all but finitely many rational points of X lie in S?

The dim $X = 1$ case is equivalent to the Mordell conjecture [56], now Faltings' Theorem [26]: it states that a curve of genus at least 2 has at most finitely many rational points. Faltings [27] used diophantine approximation methods of Vojta to prove more generally that the answer is yes whenever X can be embedded in an abelian variety. For more conjectures along these lines, see [42, Chapter I, §3].

3.2   Colliot-Thélène and Sansuc

If X is smooth and projective, and X is birational to $\mathbf{P}^d$ over $\overline{\mathbf{Q}}$ (for instance, X could be a smooth cubic surface in $\mathbf{P}^3$), is the Brauer-Manin obstruction to the Hasse principle the only one? Actually, this was posed originally only for surfaces, as question (k1) in [25] but as Colliot-Thélène has pointed out, the answer could be yes in higher dimensions as well.

The Hasse principle is the statement that $X(\mathbf{Q}) \neq \emptyset$ if and only if $X(\mathbf{Q}_p) \neq \emptyset$ for all primes $p \leq \infty$. (By convention, $\mathbf{Q}_\infty = \mathbf{R}$.) This statement is proven for some varieties X (e.g., all degree 2 hypersurfaces in $\mathbf{P}^n$) and is known to be false for others (e.g., certain genus 1 curves). In 1970, Manin [48] discovered a possible obstruction to the Hasse principle coming from elements of the Brauer group of X, and he and others subsequently showed that this obstruction accounted for all violations of the Hasse principle known at the time. Much later, Skorobogatov [75] constructed an example of a surface X with no rational points, even though there was no Brauer-Manin obstruction; in other words, one could say that other nontrivial obstructions exist. Nevertheless, it is conceivable, and this is the point of the question of Colliot-Thélène and Sansuc, that for geometrically rational varieties, the nonexistence of a Brauer-Manin obstruction is a necessary and sufficient condition for the existence of rational points.

3.3   Mazur

Does the topological closure of $X(\mathbf{Q})$ in $X(\mathbf{R})$ have at most finitely many connected components?

See [51] and [53] for this and related questions. In [16] a counterexample is given to the following stronger version: If X is a smooth integral variety over $\mathbf{Q}$ such that $X(\mathbf{Q})$ is Zariski dense in X, then the topological closure of $X(\mathbf{Q})$ in $X(\mathbf{R})$ is a union of connected components of $X(\mathbf{R})$. The end of the paper [16] also suggests some other variants.

Even if the three questions above were answered tomorrow, we still would not know whether there exists an algorithm for deciding the existence of rational points on varieties.

4 Rational Points on Curves

For the rest of this article, we consider the problem of determining $X(\mathbf{Q})$ in the case dim $X = 1$, i.e., the case of curves. We begin with a few reductions, so that in the future we need consider only "nice" curves. Computational algebraic geometry provides algorithms for decomposing X into irreducible components over $\mathbf{Q}$ and over $\overline{\mathbf{Q}}$. Clearly then we can reduce to the case that X is irreducible over $\mathbf{Q}$. If X is irreducible over $\mathbf{Q}$, but not over $\overline{\mathbf{Q}}$, then the action of Galois acts transitively on the $\overline{\mathbf{Q}}$-irreducible components, but rational points are fixed by Galois, so $X(\mathbf{Q})$ is contained in the intersection of the $\overline{\mathbf{Q}}$-irreducible components; thus in this case we reduce to the 0-dimensional problem, which according to Section 3 is easily solved. Hence from now on, we will assume that X is geometrically integral.

Taking a projective closure and blowing up singularities changes X only by 0-dimensional sets, whose rational points we understand. Therefore, from now on, all our curves will be assumed to be smooth, projective, and geometrically integral. (Alternatively, at the expense of introducing nodes (violating smoothness), we could project X to a curve in $\mathbf{P}^2$ so that X would be described by a single polynomial equation. But in this article we prefer to talk about the smooth curves, except when presenting a curve explicitly, in which case we sometimes give an equation for an affine plane curve birational to the smooth projective curve that we are really interested in.)

The most important geometric invariant of a (smooth, projective, and geometrically integral) curve over $\mathbf{Q}$ is its genus g, which has several equivalent definitions:

(1) $g = \dim_\mathbf{Q} \Omega$ where Ω is the vector space of everywhere regular differentials on X. (Here regular means "no poles.") See [34, Chapter II, §8, p. 181] or [73, Chapter II, §5, p. 39] for more details.

(2) g is the topological genus of the compact Riemann surface $X(\mathbf{C})$.

(3) g is the dimension of the sheaf cohomology group $H^1(X, \mathcal{O}_X)$. See [34, Chapter III] for definitions.

(4) $g = \dfrac{(d-1)(d-2)}{2} -$ (terms for singularities), where Y is a (possibly singular) plane curve of degree d birational to X (e.g., the image of X under a sufficiently generic projection to $\mathbf{P}^2$). In this formula one subtracts a computable positive integer for each singularity. The integer depends on the complexity of the singularity: for nodes (ordinary double points), the integer is 1. See [34, Chapter V, Example 3.9.2, p. 393] for more details.

(The equivalence of these is certainly not obvious.)

Although the genus is a measure of geometric complexity, it has been discovered over the years that the geometry also influences the rational points. Hence we subdivide the problem of determining $X(\mathbf{Q})$ according to the genus. We summarize the situation in the following table:

genus g	$\exists$ algorithm to determine $X(\mathbf{Q})$?
0	YES
1	YES, if $\text{Ш}(\text{Jac } X)$ is finite
≥ 2	Not known, but probably YES

4.1 Genus Zero

In this case, one shows using the Riemann-Roch Theorem [34, Chapter IV, §1] that the anticanonical divisor class on X induces an embedding of X as a degree 2 curve in $\mathbf{P}^2$. In other words, X is isomorphic to a conic, the zero locus in $\mathbf{P}^2$ of an absolutely irreducible homogeneous polynomial $f \in \mathbf{Q}[x, y, z]$ of degree 2. By a linear change of variables, we may assume that f has the form $ax^2 + by^2 + cz^2$ for some nonzero $a, b, c \in \mathbf{Z}$. Conversely, nonsingular degree 2 curves in $\mathbf{P}^2$ are curves of genus zero, by the fourth definition of g above. Over an algebraically closed field, we could say further that X is isomorphic to the projective line $\mathbf{P}^1$, but this is not necessarily the case for genus zero curves over $\mathbf{Q}$: for instance, the conic defined by $x^2 + y^2 + z^2 = 0$ in $\mathbf{P}^2$ is not isomorphic to $\mathbf{P}^1$ over $\mathbf{Q}$, because the latter has rational points, whereas the former does not.

As mentioned in Section 3.2, degree 2 curves in $\mathbf{P}^2$ (and more generally degree 2 hypersurfaces in $\mathbf{P}^2$) satisfy the Hasse principle, so we can decide the existence of a rational point by checking the existence of a $\mathbf{Q}_p$-point for each prime $p \leq \infty$. And the latter is in fact a finite problem since one can show a priori that $ax^2 + by^2 + cz^2 = 0$ has $\mathbf{Q}_p$-points for p not dividing $2abc$, and for each of the finitely many remaining p (including ∞) one has an algorithm for deciding the existence of a $\mathbf{Q}_p$-point, as explained in Section 2. For a more explicit criterion, see [57], for instance.

In the case where X does have a rational point P, there is an isomorphism $\mathbf{P}^1 \to X$ defined as follows: thinking of $\mathbf{P}^1$ as the set of lines in $\mathbf{P}^2$ through P, map the line L to the point $Q \in L \cap X$ not equal to P. (If L is the tangent line to X at P, take $Q = P$.) Hence we obtain an explicit parameterization of $X(\mathbf{Q})$.

4.2 Genus One

A genus one curve over $\mathbf{Q}$ with a rational point is called an *elliptic curve over* $\mathbf{Q}$. One shows using the Riemann-Roch Theorem that an elliptic curve E over $\mathbf{Q}$ is isomorphic to the projective closure of the affine curve $y^2 = x^3 + Ax + B$ for some $A, B \in \mathbf{Z}$ with $4A^3 + 27B^2 \neq 0$. More importantly, it can be shown that E can be given the structure of an algebraic group [73, Chapter 3]. Roughly, this means that there are rational functions that induce a group structure on the set $E(k)$ for any field k containing $\mathbf{Q}$. The Mordell-Weil Theorem[7] states that the set $E(\mathbf{Q})$ of rational points on an elliptic curve form a finitely generated abelian group [73, Chapter 8]. The group $E(\mathbf{Q})$ is called the Mordell-Weil group, and its rank is called the Mordell-Weil rank or simply the rank of E. The torsion subgroup of $E(\mathbf{Q})$ is easy to compute. But the equivalent problems of determining the rank of $E(\mathbf{Q})$ and of determining a list of generators have not yet been solved. There is a proposed method, "descent," for solving these problems, which is a generalization of the infinite descent method used by Fermat. Descent usually works well in practice when A and B are not too large (see [19]), but its success in general relies upon the conjecture that the Shafarevich-Tate group $\text{III}(E)$, a certain abelian group associated to E, is finite, or at least that the p-primary part of $\text{III}(E)$ is finite for some prime p. Using the modularity of elliptic curves over $\mathbf{Q}$ [7] and the work of Kolyvagin [40] supplemented by [12] or [58], we know $\#\text{III}(E) < \infty$ for infinitely many E, namely, those for which $\text{ord}_{s=1} L_E(s) \leq 1$, where $L_E(s)$ is the L-function of E. (See [73, Appendix C, §16] for a definition of $L_E(s)$.)

A general genus one curve X over $\mathbf{Q}$ need not have a rational point. In fact, Lind [44] and Reichardt [64] independently discovered examples where X does not satisfy the Hasse principle. Explicitly, the smooth projective models of the affine curves $2y^2 = 1 - 17x^4$ and $3x^3 + 4y^3 = 5$ (from [64] and [68], respectively) are curves having $\mathbf{Q}_p$-points for all $p \leq \infty$, but no rational points. To any genus one curve X one can associate an elliptic curve E, namely the *Jacobian* of X. The Jacobian $\text{Jac}\,X$ of a curve X of genus g is an abelian variety (irreducible projective algebraic group) of dimension g, whose geometric points correspond to elements of $\text{Pic}^0(X_{\overline{\mathbf{Q}}})$, i.e., to divisor classes of degree zero on $X_{\overline{\mathbf{Q}}}$. (Note: $X_{\overline{\mathbf{Q}}}$ denotes the same variety as X except where the defining polynomials are viewed as having coefficients in $\overline{\mathbf{Q}}$, even though the coefficients actually are in the subfield $\mathbf{Q}$. See [73, Chapter 2] for a definition of Pic^0, and see [55] for details about Jacobians.) In the case where $g = 1$, X is a *principal homogeneous*

---

[7] Actually Mordell alone proved this fact. Weil generalized Mordell's result in two directions: elliptic curves were replaced by abelian varieties of arbitrary dimension, and $\mathbf{Q}$ was replaced by an arbitrary number field.

space [73, Chapter 10], or *torsor*, of its Jacobian E. This means that there is an isomorphism $E_{\overline{\mathbf{Q}}} \simeq X_{\overline{\mathbf{Q}}}$ over $\overline{\mathbf{Q}}$, and a morphism of varieties $E \times X \to X$ over $\mathbf{Q}$, which when considered over $\overline{\mathbf{Q}}$ becomes equivalent (after identifying $X_{\overline{\mathbf{Q}}}$ with $E_{\overline{\mathbf{Q}}}$) to the addition morphism $E_{\overline{\mathbf{Q}}} \times E_{\overline{\mathbf{Q}}} \to E_{\overline{\mathbf{Q}}}$. Then $\text{III}(E)$ is defined as the set of principal homogeneous spaces for E that have $\mathbf{Q}_p$-points for all $p \leq \infty$, up to isomorphism as principal homogeneous spaces of E over $\mathbf{Q}$. It is known that if $\text{III}(E)$ is finite, then in principle, there is an algorithm for determining whether $X(\mathbf{Q})$ is nonempty: if $X(\mathbf{Q})$ is nonempty, a rational point can be found by search; if $X(\mathbf{Q}_p) = \emptyset$ for some $p \leq \infty$, then $X(\mathbf{Q}) = \emptyset$; if neither holds, then X represents a nonzero element of $\text{III}(E)$, and this can proved by finding another element Y of $\text{III}(E)$ such that the Cassels-Tate pairing of X and Y is nonzero in $\mathbf{Q}/\mathbf{Z}$. See [63] for some definitions of the Cassels-Tate pairing.

4.3   Genus At Least Two

Let X be a curve over $\mathbf{Q}$ of genus at least 2. Mordell [56] conjectured in 1922 that $X(\mathbf{Q})$ is finite, and this was finally proved in 1983 by Faltings [26]. A new proof based on diophantine approximation was found by Vojta [79]. Simplifications of Vojta's argument were found by Faltings and by Bombieri, who presented a relatively elementary proof in [5].

These proofs let one calculate a bound on the *number* of rational points on X given the equations defining X. But they are ineffective in that they do not provide an upper bound on the numerators and denominators of the coordinates of the rational points, so they cannot be used to determine $X(\mathbf{Q})$ rigorously. They are unable to decide even whether $X(\mathbf{Q})$ is empty.

Ironically, certain other methods (mostly older), which so far have failed to prove the Mordell conjecture in full generality, are the ones that have succeeded in determining $X(\mathbf{Q})$ in many examples. In the next section, we discuss these methods, and the ways in which they have been developed recently into practical algorithms.

5   Methods for Curves of Genus At Least Two

We use the following brief names to refer to the various methods that are used to determine $X(\mathbf{Q})$ for a curve X over $\mathbf{Q}$ of genus at least 2:

    (1) Local points
    (2) Dem'yanenko-Manin
    (3) Chabauty
    (4) Going-down
    (5) Going-up (Chevalley-Weil)

The last two are transitional in the sense that they by themselves do not determine $X(\mathbf{Q})$ directly, but instead reduce the problem of determining $X(\mathbf{Q})$ to the problem of determining the rational points on certain auxiliary varieties. In addition to the methods listed, there is a method based on the modularity of elliptic curves, and another method called "elliptic Chabauty." We will discuss these too, but in fact they can be interpreted as combinations of the methods already listed.

5.1   Local Points

This method attempts to prove that $X(\mathbf{Q})$ is empty without much work, by showing that $X(\mathbf{Q}_p)$ is empty for some $p \le \infty$.

For a curve X over $\mathbf{Q}$ of any genus g, it is possible to compute a finite set S of primes such that $X(\mathbf{Q}_p) \ne \emptyset$ for all $p \notin S$. (Because of Hensel's lemma, S can be taken as the set of primes of bad reduction, together with ∞ and the primes p for which the Weil lower bound $p + 1 - 2g\sqrt{p}$ for $X(\mathbf{F}_p)$ is nonpositive.) Then for each $p \in S$, it is possible to check whether $X(\mathbf{Q}_p)$ is empty, since according to Section 2 this problem is decidable for fixed p.

If we find p for which $X(\mathbf{Q}_p)$ is empty, then we know that $X(\mathbf{Q})$ is empty, and we are done. Otherwise we have learned nothing: the Hasse principle, the statement that $X(\mathbf{Q}_p) \ne \emptyset$ for all p implies $X(\mathbf{Q}) \ne \emptyset$, often fails for curves of positive genus.

5.2   Dem'yanenko-Manin

This method applies to certain special curves X. If A is an abelian variety, the group structure on A induces a group structure on the set of morphisms $X \to A$ over $\mathbf{Q}$. Let $\mathrm{Mor}(X, A)$ denote the quotient of this group by the subgroup of constant morphisms. If there is an abelian variety A over $\mathbf{Q}$ such that we can prove $\mathrm{rank}\,\mathrm{Mor}(X, A) > \mathrm{rank}\,A(\mathbf{Q})$, then the Dem'yanenko-Manin method [23], [47] provides an explicit upper bound on the sizes of the numerators and denominators of the coordinates of the rational points on X, so that $X(\mathbf{Q})$ can be computed by a finite search. Note that it is not necessary to know $\mathrm{rank}\,A(\mathbf{Q})$ exactly. For $\mathbf{Q}$-simple abelian varieties A, the needed inequality is equivalent to the condition that A^m appears in the decomposition of $\mathrm{Jac}\,X$ into $\mathbf{Q}$-simple abelian varieties up to isogeny, with

$$m > \frac{\mathrm{rank}\,A(\mathbf{Q})}{\mathrm{rank}\,\mathrm{End}_{\mathbf{Q}}\,A},$$

where $\mathrm{End}_{\mathbf{Q}}\,A$ denotes the ring of endomorphisms of A that are defined over $\mathbf{Q}$. For a fuller exposition of this method, see [70], and for some explicit applications of it; see [74], [41], and [33]. Its main disadvantage is that the condition necessary for its application fails for most curves.

5.3 Chabauty

This is a method based on p-adic geometry. Suppose that X embeds in an abelian variety A such that $\operatorname{rank} A(\mathbf{Q}) < \dim A$. Suppose also that X generates A in the sense that the differences of points $P - Q$ of $X(\overline{\mathbf{Q}})$ generate the group $A(\overline{\mathbf{Q}})$. For example, A might be $\operatorname{Jac} X$, in which case the condition becomes $\operatorname{rank} A(\mathbf{Q}) < g$, where g is the genus of X. Then the p-adic closure $\overline{A(\mathbf{Q})}$ of $A(\mathbf{Q})$ in $A(\mathbf{Q}_p)$ can be shown to be an "analytic subvariety" of dimension at most $\operatorname{rank} A(\mathbf{Q})$; as a topological group, it is an extension of a finite abelian group by a free $\mathbf{Z}_p$-module of finite rank. The inequality hypothesis guarantees that $\overline{A(\mathbf{Q})}$ has positive codimension in $A(\mathbf{Q}_p)$. Hence by dimension counting, one expects that $X(\mathbf{Q}_p) \cap \overline{A(\mathbf{Q})}$ is at most a zero-dimensional closed subset of the compact group $A(\mathbf{Q}_p)$, hence finite. Chabauty [14] proved this finiteness statement. But $X(\mathbf{Q}) \subseteq X(\mathbf{Q}_p) \cap \overline{A(\mathbf{Q})}$, and by computing the intersection to a given p-adic precision, one can bound the number of rational points on $X(\mathbf{Q})$, and obtain p-adic approximations to their possible locations. Coleman [15] gave an explicit upper bound on the size of this intersection. The intersection can be computed to some p-adic precision either by working with the formal group of J, or by looking at the p-adic integrals of regular differentials on X. The latter seems to be easier, especially for higher genus curves.

Unfortunately, $X(\mathbf{Q})$ may be strictly smaller than $X(\mathbf{Q}_p) \cap \overline{A(\mathbf{Q})}$, although heuristically this may be rare when $\operatorname{rank} A(\mathbf{Q}) \le \dim A - 2$: in that case the naive dimension count suggests that the intersection is empty so perhaps, if there are points in the intersection, they are there for a reason! Because of the possibility that $X(\mathbf{Q}) \ne X(\mathbf{Q}_p) \cap \overline{A(\mathbf{Q})}$, the condition $\operatorname{rank} A(\mathbf{Q}) < \dim A$ alone is not sufficient for success. In practice, however, Chabauty's method has proved successful, especially in conjunction with the transitional methods to be discussed below. See for example, [32], [54], [28], [3], and [45].

5.4 Going-Down

If X admits a nonconstant morphism $X \to Y$ over $\mathbf{Q}$ to another variety Y over $\mathbf{Q}$, where $Y(\mathbf{Q})$ is finite and computable, then one can determine $X(\mathbf{Q})$, since it suffices to examine the finitely many points in $X(\overline{\mathbf{Q}})$ mapping to the points in $Y(\mathbf{Q})$: every point in $X(\mathbf{Q})$ must map to some point in $Y(\mathbf{Q})$. In practice, Y is usually an abelian variety with $Y(\mathbf{Q})$ finite, or Y is another curve.

5.5 Going-Up

If $f: Y \to X$ is an *unramified* morphism of curves over $\mathbf{Q}$, Chevalley and Weil proved that there is a computable finite extension k of $\mathbf{Q}$ such that $f^{-1}(X(\mathbf{Q})) \subseteq Y(k)$. If X has genus at least 2, then Y does too, so $Y(k)$ is known to be finite. Hence one can reduce the problem of computing $X(\mathbf{Q})$ to that of computing $Y(k)$.

One difficulty with this method is that k may be much larger than $\mathbf{Q}$. Fortunately, Coombes and Grant [17] and Wetherell [80] found variants of the method that instead gave a finite set of unramified covering curves $Y_i \to X$ over $\mathbf{Q}$, $1 \leq i \leq n$, all isomorphic over $\overline{\mathbf{Q}}$, such that $X(\mathbf{Q}) \subseteq \bigcup_{i=1}^{n} f_i(Y_i(\mathbf{Q}))$. Given such a covering collection, one can determine $X(\mathbf{Q})$ if one can determine $Y_i(\mathbf{Q})$ for all i. Specialized to the case where X is an elliptic curve, this becomes the method of descent used to prove the Mordell-Weil Theorem.

Let us give one example of this method, taken from [8]. Suppose that we want to find $X(\mathbf{Q})$, where X is the curve of genus 2 that is the smooth projective model of the affine hyperelliptic curve $y^2 = 6x(x^4 + 12)$. (This is one of the curves that comes up when one studies the integer solutions to the equation $x^8 + y^3 = z^2$.) Suppose we have an affine point $(x_0, y_0) \in X(\mathbf{Q})$ with $x_0, y_0 \neq 0$. If we write $x_0 = X/Z$ in lowest terms with $X, Z \in \mathbf{Z}$, $Z \neq 0$, we obtain $y_0^2 Z^6 = 6XZ(X^4 + 12Z^4)$. Setting $Y = y_0 Z^3$, which must be an integer, since its square equals the right hand side, we obtain $Y^2 = 6XZ(X^4 + 12Z^4)$. If a prime $p \geq 5$ divides both $6XZ$ and $X^4 + 12Z^4$, then it divides either X or Z, and then in order to divide $X^4 + 12Z^4$ it must divide *both* X and Z, contradicting the assumption that X/Z is in lowest terms. (More generally, one would argue using primes not dividing the resultant of the two homogeneous polynomial factors.) Thus each prime $p \geq 5$ divides at most one of $6XZ$ and $X^4 + 12Z^4$. But their product is a square, so if p divides one of $6XZ$ and $X^4 + 12Z^4$, the exponent of p in that factor is even. Since this holds for all $p \geq 5$, we have $X^4 + 12Z^4 = \delta W^2$ for some $\delta \in \{\pm 1, \pm 2, \pm 3, \pm 6\}$ and $W \in \mathbf{Z}$. Dividing by Z^4, we obtain a rational point (u, v) on the curve $E_\delta: \delta v^2 = u^4 + 12$. We may assume $\delta > 0$ (since otherwise $E_\delta(\mathbf{R}) = \emptyset$) and $2|\delta$ (since otherwise $E_\delta(\mathbf{Q}_2) = \emptyset$). Therefore we need only search for rational points on

$$E_1: v^2 = u^4 + 12, \quad \text{and} \quad E_3: 3v^2 = u^4 + 12.$$

These are curves of genus one, and with a little work, using descent, one can show that $E_1(\mathbf{Q})$ and $E_3(\mathbf{Q})$ are finite, each of size 2, counting points on the nonsingular projective models. Checking to see what points on X these give rise to, we find that $X(\mathbf{Q})$ consists of $(0,0)$ and a point at infinity on the projective model.

Geometrically what has happened here is that for each $\delta \in \mathbf{Q}^*$, the genus 3 curve Y_δ defined by the *system* of equations

$$y^2 = 6x(x^4 + 12), \quad \delta z^2 = x^4 + 12$$

in (x, y, z)-space maps to X via $(x, y, z) \mapsto (x, y)$, and Y_δ is an unramified cover of X. Moreover, the union of the images of $Y_\delta(\mathbf{Q})$ in X equals $X(\mathbf{Q})$. Since the isomorphism class of Y_δ depends only on the image of δ in $\mathbf{Q}^*/\mathbf{Q}^{*2}$, we may assume that $\delta \in \mathbf{Z}$ is nonzero and squarefree. If a prime $p \geq 5$ divides δ, then $Y_\delta(\mathbf{Q}_p) = \emptyset$, so we may discard Y_δ. By also demanding the existence of local points over $\mathbf{R}$ and $\mathbf{Q}_2$, we may discard all but Y_1 and Y_3. (As it turns out, $\mathbf{Q}_3$ gives no further restriction.) Finally, $(x, y, z) \mapsto (x, z)$ gives a nonconstant morphism $Y_\delta \to E_\delta$, so by "going down" it suffices to find $E_1(\mathbf{Q})$ and $E_3(\mathbf{Q})$, if these are finite. We are lucky: both E_1 and E_3 turn out to be elliptic curves of rank zero.

In general, covering collections by geometrically *abelian* covers are described by geometric class field theory. They arise as follows. Suppose that X is embedded in its Jacobian J using a basepoint $P_0 \in X(\mathbf{Q})$. Choose an isogeny $\phi \colon A \to J$, i.e., a surjective homomorphism between abelian varieties with finite kernel. Choose a representative $R \in J(\mathbf{Q})$ of each element of $J(\mathbf{Q})/\phi(A(\mathbf{Q}))$, let $\phi_R \colon A \to J$ be the composition ϕ followed by translation-by-R on J, and let Y_R denote the following fiber product:

$$
\begin{array}{ccc}
Y_R & \longrightarrow & X \\
\downarrow & & \downarrow \\
A & \xrightarrow{\ \phi_R\ } & J.
\end{array}
$$

In other words, Y_R is the inverse image $\phi_R^{-1}(X)$ in A. Then the Y_R form a finite set of unramified covers of X, and the union of the images of $Y_R(\mathbf{Q})$ in X equals $X(\mathbf{Q})$, since $\bigcup_R \phi_R(A(\mathbf{Q})) = J(\mathbf{Q})$.

Note that given X and hence J, there are many pairs (A, ϕ) where $\phi \colon A \to J$ is an isogeny over $\mathbf{Q}$: if nothing else is available, one can take ϕ as the multiplication-by-m map $[m] \colon J \to J$ for some $m \geq 2$. Hence going up is always possible. On the other hand, for $\phi = [m]$, the genus of each Y_R equals $m^{2g}(g - 1) + 1$ by the Riemann-Hurwitz formula [34, IV.2.4] so if m is too large, the Y_R may be difficult to work with. In fact, it might be preferable to use an isogeny ϕ of degree lower than $\deg[2] = 2^{2g}$ if one exists.

5.6 Modularity

Through the work of Wiles, Taylor, Breuil, Conrad, and Diamond [81], [77], [7], it is now known that every elliptic curve E over $\mathbf{Q}$ is modular, meaning that there exists a nonconstant morphism from the modular curve $X_0(N)$ to E, for some integer $N \geq 1$. See [42, Chapter V] for an introduction to this concept, and see [71] and [38] for more details on modular functions and curves. On the other hand, work of Frey, Serre, and Ribet [65] showed that a nontrivial rational point on $x^p + y^p = 1$ for prime $p > 2$ would give rise to an elliptic curve over $\mathbf{Q}$ that could not be modular. Together, these results proved Fermat's Last Theorem. For an overview of the whole proof, see [20] or the book [18]. In the past few years, methods based on modularity have been adapted to solve certain other Fermat-like diophantine equations. See [22] and [66], for instance.

Darmon has pointed out that these proofs can be interpreted as instances of the going-up method. Recall that the geometric class field theory construction at the end of Section 5.5 produces only unramified covers Y that when considered over $\overline{\mathbf{Q}}$ are Galois and abelian over the base curve X. These abelian covers are the easiest unramified covers to work with, but they form only a small subset of *all* the unramified covers. Kummer's partial results on Fermat's Last Theorem can be reinterpreted as a study of the abelian covers of the Fermat curve $X : x^p + y^p = 1$ with Galois group $\mathbf{Z}/p$. The modularity proof of Fermat's Last Theorem is equivalent to a study of certain unramified covers with nonabelian Galois group. More precisely, the map $(x, y) \mapsto x^p$ from $X \to \mathbf{P}^1$ exhibits X as a cover of $\mathbf{P}^1$ ramified above $\{0, 1, \infty\}$, and the latter can be thought of as the modular curve $X(2)$ with its three cusps. The fiber product

$$
\begin{array}{ccc}
Y & \longrightarrow & X \\
\downarrow & & \downarrow \\
X(2p) & \longrightarrow & X(2).
\end{array}
$$

is an unramified cover Y of X with Galois group equal to that of $X(2p)$ over $X(2)$, namely the nonabelian group $\mathrm{PSL}_2(\mathbf{F}_p)$. One can then study the rational points on Y and its relevant twists by going down to $X(2p)$ and its twists. Thus one reduces to questions about rational points on modular curves, to which one can apply the work of Mazur.

5.7 Elliptic Chabauty

This is not so much a separate method as it is a clever way to combine the going-up and Chabauty methods. It was discovered independently by

Bruin [9] and by Flynn and Wetherell [29].

Suppose that X is a curve of genus 2 over $\mathbf{Q}$ embedded in its Jacobian J_X using a basepoint $P_0 \in X(\mathbf{Q})$. Consider the curves Y obtained as in Section 5.5 by pulling back X under the multiplication-by-2 isogeny $J_X \to J_X$, or its translates. Then Y is an unramified cover of X, and when considered over $\overline{\mathbf{Q}}$, it is an abelian cover with Galois group $J_X[2](\overline{\mathbf{Q}}) \simeq (\mathbf{Z}/2)^4$. As mentioned in Section 5.5, the Riemann-Hurwitz formula shows that the genus of Y is 17. By Galois theory, there are 15 intermediate covers Z_i of $X_{\overline{\mathbf{Q}}}$ of degree 2, and these have genus 3. Hence each Jacobian J_{Z_i} is isogenous over $\overline{\mathbf{Q}}$ to $J_X \times E_i$ for some elliptic curve E_i, and it follows that J_Y is isogenous to $J_X \times A$ where A is a 15-dimensional abelian variety isomorphic to $\prod_{i=1}^{15} E_i$ over $\overline{\mathbf{Q}}$. More precisely, one can show that one can take A to the Weil restriction of scalars $\mathrm{Res}_{K/\mathbf{Q}} E$ of an elliptic curve E over the 15-dimensional $\mathbf{Q}$-algebra K of global sections of the structure sheaf on $J_X[2] - \{0\}$. (See [6, Section 7.6] for the definition of the Weil restriction of scalars.) More concretely, K is the product of the fields of definition of representatives for the Galois orbits of nontrivial 2-torsion points of J; often there is just one orbit and K is a number field of degree 15 over $\mathbf{Q}$. One then can attempt to apply Chabauty to $Y \to A$ for each Y.

Whereas success of the direct Chabauty method required rank $J_X(\mathbf{Q}) < 2$, elliptic Chabauty requires the (independent?) condition rank $A(\mathbf{Q}) < \dim A$ for each A that arises. One of the properties of restriction of scalars is that $A(\mathbf{Q}) \simeq E(K)$, and we also know $\dim A = 15$, so the latter condition is equivalent to rank $E(K) < 15$. Perhaps this is more likely than rank $J_X(\mathbf{Q}) < 2$. In the worst case, when K is a number field and each nonzero 2-torsion point of E is defined only over a cubic extension of K, we apparently need to compute the 2-part of the class group of a degree 45 number field to complete the descent to compute rank $E(K)$. But in favorable cases, the number fields are much smaller, and the method is practical.

Elliptic Chabauty has the advantage over the original Chabauty method that one can do all the computations with the group law of E over K, instead of a Jacobian over $\mathbf{Q}$. The former is usually easier from the computational point of view: as Bruin says, "simple geometry over a field with complicated arithmetic is to be preferred over complicated geometry over a field with simple arithmetic."

6   The Mordell-Weil Race

If faced with the problem of determining $X(\mathbf{Q})$ for a specific curve X over $\mathbf{Q}$ of genus $g \geq 2$, it is probably best first to try the method of local points. If this fails, next one can try to see if X admits a nonconstant morphism to a curve Y where Y has genus at least 2, or where Y is a curve of genus 1 for which $Y(\mathbf{Q})$ is finite and can be determined. If not, one can attempt the Dem'yanenko-Manin method, and the method of Chabauty.

If all of these fail, one can go up to a set of unramified covering curves, and then recursively apply the above methods to each of these.

It seems plausible that iteration of going-up and Chabauty alone are sufficient to resolve $X(\mathbf{Q})$ for every curve X of genus $g \geq 2$ over $\mathbf{Q}$! Starting from X, one replaces the problem on X with the problem on a finite set of covering curves Y_R. For each Y_R whose rational points are not resolved by Chabauty, one must replace Y_R by a finite set of its covering curves, and so on. We obtain a tree and hope that all the branches will eventually be terminated by Chabauty. If one applies Chabauty to the Jacobians alone, the issue is whether the genus eventually outpaces the rank of the Jacobian as one goes up along any branch of the tree. If so, Chabauty is likely to succeed in terminating all the branches.

In practice, one can apply Chabauty to morphisms from the curve into *quotients* of its Jacobian at each node of the tree, as in the elliptic Chabauty method of Section 5.7. This is preferable especially if the original X has large rank: if Y covers X, the Jacobian J_Y is isogenous over $\mathbf{Q}$ to $J_X \times A$ for some abelian variety A over $\mathbf{Q}$, so $\operatorname{rank} J_Y(\mathbf{Q}) = \operatorname{rank} J_X(\mathbf{Q}) + \operatorname{rank} A(\mathbf{Q})$, which will still be large, if $\operatorname{rank} J_X(\mathbf{Q})$ was large to begin with. On the other hand, there seems to be no direct correlation between $\operatorname{rank} A(\mathbf{Q})$ and $\operatorname{rank} J_X(\mathbf{Q})$.

Unfortunately, virtually nothing is known about the growth of Mordell-Weil ranks as one ascends an unramified tower of curves, and hence it seems impossible to prove anything about the success of this tree method. All we have now are a few isolated examples in which the method has been successful. For a genus g curve X, is $\operatorname{rank} J_X(\mathbf{Q})$ typically of size around g? Or is it typically $O(1)$ as $g \to \infty$? It seems difficult even to find a heuristic that predicts the answers. Perhaps analytic methods generalizing [11] will provide hints.

Even if the truth is that the going-up and Chabauty methods are *not* always enough to determine $X(\mathbf{Q})$, there are several other conjectural approaches towards an effective algorithm. See Section F.4.2 of [35] for a survey of some of these.

7 Some Success Stories

7.1 Diophantus

About 1700 years ago, Diophantus challenged his readers to find a solution to

$$y^2 = x^8 + x^4 + x^2$$

in positive rational numbers. This is not hard, but these days one wants to know *all* the solutions. Wetherell [80] combined going-up, going-down, and Chabauty to prove that $(1/2, 9/16)$ is the only positive rational solution.

7.2 Serre

Over 15 years ago, Serre challenged the mathematical community to find the rational points on $x^4 + y^4 = 17$, a genus 3 curve whose Jacobian is isogenous to $E \times E \times E'$ where E, E' are elliptic curves over $\mathbf{Q}$, each of rank 2. This past year, Flynn and Wetherell [30] found suitable unramified covers and applied a version of elliptic Chabauty to them to prove that the only rational points are the obvious eight with

$$\{|x|, |y|\} = \{1, 2\}.$$

7.3 Generalized Fermat

Work of Beukers [4], and of Darmon and Granville [21] reduces solving $x^p + y^q = z^r$ in relatively prime integers x, y, z for fixed $p, q, r > 1$ to computing $X(\mathbf{Q})$ for a finite set of curves X over $\mathbf{Q}$. As has already been mentioned, methods based on modularity of elliptic curves solve the equation for many (p, q, r). Some of the small exponent cases, too small for modularity methods to work easily, are worked out in [9], [8], [10], and [62]. For example, [8] applies elliptic Chabauty and other methods to prove that the only solutions to

$$x^8 + y^3 = z^2$$

in nonzero relatively prime integers are

$$(\pm 1, 2, \pm 3) \quad \text{and} \quad (\pm 43, 96222, \pm 30042907).$$

References

[1]   J. Ax and S. Kochen, *Diophantine problems over local fields. II. A complete set of axioms for p-adic number theory*, Amer. J. Math. **87** (1965), 631–648.

[2]   ———, *Diophantine problems over local fields. III. Decidable fields*, Annals of Math. **83** (1966), 437–456.

[3]   M. H. Baker, *Kamienny's criterion and the method of Coleman and Chabauty*, Proc. Amer. Math. Soc. **127** (1999), 2851–2856.

[4] F. Beukers, *The Diophantine equation $Ax^p + By^q = Cz^r$*, Duke Math. J. **91** (1998), 61–88.

[5] E. Bombieri, *The Mordell conjecture revisited*, Ann. Scuola Norm. Sup. Pisa Cl. Sci. (4) **17** (1990), 615–640, Errata-corrige: ibid. **18** (1991), 473.

[6] S. Bosch, W. Lütkebohmert, and M. Raynaud, *Néron models*, Springer-Verlag, Berlin, 1990.

[7] C. Breuil, B. Conrad, F. Diamond, and R. Taylor, *On the modularity of elliptic curves over* **Q**, J. Amer. Math. Soc. **14** (2001), 843–939.

[8] N. Bruin, *Chabauty methods using covers on curves of genus 2*, preprint, 1999.

[9] _____, *Chabauty methods using elliptic curves*, preprint, 1999.

[10] _____, *The diophantine equations $x^2 \pm y^4 = \pm z^6$ and $x^2 + y^8 = z^3$*, Compositio Math. **118** (1999), 305–321.

[11] A. Brumer, *The average rank of elliptic curves. I*, Invent. Math. **109** (1992), 445–472.

[12] D. Bump, S. Friedberg, and J. Hoffstein, *Nonvanishing theorems for L-functions of modular forms and their derivatives*, Invent. Math. **102** (1990), 543–618.

[13] H. Cartan and C. Chevalley, *Géometrie algébrique*, Séminaire Cartan-Chevalley (Paris), Secrétariat Math., 1955–1956.

[14] C. Chabauty, *Sur les points rationnels des courbes algébriques de genre supérieur à l'unité*, C.R. Acad. Sci. Paris **212** (1941), 882–885.

[15] R. F. Coleman, *Effective Chabauty*, Duke Math. J. **52** (1985), 765–780.

[16] J.-L. Colliot-Thélène, A. N. Skorobogatov, and P. Swinnerton-Dyer, *Double fibres and double covers: paucity of rational points*, Acta Arith. **79** (1997), 113–135.

[17] K. R. Coombes and D. R. Grant, *On heterogeneous spaces*, J. London Math. Soc. (2) **40** (1989), 385–397.

[18] G. Cornell, J. H. Silverman, and G. Stevens, eds., *Modular forms and Fermat's last theorem*, Springer-Verlag, New York, 1997, Papers from the Instructional Conference on Number Theory and Arithmetic Geometry held at Boston University, Boston, MA, August 9–18, 1995.

[19] J. E. Cremona, *Algorithms for modular elliptic curves*, second ed., Cambridge University Press, Cambridge, 1997.

[20] H. Darmon, F. Diamond, and R. Taylor, *Fermat's last theorem*, Elliptic curves, modular forms, & Fermat's last theorem (Hong Kong, 1993) (John Coates and S. T. Yau, eds.), International Press, Cambridge, MA, second ed., 1997, pp. 2–140.

[21] H. Darmon and A. Granville, *On the equations $z^m = F(x, y)$ and $Ax^p + By^q = Cz^r$*, Bull. London Math. Soc. **27** (1995), 513–543.

[22] H. Darmon and L. Merel, *Winding quotients and some variants of Fermat's last theorem*, J. Reine Angew. Math. **490** (1997), 81–100.

[23] V. Dem'yanenko, *Rational points on a class of algebraic curves*, Amer. Math. Soc. Transl. **66** (1968), 246–272.

[24] Yu. Ershov, *Undecidability of certain fields (Russian)*, Dokl. Akad. Nauk SSSR **161** (1965), 349–352.

[25] J.-L. Colliot-Thélène et J.-J. Sansuc, *La descente sur les variétés rationnelles*, Journées de Géométrie Algébrique d'Angers (Angers, 1979), Sijthoff & Noordhoff, Alphen aan den Rijn, 1980, pp. 223–237.

[26] G. Faltings, *Endlichkeitssätze für abelsche Varietäten über Zahlkörpern*, Invent. Math. **73** (1983), 349–366, English translation: Arithmetic geometry (G. Cornell and J. Silverman, eds.), Springer-Verlag, New York-Berlin, 1986, pp. 2–27.

[27] ———, *The general case of S. Lang's conjecture*, Barsotti Symposium in Algebraic Geometry (Abano Terme, 1991), Perspect. Math., vol. 15, Academic Press, San Diego, CA, 1994, pp. 175–182.

[28] E. V. Flynn, *A flexible method for applying Chabauty's theorem*, Compositio Math. **105** (1997), 79–94.

[29] E. V. Flynn and J. L. Wetherell, *Finding rational points on bielliptic genus 2 curves*, Manuscripta Math. **100** (1999), 519–533.

[30] ———, *Covering collections and a challenge problem of Serre*, Acta Arith. **98** (2001), 197–205.

[31] K. Gödel, *Über formal unentscheidbare Sätze der Principia Mathematica und verwandter System I*, Monatshefte für Math. und Physik **38** (1931), 173–198, English translation by Elliot Mendelson: "On formally undecidable propositions of Principia Mathematica and related systems I" in M. Davis, The undecidable, Raven Press, 1965.

[32] D. Grant, *A curve for which Coleman's effective Chabauty bound is sharp*, Proc. Amer. Math. Soc. **122** (1994), 317–319.

[33] G. Grigorov and J. Rizov, *Heights on elliptic curves and the equation* $x^4 + y^4 = cz^4$, preprint, 1998.

[34] R. Hartshorne, *Algebraic geometry*, Graduate Texts in Mathematics, vol. 52, Springer-Verlag, New York-Heidelberg, 1977.

[35] M. Hindry and J. H. Silverman, *Diophantine geometry. An introduction*, Graduate Texts in Mathematics, vol. 201, Springer-Verlag, New York, 2000.

[36] J. E. Hopcroft and J. D. Ullman, *Formal languages and their relation to automata*, Addison-Wesley Publishing Co., Reading, Mass, 1969.

[37] N. Jacobson, *Basic algebra I*, second ed., W. H. Freeman and Company, New York, 1985.

[38] N. M. Katz and B. Mazur, *Arithmetic moduli of elliptic curves*, Annals of Mathematics Studies, vol. 108, Princeton University Press, Princeton, N.J., 1985.

[39] N. Koblitz, *p-adic numbers, p-adic analysis, and zeta-functions*, Graduate Texts in Mathematics, vol. 58, Springer-Verlag, New York-Berlin, second ed., 1984.

[40] V. Kolyvagin, *Finiteness of $E(\mathbf{Q})$ and $\text{III}(E, \mathbf{Q})$ for a subclass of Weil curves (Russian)*, Izv. Akad. Nauk SSSR Ser. Mat. **52** (1988), 522–540, 670–671, English translation: Math. USSR-Izv. **32** (1989), no. 3, 523–541.

[41] L. Kulesz, *Application de la méthode de Dem'janenko-Manin à certaines familles de courbes de genre 2 et 3*, J. Number Theory **76** (1999), 130–146.

[42] S. Lang, *Number theory. III. Diophantine geometry*, Encyclopaedia of Mathematical Sciences, vol. 60, Springer-Verlag, Berlin, 1991.

[43] A. K. Lenstra, H. W. Lenstra Jr., and L. Lovász, *Factoring polynomials with rational coefficients*, Math. Ann. **261** (1982), 515–534.

[44] C.-E. Lind, *Untersuchungen über die rationalen Punkte der ebenen kubischen Kurven vom Geschlecht Eins*, Ph.D. thesis, University of Uppsala, 1940.

[45] D. Lorenzini and T. J. Tucker, *Thue equations and the method of Chabauty-Coleman*, preprint, May 2000.

[46] A. Macintyre, *On definable subsets of p-adic fields*, J. Symbolic Logic **41** (1976), 605–610.

[47] Yu. I. Manin, *The p-torsion of elliptic curves is uniformly bounded (Russian)*, Izv. Akad. Nauk. SSSR Ser. Mat. **33** (1969), 459–465, English translation: *Amer. Math. Soc. Transl.*, 433–438.

[48] ———, *Le groupe de Brauer-Grothendieck en géométrie diophantienne*, Actes du Congrès International des Mathématiciens (Nice, 1970), vol. 1, Gauthier-Villars, Paris, 1971, pp. 401–411.

[49] Yu. Matiyasevich, *The Diophantineness of enumerable sets (Russian)*, Dokl. Akad. Nauk SSSR **191** (1970), 279–282.

[50] H. Matsumura, *Commutative algebra*, W. A. Benjamin, Inc., New York, 1970.

[51] B. Mazur, *The topology of rational points*, Experiment. Math. **1** (1992), 35–45.

[52] ———, *Questions of decidability and undecidability in number theory*, J. Symbolic Logic **59** (1994), 353–371.

[53] ———, *Speculations about the topology of rational points: an update*, Astérisque **4** (1995), 165–182, Columbia University Number Theory Seminar (New York, 1992).

[54] W. McCallum, *On the method of Coleman and Chabauty*, Math. Ann. **299** (1994), 565–596.

[55] J. S. Milne, *Jacobian varieties*, Arithmetic geometry (G. Cornell and J. H. Silverman, eds.), Springer-Verlag, New York, 1986, pp. 167–212.

[56] L. J. Mordell, *On the rational solutions of the indeterminate equations of the third and fourth degrees*, Proc. Cambridge Phil. Soc. **21** (1922), 179–192.

[57] ———, *On the magnitude of the integer solutions of the equation $ax^2 + by^2 + cz^2 = 0$*, J. Number Theory **1** (1969), 1–3.

[58] M. Ram Murty and V. Kumar Murty, *Mean values of derivatives of modular L-series*, Annals of Math. (2) **133** (1991), 447–475.

[59] Yu. Penzin, *Undecidability of fields of rational functions over fields of characteristic 2 (Russian)*, Algebra i Logika **12** (1973), 205–210, 244.

[60] T. Pheidas, *Hilbert's tenth problem for fields of rational functions over finite fields*, Invent. Math. **103** (1991), 1–8.

[61] T. Pheidas and K. Zahidi, *Undecidability of existential theories of rings and fields: a survey*, Hilbert's tenth problem: relations with arithmetic and algebraic geometry (Ghent, 1999), Contemporary Mathematics, vol. 270, Amer. Math. Soc., 2000, pp. 49–105.

[62] B. Poonen, *Some Diophantine equations of the form $x^n + y^n = z^m$*, Acta Arith. **86** (1998), 193–205.

[63] B. Poonen and M. Stoll, *The Cassels-Tate pairing on polarized abelian varieties*, Annals of Math. (2) **150** (1999), 1109–1149.

[64] H. Reichardt, *Einige im Kleinen überall lösbare im Grossen unlösbare diophantische Gleichungen*, J. Reine Angew. Math. **184** (1942), 12–18.

[65] K. A. Ribet, *On modular representations of* $\mathrm{Gal}(\overline{\mathbf{Q}}/\mathbf{Q})$ *arising from modular forms*, Invent. Math. **100** (1990), 431–476.

[66] _____, *On the equation $a^p + 2^\alpha b^p + c^p = 0$*, Acta Arith. **79** (1997), 7–16.

[67] J. Robinson, *Definability and decision problems in arithmetic*, J. Symbolic Logic **14** (1949), 98–114.

[68] E. Selmer, *The diophantine equation $ax^3 + by^3 + cz^3 = 0$*, Acta Math. **85** (1951), 203–362, and **92** (1954) 191–197.

[69] J.-P. Serre, *A course in arithmetic*, Graduate Texts in Mathematics, vol. 7, Springer-Verlag, New York-Heidelberg, 1973.

[70] _____, *Lectures on the Mordell-Weil theorem*, Aspects of Mathematics, vol. E15, Friedr. Vieweg & Sohn, Braunschweig, 1989.

[71] G. Shimura, *Introduction to the arithmetic theory of automorphic functions*, Publications of the Mathematical Society of Japan, vol. 11, Princeton University Press, Princeton, NJ, 1994, Reprint of the 1971 original.

[72] A. Shlapentokh, *Hilbert's Tenth Problem over number fields, a survey*, Hilbert's tenth problem: relations with arithmetic and algebraic geometry (Ghent, 1999), Contemporary Mathematics, vol. 270, Amer. Math. Soc., 2000, pp. 107–137.

[73] J. H. Silverman, *The arithmetic of elliptic curves*, Graduate Texts in Mathematics, vol. 106, Springer-Verlag, New York-Berlin, 1986.

[74] _____, *Rational points on certain families of curves of genus at least 2*, Proc. London Math. Soc. (3) **55** (1987), 465–481.

[75] A. N. Skorobogatov, *Beyond the Manin obstruction*, Invent. Math. **135** (1999), 399–424.

[76] A. Tarski, *A decision method for elementary algebra and geometry*, 2nd ed., University of California Press, Berkeley and Los Angeles, Calif., 1951.

[77] R. Taylor and A. Wiles, *Ring-theoretic properties of certain Hecke algebras*, Annals of Math. (2) **141** (1995), 553–572.

[78] C. Videla, *Hilbert's tenth problem for rational function fields in characteristic 2*, Proc. Amer. Math. Soc. **120** (1994), 249–253.

[79] P. Vojta, *Siegel's theorem in the compact case*, Annals of Math. (2) **133** (1991), 509–548.

[80] J. L. Wetherell, *Bounding the number of rational points on certain curves of high rank*, Ph.D. thesis, University of California at Berkeley, 1997.

[81] A. Wiles, *Modular elliptic curves and Fermat's last theorem*, Annals of Math. (2) **141** (1995), 443–551.

G. H. Hardy As I Knew Him

Robert A. Rankin[1]

1 Introduction

This article stems from my desire to make a complete list of all the research students of G. H. Hardy (1877–1947) and that is still one of its objectives; see §§3–5. However, it occurs to me that it may be of some interest to set down some reminiscences of my association with Hardy during the last ten years of his life, and to add information about his family and ancestry.

After completing Part III of the Mathematical Tripos in 1937 I enrolled as a research student of A. E. Ingham (1900–1967). Ingham had done distinguished work on number theory, and, in particular, on Riemann's zeta-function, and was a very kind and considerate supervisor. I owe a lot to him; in particular, what is known as 'Rankin's trick' for the estimate of a Dirichlet series is really due to him. However, I was a very shy young man and he too was shy, so that I did not have the same rapport with him as I like to think I had later with Hardy, to whom in 1938 Ingham transferred me when he went on sabbatical leave as a Leverhulme Fellow. Hardy talked to one as an equal, which was flattering, although somewhat frightening. Also, although Hardy disliked small talk, he was at times quite chatty and his conversation was always interesting. He remains one of my heroes.

Like other research students in pure mathematics I attended Professor Hardy's seminar on Tuesday afternoons. This was advertised in the Cambridge University Reporter as the 'Conversation Class of Professors Hardy and Littlewood'; my memory is that in 1937 the word seminar was not as widely used as it is today, and that it tended to be restricted to its original German meaning of a course of lectures or study of an advanced topic. During my time Littlewood (1885–1977) never turned up at these meetings; the explanation of this I owe to Dame Mary Cartwright.

While Hardy was in Oxford, Littlewood held a regular seminar in his rooms in Trinity College and it was agreed that this should continue when Hardy returned to Cambridge and that the meetings should be chaired by them alternately. The first such meeting was held in Trinity as usual, with Littlewood giving the lecture. Hardy was present and asked a number

---

[1]The author died on January 27, 2001.

of questions. At this Littlewood took umbrage and said "I refuse to be heckled". The result was that thereafter they never appeared together at any of the meetings and eventually only Hardy came and the meetings were held elsewhere.

Both Hardy and Littlewood were admirable lecturers and were well prepared. Hardy had a curious mannerism of repeating the last six or so words in a sentence, then the last three, and ultimately converging to the last one. I think that Littlewood was marginally more stimulating than Hardy but foreign students could occasionally find him difficult to understand. I illustrate by referring to the English word *qua*, which is pronounced to rhyme with *way* but is originally the feminine ablative of the Latin relative pronoun *qui*; it is used to mean *in the capacity of*. Littlewood was lecturing on function theory and introduced a function f of two variables s and z. He was laying forth in his usual extravert style about f *qua* function of s and f *qua* function of z. A friend of mine, an Indian student called R. K. Rubugunday, who was classed in 1938 as Wrangler in the Mathematical Tripos, Part II, put up his hand and said "Please, sir, what is *qua*?" I do not recall that Littlewood gave any satisfactory reply, but shortly after poor Rubugunday was admitted to Fulbourn Mental Hospital, suffering, no doubt, from cultural shock at the strangeness of English *ways* (if not *quas*).

Hardy and Littlewood normally lectured in flannel trousers and sports jackets. I do not remember that their dress was particularly scruffy, but Hardy's sister Gertrude (1879–1963), who adored her older brother, told me that he was once taken for a beggar in the street and given sixpence. He was fond of wearing several woolen cricketing sweaters and would peel off one or two when he got too hot. However, once a month he appeared at his lecture wearing a beautiful dark crimson suit, and we knew it was the day when he went to London for the meeting of the London Mathematical Society. On these days the lecture finished early and he had a taxi waiting for him outside the Arts School to take him to the station.

He had a dislike of all formalities, including shaking hands. He told me once that when he was in Copenhagen at the invitation of Harald Bohr he was taken for a walk in the town and met several members of the mathematics staff, who all expected to shake hands; however, he kept his hands firmly in his pockets.

Hardy knew E. T. Whittaker (1873–1956) well as they were both Trinity men. Whittaker, who was knighted in 1945, was Professor of Mathematics in Edinburgh University from 1912 till 1946 and was, no doubt, responsible for Hardy's honorary LL.D. in Edinburgh. My aunt Dr. Mary Rankin, who was a Reader in Political Economy at that university and lived near the Whittakers in George Square, told me that Hardy created a bad impression by not attending one of the honorary degree parties, and going instead for a walk in the Pentland Hills with Whittaker.

In 1938 or 1939 Hardy told me that he had received an invitation from Whittaker to come to Edinburgh to give a course of lectures and was doubtful whether he should accept. I do not know why he consulted me on the subject, but he may have known that several members of my family had a strong connection with that university. He suspected that his conversation with Whitttker would involve too much discussion of religious matters, which he wanted to avoid. As a voting Fellow of Trinity Whittaker had been an elder in the Presbyterian church in Cambridge, but, over the years, he had turned to various other denominations, finally ending up as a convert to Roman Catholicism.

One of the most enjoyable occasions of my life was, surprisingly, my Ph.D. oral examination in June, 1940. My two examiners were supposed to be Hardy and Heilbronn, but Heilbronn had been interned in the Isle of Man, so that Littlewood took his place. Hardy invited me to dinner in Trinity, with strict instructions to meet him at 7:45 p.m. outside the door to the dining hall. As he would have been required to read the grace, if he had been the Senior Fellow present, we waited outside until grace had been said and then went in and had our dinner. After dinner Hardy, not being a drinking man, went back to his room, but I accompanied Littlewood to the Combination Room where we drank port and ate charcoal biscuits and salted almonds. We then went to Littlewood's room, Hardy joined us and we had, so far as I can remember, a very pleasant conversation; I suspected that Littlewood had not read the dissertation in detail, but had only glanced through it.

When I returned to Cambridge in 1945 after the war I served for a period of years as editor of the *Proceedings of the Cambridge Philosophical Society*. After reading Dorothy Sayers' book *The Nine Tailors*, I had written a paper [4] on campanology, which improved upon a result of W. H. Thomson [5] who had shown in 1886 that a complete peel of Grandshire Triples could not be rung using only plain and bob leads; this he had done without any knowledge of group theory. I was anxious that my paper should be published in Cambridge in the *Proceedings* and gave it to Hardy to send to a referee. Hardy, as I guessed, sent it to Philip Hall, who, fortunately, recommended publication. Hardy returned the paper to me with Hall's comments and his own letter to Hall. I was interested to see that he had written to Hall: "Fortunately the conclusions are negative, since anyone who had proved such feats possible would probably and deservedly be shot". Perhaps I should comment that at that time in Cambridge on Monday evenings those who lived in the centre of Cambridge had to endure a two hour practice by the change ringers in Great St. Mary's church.

I was greatly honoured, when sometime before the outbreak of war, Hardy asked me to read the manuscript and proofs of his book of lectures on the work of Ramanujan [2].

As is well known Hardy was very fond of intellectual games of various kinds. I remember his telling me that once, during a dull Royal Society Council meeting, he had before him a list of fellows divided into their different subjects. He wondered which group was the most aristocratic and assigned to the names 5 marks for a duke, 4 for an earl, etc. He told me that Mathematics came out better than he had expected, but that oceanography was the clear winner owing to the large number of admirals.

I very much regretted that I was unable to go and see him during his last illness when he had lost his interest in mathematics but retained his love of cricket. His sister used to visit him in the Evelyn Nursing Home and read to him out of the history of English county cricket, which to me seemed one of the most boring subjects imaginable. Unfortunately, I was no good at cricket at school, the summit of my achievements being made captain of a team below any eleven named Remnants, which ranked below all the other cricket teams fielded by my school at that time.

2   Family History

Most of the information in this section was obtained by my wife and me from searches in the Public Record Office and elsewhere. Some of it has been published in Kanigel's book [3], where further information can be found.

Hardy's father, Isaac Hardy (1842–1901), was born in Pinchbeck, a couple of miles north of Spalding in Lincolnshire. He taught Geography and Drawing at Cranleigh School in Surrey and later became bursar; he had a fine tenor voice and had been a keen footballer. Both his father, who was a labourer and foundryman, and his paternal grandfather were called Isaac. This must have been a popular name in Lincolnshire at the time, possibly deriving from Isaac Newton, who was a Lincolnshire man.

Hardy's mother, née Sophia Hall (1845–1917), was First Governess at Lincoln Diocesan College and was a remarkable woman; see [6]. At this school the teachers were called governesses and the pupils mistresses. She was born in Northampton. Her father Edward Hall was turnkey at Northampton County Gaol, and lived in the appropriately named Fetter Street nearby. He later became a baker in the village of Spratton, north of Northampton. Edward's wife was Charlotte Penn. Sophia Hall took up her employment at the Diocesan College in 1870. When she resigned her post in 1874 to get married, the Management Committee recorded their appreciation of the uniform excellence with which the responsible duties entrusted to her had been carried out and her wise combination of firmness and kindness.

Hardy's sister Gertrude Edith Hardy was unmarried and taught art and classics at St. Catherine's School, Bromley, a kind of sister school to Cranleigh. When her brother became ill after the war she came to Cambridge to look after him and lived in a Guest House at 5 West Road. It is possible that later on she was allowed by the College authorities to live in Hardy's rooms in Trinity in order to look after him better, and before he moved to the Evelyn Nursing home where he died. She later moved to Meadowcroft in Trumpington Road, where she died in 1963.

It is known that Hardy, possibly because of his atheism, never used his first name Godfrey. To his family and very close friends, such as Professor Donald Robertson, the Professor of Greek, he was known as Harold, and in the newspaper report of the death of his father he is listed among the mourners as Mr. Harold Hardy.

Hardy's atheism appears to have been partly a reaction against the strong religious views of his parents. At the time Cranleigh was known for its 'churchiness'. Gertrude's views were similar to her brother's, but probably not quite so strong. When she lived in Meadowcroft she was annoyed because the other old ladies living there kept on asking her where she went to church on Sundays. Her solution was to say that she was a Mohammedan, but that there was no mosque in Cambridge, and she asked her friend Marjorie Dibden, whose husband Kenneth was Secretary of the Cavendish Laboratory, to accompany her into town to purchase a prayer mat.

3 Early Days of Mathematical Research

The degree of Doctor of Philosophy was not introduced in the University of Cambridge until May 1920. A Board of Research Studies was then set up, which issued an Annual Report in the Cambridge University Reporter. In this report were listed the names and colleges of all research students, together with those of their supervisors, and a brief statement of the subjects of their research. Before that date arrangements were somewhat informal and supervisors' names can only occasionally be found in the records. From 1880 to 1913 Research Students were known as 'Advanced Students'. Of these there were two kinds, one proceeding to the Degree of B.A. or LL.B. by means of Tripos Examinations and the other by means of Certificates of Research. From 1913 to 1920 Research Students proceeded to the Degree of B.A. or LL.B. by Certificates of Research only.

1910	E. F. Clark, (b. 1887), Trin.
1900	B. Cookson, (1874–1909), Trin.
1910	W. T. David (1886–1948), Trin.
1912	G. G. Davidson, (1886–1959), Caius
1900	L. N. G. Filon (1875–1937), King's
1907	J. B. Hubrecht (1883–1978), Christ's
1916–17	E. L. Ince, (1891–1941), Trin.
1910	C. McNeil, (1886–1959), Jesus
1919–20	G. A. Newgass, (1889–1948), Trin.
1915–16	S. Ramanujan, (1887–1920), Trin.
1916–17	F. W. Richards, (b. 1890), Caius
1913–14	R. Rossi, (1888–1920), Trin.
1913–14	F. E. Rowett, (1889–1935), St. John's
1913–14	L. F. G. Simmons, (1890–1954), St. John's
1911	H. J. Swain, Emma.

Table 1. Holders of Research Certificates

In [1] a complete list is given of the colleges and Faculties of the 214 students who received Certificates of Research from 1899 to 1920. From this list I have excerpted, in Table 1, the names of the 15 students who gave Mathematics as their Faculty. Of these only three, namely, Louis Napoleon George Filon, a distinguished applied mathematician, Edward Lindsay Ince and Srinivasa Ramanujan were known to me because of their work. In this and the following tables I give the years of birth and death, where I have been able to find them, together with occasional comments on the students listed. This may enable interested readers to find biographical information from the *Dictionary of National Biography*, *Who was Who*, or other sources.

However, not every student who did research in Mathematics registered as a Research Student during this time. In Table 2 I list the names of students who, between 1911 and 1920, won a Smith's or Rayleigh Prize, and those who failed to do so, but were commended for their essays. With the exception of E. L. Ince, each person in the table had taken the Mathematical Tripos. It will be noticed that this is, on average, a much more distinguished list of students. The letters S, and R and C after the names denote Smith's Prize, Rayleigh Prize and Commended.

I now come to G. H. Hardy. From the time that he was elected a Fellow of Trinity College in 1900, his adult university career falls into three parts: (i) Cambridge 1900–1919, (ii) Oxford 1919–1931, (iii) Cambridge 1931–1947. During the first period it is known that he gave guidance and encouragement to K. Ananda Rau and S. Ramanujan, but there were, almost certainly others. The former had taken the Tripos, obtaining First Class Honours in both parts, and was a Smith's Prize winner, as may be seen from Table 2. As regards the latter, Hardy stated that he learnt more from Ramanujan than Ramanujan did from him. To illustrate that supervision arrangements were not as formalised as they became after 1920, it may be mentioned that E. L. Ince (an Edinburgh graduate) paid tribute to E. T. Whittaker, his Edinburgh professor, for help and advice while he was doing research in Cambridge; he may have had a nominal Cambridge supervisor, although that was, perhaps, unlikely to have been Hardy. I had hoped that the Minutes of the Degree Committee of the Faculty of Mathematics for the years before 1920 might have revealed the name of supervisors. These Minutes, which were lost, have only recently been found, but unfortunately they do not do so.

4   The Oxford Years

In the 1920's research students at the University of Oxford were admitted either as Advanced Students studying for the D.Phil. or as B.Sc. students. This second category was of a lower level, being approximately equivalent to a present University M.Sc. course. Table 3 lists the names of students supervised by Hardy who were B.Sc. students. The first column gives the year of admission and the last year of graduation, where relevant. It may be noted that T. Vijayaraghavan did not graduate, but received a Certificate entitling him to do so. Also, W. L. Ferrar (1893–1990) succeeded Hardy as supervisor for J. G. Nicholas.

Duminy later became Vice-Chancellor of the University of Capetown. Sutton was a distinguished meteorologist and Gertrude Stanley became head of the Mathematics Department at Westfield College, London.

I now turn to Hardy's Advanced Students reading for the Degree of D.Phil. They are listed in Table 4, arranged similarly to Table 3. It is interesting to note that E. C. Titchmarsh, although qualified to graduate, did not do so, and the same is true again for T. Vijayaraghavan. The asterisk against the last four names indicates that they were transferred to Titchmarsh as supervisor on Hardy's return to Cambridge.

1911	W. E. H. Berwick (S) (1888–1944)	1912	L. J. Mordell (S) (1888–1972)
1913	S. Chapman (S) (1888–1970)	1912	E. H. Neville (S) (1889–1961)
1912	P. J. Daniell (R) (1889–1946)	1920	S. Pollard (S) (1894–1925)
1911	C. G. Darwin (C) (1887–1962)	1915	J. Proudman (S) (1888–1966)
1913	R. H. Fowler (R) (1889–1944)	1918	K. A. Rau (S) (1893–1975)
1914	R. A. Frazer (R) (1891–1959)	1919	S. R. U. Savoor (S) (b. 1893)
1916	H. M. Garner (S) (1891–1977)	1914	B. P. Sen (S)
1913	A. H. S. Gillson (C) (1889–1954)	1916	W. M. Smart (R) (1889–1975)
1915	H. Glauert (R) (1892–1934)	1913	H. Spencer Jones (S) (1890–1967)
1913	A. R. Grieve (C) (1886–1952)	1914	C. A. Stewart (C) (1888–1959)
1918	E. L. Ince (S) (1891–1941)	1913	R. O. Street (R) (1890–1967)
1914	J. Jackson (S) (1887–1958)	1911	A. W. H. Thomson (C) (b. 1888)
1915	H. Jeffreys (S) (1891–1989)	1916	G. P. Thomson (S) (1892–1972)
1911	S. Lees (R) (1885–1940)	1917	H. Todd (S) (b. 1894)
1911	G. H. Livens (S) (1886–1950)	1918	H. W. Unthank (C) (1893–1979)
1919	C. N. H. Lock (S) (1894–1949)	1913	T. L. Wren (R) (1889–1972)

Table 2. Smith and Rayleigh Prize Students

1922	J. P. Duminy	(1897–1980)	Univ. 1923
1923	E. H.Saayman	(b. 1897)	New Coll.
1924	O. G. Sutton	(1903–1977)	Jesus 1927
1925	T. Vijayaraghavan	(1902–1955)	New Coll.
1925	G. K. Stanley	(1897–1974)	Home Stud. 1927
1925	N. L. Clapton	(b. 1903)	Hertford
1926	J. G. Nicholas	(b. 1908)	Jesus 1932

Table 3. Oxford B.Sc. Students

1919	W. R. Burwell	(1894–1971)	Merton
1921	F. V. Morley	(1899–1980)	New Coll. 1923
1922	E. C. Titchmarsh	(1899–1963)	Balliol
1924	G. L. Frewin	(b. 1902)	New Coll.
1925	P. L. Srivastava	(b. 1898)	New Coll. 1927
1926	U. S. Haslam-Jones	(1903–1962)	Queen's 1928
1926	E. H. Linfoot	(1905–1982)	Balliol 1928
1926	F. J. Brand	(1905–1995)	Jesus, Grad. B.Sc. 1929
1926	T. Vijayaraghavan	(1902–1955)	New Coll.
1927	L. S. Bosanquet	(1903–1984)	Balliol 1929
1928	M. L. Cartwright	(1900–1998)	St. Hugh's 1930
1929	E. M. Wright	(b. 1906)	Jesus 1932
1929	P. M. Owen*	(1906–1962)	Jesus 1933
1930	R. Profitt*	(1907–1974)	Exeter 1936
1930	E. G. Phillips*	(1909–1984)	Christ Ch. 1932
1930	A. C. Bassett*	(1908–1989)	Jesus 1936

Table 4. Oxford D.Phil. Students

	Name		University	B.A.	Ph.D.
1931	M. M. Ahmed	(b. 1908)	Edinburgh		
1932	M. Hall	(1910–1989)	Yale		
1933	J. M. Hyslop	(1908–1984)	CU	1932	1935
1933	R. Rado	(1906–1989)	Berlin		1935
1934	G. W. Morgan	(1911–1989)	CU	1932	1935
1934	A. C. Offord	(b. 1906)	London		1936
1934	F. Smithies	(b. 1912)	CU	1933	1937
1937	H. R. Pitt	(b. 1914)	CU	1935	1939
1938	F. M. C. Goodspeed	(b. 1914)	Winnipeg		1942
1939	Y. C. Chow	(b. 1918)	No degree		
1940	R. A. Rankin	(b. 1915)	CU	1937	1940
1942	S. M. Edmonds	(b. 1916)	CU	1938	1942

Table 5. Hardy's later Cambridge Research Students

I am indebted to Simon Bailey and Richard Hughes of Oxford University for the information contained in Tables 3 and 4.

F. V. Morley, a London author and publisher, wrote a book on Inversive Geometry together with his father Frank Morley, who was professor of mathematics at Johns Hopkins University.

5   Return to Cambridge

In Table 5 Hardy's research students are given as listed in the Annual Reports of the Board of Research Studies. The first column gives the year when the student's name first occurs with Hardy as supervisor. However, in nearly every case the date of commencement of research was earlier. There are various reasons for this, such as transferral from a former supervisor. For example, I transferred in 1938 to Hardy, but only appear in 1940 as having done so. For this reason I include, in the second last column, the year when students who had taken the Tripos examinations were awarded their B.A.

The third column lists the previous university CU, denoting Cambridge University. The first student listed is better known as M. Mursi. He was a student of E. T. Whittaker and had just obtained his Edinburgh Ph.D. for a dissertation on automorphic functions, and, in particular uniformization. It is very likely that both he and Marshall Hall intended, at the outset, to stay no more than one year in Cambridge. Y. C. Chow was an able young Chinese mathematician, with no University degree, and was an expert on inequalities on which he published two papers in 1939 in the Journal of the London Mathematical Society. He had to leave Cambridge when he got into financial difficulties and was, as I was informed, killed in an air crash. I knew him well and taught him how to ride a bicycle on Midsummer Common. From the tables it appears that Hardy had three female research students, Gertrude Stanley, Mary Cartwright and Sheila Edmonds.

Littlewood's Students

In view of the close connection between Hardy and Littlewood, I conclude by giving a list of the latter's research students. Apart from the years 1907–1910 when he was at Manchester University, Littlewood spent all his academic life in Cambridge. I have not been able to discover the names of any research students he may have supervised before 1920, but, with one exception that I mention later, Table 6 lists the names of his students who appear after that date in the reports of the Board of Research Studies. The second last column gives the date of obtaining the B.A. degree for those students who had been Cambridge undergraduates. As before, the last column gives the date of the award of the Ph.D. The exception mentioned above is Sir Peter Swinnerton-Dyer, who was a research student of Littlewood, but did not register with the Board of Research Studies. This was possible at that time, and I have discovered a number of additional Cambridge research students who did not enrol with the Board of Research Studies, but made private arrangements with their supervisors. One of those was Douglas C. Noerthcott, who was probably Hardy's last research student. There were others such as Norbert Wiener and Norman Levinson who, during their time in Cambridge, regarded Hardy as their teacher. Freeman Dyson was another who knew Hardy and Littlewood well and had a private arrangement with his supervisor A. S. Besicovitch.

The dates given for V. J. Levin are under the assumption that he was the Viktor Josifovich Levin who worked on complex function theory. His name only occurs once in 1933 in the Report of the Board of Research Studies. He was admitted 'conditional of being accepted by a college, or as a Non-Collegiate Student'. It is likely that he never came to Cambridge because no college wanted him or, more probably, he changed his mind.

Dr. Jeremy Bray was Member of Parliament for Motherwell South from 1983 until he retired in 1997.

1924	T. A. A. Broadbent	(1903–1973)	1924	
1924	R. Cooper	(1903–1979)	1924	1927
1925	F. W. Bradley	(1904–1953)		1931
1925	E. F. Collingwood	(1900–1970)	1921	1929
1925	G. A. A. H. Gyllensvard	(b. 1898)		
1926	F. G. Maunsell	(1898–1956)	1923	1928
1927	H. P. Mulholland	(1906–1977)	1926	1930
1929	S. Verblunsky	(1906–1996)	1927	1930
1930	S. D. S. Chowla	(1907–1995)		1931
1931	J. Cossar	(b. 1907)	1930	1933
1931	H. Davenport	(1907–1969)	1929 Sc.D.	1938
1932	S. Skewes	(1899–1988)	1925	1938
1933	A. E. Gwilliam	(1912–1984)		1935
1933	V. J. Levin	(1909–1986)		
1936	D. C. Spencer	(b. 1912)		1939
1947	A. O. L. Atkin	(b. 1925)	1946	1952
1947	T. M. Flett	(1923–1976)		1950
1948	A. C. Allan	(b. 1928)		
1948	N. DuPlessis	(1921–1983)	1951	
1948	G. R. Morris	(b. 1922)	1948	1952
1948	H. P. F. Swinnerton-Dyer	(b. 1927)	1948	
1949	S. R. Tims	(1926–1971)		1952
1949	E. J. Watson	(b. 1924)	1945	
1949	M. N. Ghabour	(b. 1915)		
1951	P. S. Bullen	(b. 1928)		1955
1951	F. R. Keogh	(1923–1991)		1954
1953	M. F. C. Woollett	(b. 1925)		
1954	J. W. Bray	(b. 1930)	1952	1957

Table 6. Littlewood's Research Students

I should be grateful for any information enabling me to supply dates missing from the above tables. I acknowledge with thanks the information on research students supplied by the archivists of Oxford and Cambridge Colleges. I am also grateful for help and advice received from Dr. Elizabeth Leedham-Green of Cambridge University Archives and Professor J. Milne Anderson of University College, London.

References

[1] *Cambridge Historical Register Supplement 1911–1920*, Cambridge, 1922.

[2] G. H. Hardy, *Ramanujan: Twelve Lectures on subjects suggested by his life and work*, University Press, Cambridge, 1940.

[3] R. Kanigel, *The man who knew infinity*, Charles Scribner's Sons, New York, 1991.

[4] R. A. Rankin, *A campanological problem in group theory*, Proc. Cambridge Phil. Soc. **44** (1948), 17–25.

[5] W. H. Thompson, *A note on Grandsire Triples*, Cambridge, 1886.

[6] D. H. J. Zebedee, *Lincoln Diocesan Training College 1862–1962*, Lincoln, 1962.

Some Applications of Diophantine Approximation

R. Tijdeman

0 Introduction

The paper gives a survey of some results on diophantine approximation (Sections 1 and 2) and their applications (Sections 3, 4 and 5). Section 1 contains an introduction to the theory of linear forms in logarithms of algebraic numbers, and Section 2 describes some results following from the Subspace Theorem. Section 3 gives applications to the local behaviour of sequences of numbers composed of small primes and of sums of two such numbers. Section 4 deals with the transcendence of infinite sums of values of a rational function and related sums, and in Section 5 some recent applications to diophantine equations and recurrence sequences are described. The Appendix contains some elaborations of Section 4. Section 3 and the Appendix contain some new results.

1 Linear Forms in Logarithms

In 1966–1968 Baker extended the Gelfond-Schneider theorem by proving the following result.

Theorem 1 ([2], p. 10). *If* $\alpha_1, \alpha_2, \ldots, \alpha_n$ *are non-zero algebraic numbers such that* $\log \alpha_1, \ldots, \log \alpha_n$ *are linearly independent over the rationals, then* $1, \log \alpha_1, \ldots, \log \alpha_n$ *are linearly independent over the field of all algebraic numbers.*

For our convenience we shall assume in the sequel that log denotes the principal value of the logarithm. I record some consequences of Theorem 1.

Corollary 1.1. $e^{\beta_0} \alpha_1^{\beta_1} \ldots \alpha_n^{\beta_n}$ *is transcendental for any non-zero algebraic numbers* $\alpha_1, \ldots, \alpha_n$, $\beta_0, \beta_1, \ldots, \beta_n$.

The fact that $\beta_0 \neq 0$ says that we are dealing with the inhomogeneous case. The corollary implies that numbers like $e \cdot 2^{\sqrt{2}}$ and $\pi + \log 2$ are transcendental, but it is less useful for applications than its homogeneous counterpart which reads as follows.

Corollary 1.2. $\alpha_1^{\beta_1} \ldots \alpha_n^{\beta_n}$ *is transcendental for any algebraic numbers* $\alpha_1, \ldots, \alpha_n$, *other than* 0 *or* 1, *and any algebraic numbers* $\beta_1, \ldots, \beta_n$ *with* $1, \beta_1, \ldots, \beta_n$ *linearly independent over the rationals.*

An alternative formulation is as follows.

Corollary 1.3. *If* $\alpha_1, \ldots, \alpha_n$, $\beta_1, \ldots, \beta_n$ *are non-vanishing algebraic numbers and* $\Lambda = \beta_1 \log \alpha_1 + \cdots + \beta_n \log \alpha_n$, *then*

$$\Lambda = 0 \quad or \quad \Lambda \ is \ transcendental.$$

Baker used the effectiveness of his method to give lower bounds for $|\Lambda|$ in the case $\Lambda \neq 0$. Of course, these bounds depend on $\alpha_1, \ldots, \alpha_n$, $\beta_1, \ldots, \beta_n$. They are functions of the degrees and the heights of these numbers. The earlier bounds were expressed in terms of the classical height, which is defined as the maximal absolute value of the coefficients of the minimal defining polynomial of the number. These bounds have been successively improved, and they are now stated in terms of the Mahler height or its variants. This is more convenient for the proofs and leads to better constants. Since the different heights used are more or less equivalent, I do not want to go into this rather technical aspect here.

Let $\alpha_1, \ldots, \alpha_n$ be algebraic numbers $\neq 0, 1$. Let $\mathbb{Q}(\alpha_1, \ldots, \alpha_n)$ have degree at most d over $\mathbb{Q}$. Let the classical height of α_j be at most $A_j \geq 4$ for $j = 1, \ldots, n$. Put $A = \max A_j$, $\Omega = (\log A_1) \ldots (\log A_n)$, $\Omega' = (\log A_1) \ldots (\log A_{n-1})$. Let $\beta_1, \ldots, \beta_n$ be rational integers of absolute values at most $e^B \geq 4$. In 1977 Baker [3] proved that

$$\Lambda = 0 \ \text{or} \ \log|\Lambda| \geq -(16nd)^{200n} B\Omega \log \Omega'. \tag{1.1}$$

This bound has three weaknesses: The constants are rather large; it is expected that the factor $(16nd)^{200n}$ can be replaced by a polynomial expression in n and d; and it is conjectured that the product of the logarithms in Ω may be replaced by the sum of the logarithms. The first drawback has been largely overcome. After several improvements by Waldschmidt and others, Baker and Wüstholz [6] published in 1993 a result which implies that

$$\Lambda = 0 \ \text{or} \ \log|\Lambda| \geq -(16nd)^{2(n+2)} B\Omega.$$

In addition to having a smaller constant in the exponent, we see that the factor $\log \Omega'$ in Baker's bound has been removed. Baker and Wüstholz state that a similar improvement in the constant can be obtained in the case of arbitrary algebraic coefficients β_j, but they have not worked out the details. For bounds in the general case, see [62]. In many applications only two or three logarithms occur. In these cases bounds with better constants are available. For the case of two logarithms, see Laurent, Mignotte and Nesterenko [33] [39]; for linear forms in three logarithms see [7].

In some applications the dependence on the number n of logarithms is crucial. In this direction Matveev [36] proved the following result:

Let $\alpha_1, \ldots, \alpha_n$ be algebraic numbers distinct from 0, let $\log \alpha_j$ be fixed values of the logarithms distinct from 0, and let $h(\alpha_j)$ denote the absolute logarithmic heights. Consider the linear form

$$\Lambda := b_1 \log \alpha_1 + \cdots + b_n \log \alpha_n \neq 0, \quad b_j \in \mathbf{Z}.$$

Let K be an algebraic number field containing all numbers α_j and having degree at most D. Define $\kappa = 1$ if K is contained in the field of real numbers and $\kappa = 2$ otherwise. Put

$$B = \max(|b_1|, \ldots, |b_n|).$$

Let A_j be real numbers such that

$$A_j \geq \max(Dh(\alpha_j), |\log \alpha_j|, 0.16), \quad j = 1, \ldots, n.$$

Then

$$\log |\Lambda| \geq -C_1 D^2 A_1 \ldots A_n \log(eD) \log(eB),$$

where

$$C_1 = \min \left(\kappa^{-1} \left(\frac{1}{2} en \right)^\kappa 30^{n+3} n^{3.5}, \ 2^{6n+20} \right).$$

The dependence on n in this result is reduced to c^n; furthermore, the constants are remarkably small.

There exist p-adic analogues of the complex linear forms estimates which are important for many applications. Generally speaking, the estimates are similar and the constants are slightly better. Kunrui Yu obtained the p-adic analogues of the Baker-Wüstholz estimate [64] and of a preliminary version of Matveev's result [65]. A p-adic analogue of the result of Laurent, Mignotte and Nesterenko on linear forms in two logarithms was given by Bugeaud and Laurent [16]. In general the upper bounds for

$$\mathrm{ord}_{\mathfrak{p}}(\alpha_1^{\beta_1} \ldots \alpha_n^{\beta_n} - 1),$$

where $\mathrm{ord}_{\mathfrak{p}}(\alpha)$ denotes the exponent to which the prime ideal $\mathfrak{p}$ divides the principal fractional ideal generated by $\alpha \in \mathbb{Q}(\alpha_1, \ldots, \alpha_n)$, $\alpha \neq 0$, are similar to the estimates for $-\log |\Lambda|$ with respect to $n, \alpha_1, \ldots, \alpha_n$ and $\beta_1, \ldots, \beta_n$. There is, however, an additional factor $p^{f_{\mathfrak{p}}}$ where $f_{\mathfrak{p}}$ denotes the residue class degree of $\mathfrak{p}$. For some applications this dependence on p is an impediment.

2   The Subspace Theorem

In this section we state some consequences of the Subspace Theorem which are quite useful in arithmetical applications. The Subspace Theorem is a generalization of Roth's theorem proved by W. M. Schmidt [45] in 1972. A p-adic analogue was obtained by Schlickewei [42] in 1977. Both van der Poorten and Schlickewei [59] and Evertse [27] applied the p-adic Subspace Theorem to S-unit equations. For the general statements I refer to the original papers. Here I state the consequence for rational integers of a result of Evertse.

Theorem 2 ([27], p. 227). *Let c, d be constants with $c > 0$, $0 < d < 1$. Let S be a finite set of prime numbers and let n be a positive integer. Then there are only finitely many tuples $\boldsymbol{x} = (x_0, x_1, \ldots, x_n)$ of rational integers such that*

$$x_0 + x_1 + \cdots + x_n = 0,$$

$$\sum_{i \in I} x_i \neq 0$$

for each proper, non-empty subset I of $\{0, 1, \ldots, n\}$,

$$\gcd(x_0, x_1, \ldots, x_n) = 1,$$

and

$$\prod_{k=0}^{n} \left(|x_k| \prod_{p \in S} |x_k|_p \right) \leq c \cdot |\boldsymbol{x}|^d,$$

where $|a|_p = p^{-\operatorname{ord}_p(a)}$.

In 1989 Schmidt [46] made another breakthrough by proving a quantitative version of his Subspace Theorem. Soon afterwards Schlickewei [43] proved the p-adic analogue of this result and extended it to number fields. In 1995 Evertse [28] improved upon earlier upper bounds for the number of equivalence classes of solutions of S-unit equations. For the general formulation and related results I again refer to the original papers. Here I state the result for S-unit equations in the case of rational integers.

Theorem 3 ([28], p. 564). *Let $a_1, \ldots, a_n$ be non-zero rational numbers. Let S be a finite set of s prime numbers. Then the equation*

$$a_1 u_1 + \cdots + a_n u_n = 1$$

with

$$\sum_{i \in I} a_i u_i \neq 0$$

for each non-empty subset I of $\{0, 1, \ldots, n\}$, has at most $(2^{35}n^2)^{n^3(s+1)}$ solutions in rational numbers $u_1, \ldots, u_n$ which are entirely composed of primes from S.

It is remarkable that the bound depends only on n and s, and not on the primes in S or the coefficients a_i. The upper bound is large, but the dependence on n and s cannot be polynomial [26], [30]. For a corresponding estimate for linear equations in variables which lie in an arbitrary finitely generated multiplicative group, see [29]. The existing proofs of the Subspace Theorem do not enable us to give upper bounds for the solutions themselves. We call them therefore *ineffective.*

3 Numbers Composed of Small Primes

The results mentioned in Sections 1 and 2 have immediate consequences for the local distribution of integers composed of small primes. In this section I give a survey of such results.

Let $p_1, \ldots, p_k$ be fixed primes, each at most P. Consider all integers $p_1^{r_1} \ldots p_k^{r_k}$, where $r_1, \ldots, r_k$ are non-negative integers and order them in increasing order of magnitude as $1 = n_0 < n_1 < n_2 < \ldots$. For example, for $k = 2$, $p_1 = 2$, $p_2 = 3$ we get

$$(n_i)_{i=0}^{\infty} = 1, 2, 3, 4, 6, 8, 9, 12, 16, 18, 24, 27, \ldots.$$

Improving upon earlier lower bounds by Störmer, Thue and Erdős, I noticed in 1973 [55] that there is a computable number C_1 depending only on P such that

$$n_{i+1} - n_i > \frac{n_i}{(\log n_i)^{C_1}}$$

for $n_i \geq 3$. The result is an immediate consequence of linear forms estimates. From

$$\frac{n_{i+1}}{n_i} - 1 > \log \frac{n_{i+1}}{n_i} = t_1 \log p_1 + \cdots + t_k \log p_k > 0$$

we obtain, by (1.1),

$$n_{i+1} - n_i > n_i (\max_j t_j)^{-C_2} > \frac{n_i}{(\log n_i)^{C_1}}.$$

Apart from the value of C_1 the result is the best possible. In another paper [56] I proved that there are computable numbers C_3 and N depending only on P such that

$$n_{i+1} - n_i < \frac{n_i}{(\log n_i)^{C_3}}$$

for $n_i \geq N$. The average order of the difference is

$$\frac{n_i}{(\log n_i)^{k-1}}$$

so that $C_3 \leq k - 1 \leq C_1$. Actually we need not assume that $p_1, \ldots, p_k$ are fixed. Using the Baker-Wüstholz estimate and the inequality $t_i \leq 2 \log n_i$ we see that we can take $C_1 = (ck^2 \log P)^k$ where c is some absolute constant.

Let $1 = m_0 < m_1 < m_2 < \ldots$ be a sequence of positive integers each of which is the sum of at most two terms from the sequence $(n_i)_{i=1}^{\infty}$ such that m_i and m_{i+1} do not use the same n_i. For example, for $k = 2$, $p_1 = 2$, $p_2 = 3$ we may take

$$(m_i)_{i=0}^{\infty} = 1, 2, 3, 4, 5, 6, 7, 8, 9, 10, 11, 12, 13, 14, 15, 16, \ldots,$$

but 23 will not occur. What can be said about the differences $m_{i+1} - m_i$? By our restriction we avoid large neighbouring numbers like

$$2^k 3^l + 2^3, \quad 2^k 3^l + 3^2.$$

Baker-type estimates are of no use here. I do not know of any method giving explicit lower bounds. However, the ineffective methods from Section 2 show that $m_{i+1} - m_i \to \infty$. In fact, it follows from Theorem 2 that for every $\epsilon > 0$ there are only finitely many i such that

$$m_{i+1} - m_i < m_i^{1-\epsilon}. \tag{3.1}$$

By applying Theorem 3 to the equation

$$\frac{n_{i_1}}{n_{i_4}} + \frac{n_{i_2}}{n_{i_4}} - \frac{n_{i_3}}{n_{i_4}} = 1$$

we see that there are at most

$$(2^{35} \cdot 9)^{27(k+1)}$$

integers m which can be written as

$$m = n_{i_1} + n_{i_2} = n_{i_3} + n_{i_4}$$

with $\gcd(n_{i_1}, n_{i_2}, n_{i_3}, n_{i_4}) = 1$ and $n_{i_1} \neq n_{i_3}, n_{i_4}$. Similar results can be derived for numbers which can be written in more than one way as the sum of at most t numbers n_i, but if $t > 2$ the vanishing subsums cause complications.

It is not necessary that the n_i be entirely composed of $p_1, \ldots, p_k$. For example, let $0 \leq \delta < 1/4$. Let $1 = l_0 \leq l_1 \leq l_2 \leq \ldots$ be a sequence of positive integers which are the sum of two numbers of the form $p_1^{r_1} \ldots p_k^{r_k} q$ where q is some positive integer with $q \leq (p_1^{r_1} \ldots p_k^{r_k})^\delta$ such that l_i and l_{i+1} do not use the same summand for any i. Then, by Theorem 2, for every positive ϵ there are only finitely many i such that

$$l_{i+1} - l_i < l_i^{1-4\delta-\epsilon}.$$

The inequality reduces to (3.1) if $\delta = 0$.

4   On the Transcendence of Infinite Sums

It is well known that

$$\sum_{n=1}^{\infty} \frac{1}{n(n+1)} = 1 - \frac{1}{2} + \frac{1}{2} - \frac{1}{3} + \frac{1}{3} - \frac{1}{4} + \frac{1}{4} - \frac{1}{5} + \cdots = 1$$

is an integer, but

$$\sum_{n=0}^{\infty} \frac{1}{(2n+1)(2n+2)} = 1 - \frac{1}{2} + \frac{1}{3} - \frac{1}{4} + \frac{1}{5} - \frac{1}{6} + \cdots = \log 2$$

is transcendental. Remarkably, little attention has been given in the literature to the transcendence of sums $\sum_{n=0}^{\infty} f(n)$ where $f \in \mathbb{Z}(x)$, whereas a number of exotic series such as

$$\sum_{n=1}^{\infty} \frac{1}{2^{n!}}, \quad \sum_{n=0}^{\infty} \frac{1}{2^{2^n}}, \quad \sum_{h=0}^{\infty} \frac{1}{F_{2^h+1}}, \quad \sum_{n=1}^{\infty} \frac{(-1)^n}{F_n^2},$$

where $(F_h)_{h=0}^{\infty}$ is the Fibonacci sequence, have been studied in the literature. As to sums $\sum_{n=0}^{\infty} f(n)$ with $f \in \mathbb{Z}(x)$, Lehmer [34] gives the following examples:

$$\sum_{n=0}^{\infty} \frac{1}{(n+1)(2n+1)(4n+1)} = \frac{\pi}{3},$$

$$\sum_{n=0}^{\infty} \frac{1}{(6n+1)(6n+2)(6n+3)(6n+4)(6n+5)(6n+6)}$$

$$= \frac{1}{4320} (192 \log 2 - 81 \log 3 - 7\pi\sqrt{3}).$$

We see that both sums are transcendental. The latter example shows that Baker's theory is relevant here. If one replaces π by $\log(-1)$, the sum

becomes a linear form in logarithms of algebraic numbers with algebraic
coefficients. Corollary 1.3 tells us that such a sum is 0 or transcendental,
and a sum of positive numbers cannot be 0. Of course, there are also sums
of this form, such as $\sum_{n=1}^{\infty} n^{-3}$ and $\sum_{n=1}^{\infty} n^{-5}$, for which we do not know
whether they are transcendental.

We assume from now on that $f(x) = P(x)/Q(x)$ for $P(x), Q(x) \in \mathbb{Q}[x]$,
where the zeros of Q are simple and rational and $\deg P < \deg Q$. Hence
we can split $f(x)$ into partial fractions and write

$$f(x) = \frac{P(x)}{Q(x)} = \sum_{j=1}^{m} \frac{c_j}{k_j x + r_j} \text{ where } k_j, r_j \in \mathbb{Z}, c_j \in \mathbb{Q}.$$

An obvious simplification is possible if $k_i = k_j$ and $k_i \mid r_i - r_j$ for some i, j.
Changing the summation index we get a rational initial term and we can
combine the remaining terms. In this way we can arrange that $0 < r_j \leq k_j$
for all j (starting the summation from $n = 0$). For example,

$$\sum_{n=0}^{\infty} \left(\frac{4}{5n + 3} - \frac{3}{5n + 7} - \frac{1}{5n + 8} \right)$$

$$= \frac{3}{2} + \frac{1}{3} + \sum_{n=0}^{\infty} \left(\frac{4}{5n + 3} - \frac{3}{5n + 2} - \frac{1}{5n + 3} \right)$$

$$= \frac{11}{6} + \sum_{n=0}^{\infty} \left(\frac{3}{5n + 3} - \frac{3}{5n + 2} \right).$$

We call an infinite sum *reduced* if it can be written as

$$\sum_{n=0}^{\infty} \sum_{j=1}^{m} \frac{c_j}{k_j n + r_j} \text{ with } k_j, r_j \in \mathbb{Z}_{>0}, r_j \leq k_j, c_j \in \mathbb{Q},$$

so that all zeros of Q are in the interval $[-1, 0)$. It may happen that the
infinite sum vanishes, in which case the sum is rational; for example,

$$\sum_{n=1}^{\infty} \frac{1}{n(n + 1)} = 1 + \sum_{n=0}^{\infty} \left(\frac{1}{n + 1} - \frac{1}{n + 1} \right) = 1$$

and

$$\sum_{n=0}^{\infty} \left(\frac{1}{5n + 2} - \frac{3}{5n + 7} + \frac{2}{5n - 3} \right)$$

$$= \frac{3}{2} - \frac{2}{3} + \sum_{n=0}^{\infty} \left(\frac{1}{5n + 2} - \frac{3}{5n + 2} + \frac{2}{5n + 2} \right) = \frac{5}{6}.$$

If this is not the case, it is useful to apply formulas for $\sum_{n=0}^{N} 1/(kn+r)$ which D. H. Lehmer [34] gave in 1975. The formulas are of the form

$$\sum_{n=0}^{N} \frac{1}{kn+r} = \frac{1}{k} \log N + \gamma_{k,r} + o(1),$$

where the numbers $\gamma_{k,r}$ are generalized Euler constants. The arithmetic nature of these numbers is unknown, but fortunately they cancel in cases where $\sum_{n=0}^{\infty} P(n)/Q(n)$ converges. By using Lehmer's formulas, Adhikari, Saradha, Shorey and I proved that if the reduced sum $\sum_{n=0}^{\infty} \sum_{j=1}^{m} c_j/(k_j n + r_j)$ converges, then it can be written as a finite sum of the form

$$\sum_{j=1}^{m} \sum_{t=0}^{k_j-1} \frac{c_j}{k_j} (1 - \zeta^{-r_j t}) \log(1 - \zeta_j^t),$$

where ζ_j is a primitive k_jth root of unity. Here we recognize the linear form in logarithms of algebraic numbers with algebraic coefficients. According to Corollary 1.3 the double sum is either 0 or transcendental. Hence we obtain the following result.

Theorem 4 ([1]). *Let $P(x) \in \mathbb{Q}[x]$. Let $Q(x) \in \mathbb{Q}[x]$ have only simple rational zeros. If $S := \sum_{n=0}^{\infty} P(n)/Q(n)$ converges, then either S equals a computable rational number or S is transcendental.*

I call a polynomial $Q \in \mathbb{Q}[x]$ *reduced* if it has only simple rational zeros which are all in the interval $[-1, 0)$. If the polynomial Q in Theorem 4 is reduced, then the computable number equals 0. Hence, for example, it is obvious that the sum

$$\sum_{n=0}^{\infty} \frac{1}{(5n+1)(5n+3)(5n+5)}$$

is transcendental.

When applying Theorem 4 it may be important to exclude the case where the infinite sum S has the exceptional rational value, say q. If $S \neq q$, then this can be checked by numerical methods. If $S = q$, then the logarithms are linearly dependent over $\mathbb{Q}$ and the equality may be proved by simplifying the linear forms in logarithms to 0. For example, consider

$$\sum_{n=0}^{\infty} \left(\frac{1}{4n+1} - \frac{3}{4n+2} + \frac{1}{4n+3} + \frac{1}{4n+4} \right). \tag{4.1}$$

The corresponding linear form is

$$\log 2 - \log(1 - i) - \log(1 + i),$$

which happens to be 0. This example shows that a sum may vanish without
an obvious reason.

The same method can be applied to other infinite sums. In this way we
derived the following results. We denote here the field of algebraic numbers
by $\overline{\mathbb{Q}}$.

Theorem 5 ([1]). *Let* $f \colon \mathbb{Z}_{\geq 0} \to \overline{\mathbb{Q}}$ *be periodic mod* q. *Let* $Q(x) \in \mathbb{Q}[x]$
have only simple rational zeros. If $S = \sum_{n=0}^{\infty} f(n)/Q(n)$ *converges, then
either* S *equals a computable algebraic number or* S *is transcendental.*

Again the computable algebraic number equals 0 if Q is reduced. On
combining Theorem 5 with Dirichlet's result that $L(1, \chi) \neq 0$ for an arbitrary non-principal Dirichlet character χ, we immediately obtain the following application.

Corollary 5.1 ([1]). *Let* $q \geq 2$ *be an integer and* χ *a non-principal Dirichlet character* mod q. *Then* $L(1, \chi) = \sum_{n=1}^{\infty} \chi(n)/n$ *is transcendental.*

We call $f \colon \mathbb{Z}_{\geq 0} \to \mathbb{Q}$ *completely multiplicative* if $f(mn) = f(m) \cdot f(n)$
for all integers m, n. Dirichlet characters are completely multiplicative. In
Theorem 9, proved in the Appendix, it is shown that $\sum_{n=1}^{\infty} f(n)/n \neq 0$ if
$f \colon \mathbb{Z} \to \mathbb{Q}$ is periodic and completely multiplicative. Using this we obtain
the following variant of Corollary 5.1.

Corollary 5.2. *Let* $f \colon \mathbb{Z} \to \mathbb{Q}$ *be completely multiplicative and periodic.
Then* $\sum_{n=1}^{\infty} f(n)/n$ *is transcendental.*

Another result obtained by the method above is as follows.

Theorem 6 ([1]). *Let* $P_1(x), \ldots, P_l(x) \in \overline{\mathbb{Q}}[x]$, $\alpha_1, \ldots, \alpha_l \in \overline{\mathbb{Q}}$. *Put*
$g(x) = \sum_{\lambda=1}^{l} P_\lambda(x)\alpha_\lambda^x$. *Let* $Q(x)$ *have only simple rational zeros. If*
$S = \sum_{n=0}^{\infty} g(n)/Q(n)$ *converges, then either* S *equals a computable algebraic number or* S *is transcendental.*

Again the special value of S equals 0 if Q is reduced. It follows immediately from this observation and Theorem 6 that $\sum_{n=1}^{\infty} F_n/(n \cdot 2^n)$ is
transcendental, in view of

$$F_n = \frac{1}{\sqrt{5}} \left(\left(\frac{1 + \sqrt{5}}{2} \right)^n - \left(\frac{1 - \sqrt{5}}{2} \right)^n \right).$$

On the other hand, we have $\sum_{n=1}^{\infty} F_n/2^n = 2$. The example (4.1) also
shows that in Theorems 5 and 6 the exceptional value can be attained for
a non-trivial reason.

The question when $S = 0$ can occur in the special case that $Q(x) = x$ has been the subject of some conjectures of Chowla and Erdős. In 1952 Chowla [20, p. 300] incorrectly attributed the following problem to Erdős:

Let $f(x)$ be a number-theoretic (integer-valued) function which is periodic mod q. Suppose that not all $f(n)$ are 0. Then $\sum_{n=1}^{\infty} f(n)/n \neq 0$.

Already in 1949 Siegel had shown that this is true if q is prime (cf. [21]). Example (4.1) shows that Chowla's assertion is not true in general. Another example was given in 1973 by Baker, Birch and Wirsing ([4, p. 225]). According to Livingston [35] Erdős made the conjecture:

If q is a positive integer and $f(x)$ is a number-theoretic function mod q for which $f(n) \in \{-1, 1\}$ when $n = 1, 2, \ldots, q - 1$ and $f(q) = 0$, then $\sum_{n=1}^{\infty} f(n)/n \neq 0$ whenever the series is convergent.

In a lecture at the Stony Brook conference in 1969 Chowla raised the question whether there exists a rational-valued function $f(n)$ that is periodic with prime period p, such that $\sum_{n=1}^{\infty} f(n)/n = 0$. Chowla proved that, under some additional conditions, this is not possible. The general question, and more, was answered by Baker, Birch and Wirsing [4] in 1973, using Baker's theory.

Theorem 7 ([4]). *Suppose f is a non-vanishing function defined on the integers with algebraic values and period q such that*
  (i) $f(r) = 0$ *if* $1 < \gcd(r, q) < q$,
  (ii) *the cyclotomic polynomial Φ_q is irreducible over $\mathbb{Q}(f(1), \ldots, f(q))$.*
Then

$$\sum_{n=1}^{\infty} \frac{f(n)}{n} \neq 0.$$

If q is prime, then condition (i) is vacuous, and if f is rational-valued then (ii) holds trivially. Baker, Birch and Wirsing showed further that their theorem becomes false if (i) or (ii) is dropped. On the other hand, they showed that all functions f with algebraic values that are periodic mod q and for which (i) and $\sum_{n=1}^{\infty} f(n)/n = 0$ hold, are odd.

In 1982 Okada [41] gave a description of the functions f which satisfy condition (ii) of Theorem 7 and do not satisfy the conclusion of Theorem 7. Denoting the number of distinct prime factors of an integer m by $\omega(m)$ and Euler's function by $\phi(m)$, he gave a system of $\phi(m) + \omega(m)$ homogeneous linear equations with rational coefficients which is satisfied if and only if (ii) implies $\sum_{n=1}^{\infty} f(n)/n = 0$. The precise statement can be found in the Appendix. The non-vanishing result used in the proof of Corollary 5.2 is derived from Okada's theorem. Okada used his characterization to show that if $2\phi(q) + 1 > q$ and $f(n) \in \{1, -1\}$ when $n = 1, \ldots, q - 1$ and

$f(q) = 0$, then

$$\sum_{n=1}^{\infty} \frac{f(n)}{n} \neq 0$$

whenever the series is convergent. This gives a partial answer to Erdős' conjecture mentioned above. Okada's result covers the cases of Erdős' conjecture in which q is either a prime, a prime power or the product of two odd primes. As shown in the Appendix, it follows from Okada's criterion that Erdős' conjecture is also true if q is even. I am not able to establish Erdős' conjecture in full, let alone answer the question on the vanishing of S in general.

5 Diophantine Equations and Related Questions

There has been a tremendous stream of important results on diophantine equations in the past decades, culminating in the proof of Fermat's Last Theorem by Wiles and Taylor. In the shadow of the heavy arithmetic algebraic geometry machinery of Faltings and Wiles, which may be the subjects of other speakers, there are some remarkable new results obtained by diophantine approximation methods or combinations of these methods with other methods. Here I present a personal selection of some results which may be of general interest, but may not be well known.

5.1 Modular Curves

In 1929 Siegel proved that if a curve over a number field K has genus at least 1, then the number of integral points on the curve is finite. (In 1982 Faltings proved that if the genus is at least 2, then the number of K-points on the curve is finite.) In 1970 Baker and Coates [5] made Siegel's theorem effective for curves of genus 1 by using linear forms estimates. At the Oberwolfach meeting in April 2000 Bilu announced that using linear forms estimates he had made Siegel's theorem effective for congruence subgroups of finite index of modular curves.

5.2 Catalan's Equation

One of the first spectacular applications of Baker's theory on linear forms was the (effective) proof that the equation

$$x^p - y^q = 1 \text{ in integers } p > 1, q > 1, x > 1, y > 1$$

admits only finitely many solutions [57]. The initial upper bounds for the unknowns were huge. Mignotte and others have improved the upper and lower bound for unknown solutions. Last year Mihailescu [40] made a breakthrough by proving that, if there is a solution different from $3^2 - 2^3 = 1$ with p, q prime, then

$$p^{q-1} \equiv 1 \pmod{q^2} \quad \text{and} \quad q^{p-1} \equiv 1 \pmod{p^2}.$$

Using this result Mignotte obtained the following improvement (announced during this conference) on previous bounds (cf. [38]):

$$15 \cdot 10^6 < \min(p, q) < 8 \cdot 10^{11}.$$

5.3   The Equation $ax^n - by^n = c$

Another early application of Baker's theory was that, given a, b and $c \neq 0$, the equation $ax^n - by^n = c$ admits only finitely many solutions in integers $n \geq 3$, x, y; see [49, Ch. 5]. Bennett [8] has found a useful upper bound for the number of solutions if $c = \pm 1$ and n is fixed:

Let a, b and n be integers with $a > b \geq 1$ and $n \geq 3$. Then $|ax^n - by^n| = 1$ admits at most one solution in positive integers x, y.

5.4   Perfect Powers with Identical Decimal Digits

An old question is to determine all perfect powers which have identical digits in their decimal representation. This yields the diophantine equation

$$\frac{10^n - 1}{10 - 1} = ay^q \quad (1 \leq a \leq 9).$$

Obláth proved in 1956 that there are no solutions with $1 < a \leq 9$. In 1976 Shorey and Tijdeman showed that there are only finitely many exceptions (cf. [49, Ch. 12]). Recently Bugeaud [17] solved the problem completely by showing that there are no perfect powers with identical digits whatsoever.

More generally, Shorey and Tijdeman proved that the more general equation

$$\frac{x^n - 1}{x - 1} = y^q \text{ in integers } x > 1,\, y > 1,\, n > 2,\, q > 1 \tag{5.1}$$

admits only finitely many solutions if x or n is fixed, or if y has a fixed prime divisor. Bugeaud, Mignotte and Roy [18] have extended Bugeaud's result as follows:

If (5.1) admits a solution x, y, n, q with $n \geq 5$, then there exists a prime p such that p divides x and q divides $p - 1$. In particular, $x \geq 2q + 1$.

The proof uses Skolem's method and a slight refinement of the estimate for p-adic linear forms in two logarithms due to Bugeaud and Laurent. For more results on (5.1) and related equations I refer to [48].

5.5 The Equation of Goormaghtigh

One step further is the equation

$$\frac{x^m - 1}{x - 1} = \frac{y^n - 1}{y - 1} \text{ in integers } x > 1, y > 1, m > n > 1.$$

The only known solutions are due to Goormaghtigh:

$$\frac{2^5 - 1}{2 - 1} = \frac{5^3 - 1}{5 - 1}, \quad \frac{2^{13} - 1}{2 - 1} = \frac{90^3 - 1}{90 - 1}.$$

This says that the numbers 31 and 8191 both have identical digits in their expansions with respect to two different bases. Shorey and Tijdeman (cf. [49, Ch. 12]) showed that if x and y are fixed, then m and n are bounded in terms of x and y. A recent result of Bugeaud and Shorey [19] says that for given x and y there are at most two solutions (m, n).

5.6 The abc-Conjecture

In 1985 Masser stated the following conjecture which refines a conjecture of Oesterlé and is now known as the abc-conjecture.

Let $a, b, c \in \mathbb{Z}_{>0}$. For every $\epsilon > 0$ there exists a number $C(\epsilon) > 0$ such that if $a + b = c$ and $\gcd(a, b, c) = 1$, then

$$c < C(\epsilon) N^{1+\epsilon} \text{ where } N = \prod_{p \mid abc} p.$$

By using p-adic linear forms in logarithms, Stewart and I [51] proved that for all such a, b, c,

$$\log c < c_1 N^{15} \qquad (c_1 \text{ some constant}).$$

In the other direction, we showed that the conjecture is not true with

$$N \cdot \exp\left((4 - \epsilon)\frac{\sqrt{\log N}}{\log\log N}\right) \tag{5.2}$$

as upper bound for c, for any $\epsilon > 0$. The constants in both of these results have recently been improved. By using the p-adic version of Matveev's refinement Stewart and Yu [52], [53] derived the inequality

$$\log c < c_2 N^{1/3} (\log N)^3 \qquad (c_2 \text{ constant}).$$

Using an idea of H. W. Lenstra, van Frankenhuysen [60] replaced the constant $(4 - \epsilon)$ in the exponent of (5.2) by 6.068.

5.7 Sums of Two Powers Being a Power

Darmon and Granville [23] proved that, for given positive integers k, l, m with $(1/k) + (1/l) + (1/m) < 1$, the equation

$$x^k + y^l = z^m \text{ in coprime integers } x, y, z \qquad (5.3)$$

admits only finitely many solutions. If the *abc*-conjecture is true, then the total number of solutions k, l, m, x, y, z is finite ([58, p. 234]). The proof of Fermat's Last Theorem by Wiles and Taylor [63], [54] shows that (5.2) has no solutions if $k = l = m$. While preparing for a "Fermat day" for a broad audience in Utrecht in November 1993 on this celebrated result, Beukers and Zagier [9] found five new large solutions of (5.3). At present, the following solutions are known:

$$1^m + 2^3 = 3^2,$$
$$13^2 + 7^3 = 2^9,$$
$$2^7 + 17^3 = 71^2,$$
$$2^5 + 7^2 = 3^4,$$
$$3^5 + 11^4 = 122^2,$$
$$17^7 + 76721^3 = 21063928^2,$$
$$1414^3 + 2213459^2 = 65^7,$$
$$33^8 + 1549034^2 = 15613^3,$$
$$43^8 + 96222^3 = 30042907^2,$$
$$9262^3 + 15312283^2 = 133^7$$

When I presented these examples during this Fermat day, I noticed that in all examples a square occurs, and I formulated the following conjecture. Later, Beal attached a reward to the resolution of this problem (cf. [37]), and the conjecture has become known as the Beal Prize Problem.

Let x, y, z, k, l, m be positive integers with $k > 2$, $l > 2$, $m > 2$. If $x^k + y^l = z^m$, then x, y, z have a factor in common.

For certain triples (k, l, m) it has recently been established that there are no solutions of (5.3) other than those listed above. All of these results were obtained by geometric methods, and some of them were subject to the Taniyama-Shimura-Weil conjecture, which has recently been proved; see [12]. Darmon [22] proved that (5.3) admits no non-trivial solutions when $k = l = p > 13$ is a prime $\equiv 1 \pmod 4$ and $m = 2$. Kraus [32] proved that there are no non-trivial solutions of (5.3) if $k = l = 3$ and m is a prime with $16 < m < 10000$. Bruin [15] used Chabauty's method to extend Kraus'

result to $m = 4$ and $m = 5$. In addition, Bruin [13], [14] showed that
the above list contains all the solutions (modulo $\pm$ signs) for the exponent
triples $(2, 3, 8)$, $(2, 4, 6)$ and $(2, 4, 5)$.

5.8 Arithmetic Progressions of Powers and Binomials Which Are Powers

Fermat claimed, and Euler proved, that four positive integers in arithmetical progression cannot be all squares. There are many triples of squares in arithmetic progression; examples are $1, 25, 49$ and $1, 841, 1641$. Darmon and Merel [24] used an extension of Wiles' method to prove that there exists no integer $n > 2$ for which there are triples of n-th powers in arithmetic progression. This beautiful result was used by Győry [31] to complete the solution of another classical problem. Erdős [25] had shown that

$$\binom{n}{k} = y^l$$

has no solution in positive integers $n \geq 2k \geq 8$, $y > 1, l > 1$. Győry treated the remaining cases $k = 2, 3$ by a combination of estimates for p-adic linear forms in logarithms, Eisenstein's reciprocity theorem and the above-mentioned result of Darmon and Merel. The only solution with $kl > 4$ turns out to be

$$\binom{50}{3} = 140^2.$$

There are many related results in the literature.

5.9 Primitive Divisors of Lucas and Lehmer Sequences

A *primitive divisor* of the n-th term of the sequence $(u_n)_{n=1}^\infty$ is a prime which divides u_n, but not u_m for any $m < n$. It follows from old results of Bang, Zsigmondy, and Birkhoff and Vandiver that $a^n - b^n$ has a primitive divisor for $n > 6$ when $a > b > 0$ are integers. For example, for $a = 2$, $b = 1$ we obtain the sequence $1, 3, 7, 15, 31, 63, 127, \ldots$, and according to this result 63 is the last term which does not have new a prime factor in the sequence. *Lucas numbers* are defined as follows: Let P and Q be coprime integers and let α and β be distinct roots of the equation $x^2 - Px - Q = 0$. Then $u_n := (\alpha^n - \beta^n)/(\alpha - \beta)$ and $v_n := \alpha^n + \beta^n$. Lehmer sequences are defined in a similar way. It follows from work of Carmichael, Ward and Durst that if α and β are reals, then there are primitive divisors in Lucas and Lehmer numbers for any $n > 12$. The Fibonacci sequence shows that this result cannot be improved in general:

$1, 1, 2, 3, 5, 8, 13, 21, 34, 55, 89, 144, 233, \ldots$ and the twelfth term 144 has no primitive divisor as its prime factors 2 and 3 both occur in previous terms of the sequence. For non-real α and β upper bounds for the integers n for which the n-th term of the sequence has not a primitive divisor were given by Schinzel and Stewart in the 1970s (see e.g. [50]). Voutier [61] found the much smaller upper bound 30030. Bilu, Hanrot and Voutier [11] have now reached the best possible upper bound 30. This result has been used in quite a few applications.

5.10 Linear Recurrence Sequences

Lucas and Lehmer sequences are special cases of linear recurrence sequences. A *(homogeneous linear) recurrence sequence* $(u_n)_{n=1}^\infty$ *of order* r satisfies a relation

$$u_n = a_1 u_{n-1} + \cdots + a_r u_{n-r} \text{ for } n > r,$$

where the numbers $a_1, \ldots, a_r$ and the initial values $u_1, \ldots, u_r$ are given. A strong feature of the p-adic Subspace Theorem is that it gives upper bounds for the numbers of solutions which depend on surprisingly few parameters. A major result in this direction, due to Schlickewei [44], states that if the zeros of the companion polynomial $x^r - a_1 x^{r-1} \cdots - a_r$ lie in a number field of degree d and no two roots have a root of unity as their ratio, then a linear recurrence sequence of order r has at most

$$d^{6r^2} \cdot 2^{28r!}$$

terms equal to 0. Beukers and Schlickewei [10] found a much better upper bound for a recurrence sequence of complex numbers of order 3 with a companion polynomial having simple zeros, namely 61. Schmidt [47] proved that for any recurrence sequence of complex numbers of order r for which the companion polynomial does not have two distinct roots with a root of unity as ratio, the number of terms 0 is bounded by a bound which depends only on r. Schmidt's bound is triply exponential in r. I refer to [29] for bounds which are "only" doubly exponential in r.

Appendix: On Erdős' Conjecture on the Vanishing of Infinite Sums

As mentioned in Section 4, the following conjecture is attributed to Erdős:
Suppose q is a positive integer and $f(x)$ is a number-theoretic function that is periodic mod q with values $f(n) \in \{-1, 1\}$ when $n = 1, 2, \ldots, q - 1$ and $f(q) = 0$. Then $\sum_{n=1}^\infty f(n)/n \neq 0$ whenever the series is convergent.

It follows from Theorem 7 that this holds if q is prime. Furthermore, Okada [41] proved that the assertion is true if $2\phi(q) + 1 > q$. He derived this result from the following criterion (cf. [41], Theorem 10 and the proof of Corollary 15).

Theorem 8 ([41]). *If f satisfies condition* (ii) *of Theorem 7, then we have $\sum f(n)/n = 0$ if and only if $(f(1), \dots, f(q))$ is a solution of the following system of $\phi(q) + \omega(q)$ homogeneous linear equations with rational coefficients*

$$f(a) + \sum_{d\mid q, 1 < d < q} \prod_{p \in P(d)} \left(1 - \frac{1}{p^{\phi(q)}}\right)^{-1} \sum_{n \in S(d)} \frac{f(adn)}{dn}$$

$$+ \frac{f(q)}{\phi(q)} = 0 \ (a \in J), \tag{5.4}$$

$$\sum_{r \in L} f(r)\epsilon(r, p) = 0 \text{ for all prime divisors } p \text{ of } q, \tag{5.5}$$

where

$$J = \{a \mid 1 \le a < q, \gcd(a, q) = 1\},$$
$$L = \{r \mid 1 \le r \le q, \gcd(r, q) > 1\},$$
$$P(d) = \left\{p \text{ prime} \mid p \mid q, \operatorname{ord}_p(d) \ge \operatorname{ord}_p(q)\right\},$$
$$\epsilon(r, p) = \begin{cases} \operatorname{ord}_p(q) + \dfrac{1}{p-1} & \text{if } p \in P(r), \\ \operatorname{ord}_p(r) & \text{otherwise} \end{cases}$$

and

$$S(d) = \left\{\prod_{p \in P(d)} p^{\alpha(p)} \mid 0 \le \alpha(p) < \phi(q)\right\}.$$

Okada derived from (5.4) that $2\phi(q) + 1 \le q$. Since (ii) is satisfied, this implies that $\sum_{n=1}^{\infty} f(n)/n \ne 0$, if q is either a prime, a prime power, or the product of two odd primes. Note that it follows from the convergence of $\sum f(n)/n$ that $\sum_{n=1}^{q} f(n) = 0$. Hence Erdős' conjecture is true if q is even.

Let $f: \mathbb{Z} \to \mathbb{Q}$ be periodic mod q such that $\sum_{n=0}^{\infty} f(n)/n = 0$. From Theorem 8 we immediately obtain the result of Baker, Birch and Wirsing [4] that q is composite. Okada's criterion gives a good way to construct such functions f: the values $f(n)$ with $\gcd(n, q) > 1$ and n composite can be chosen arbitrarily; then the values $f(n)$ with n a prime divisor of q are fixed by (5.5); finally, values $f(n)$ with $\gcd(n, d) = 1$ are determined by (5.4).

The structure of (5.4) and (5.5) becomes much more transparent if f is multiplicative. Let $f \colon \mathbb{Z} \to \mathbb{Q}$ be a multiplicative function that is periodic mod q. Observe that it follows from $f(a)f(q) = f(aq) = f(q)$ for every $a \in J$ that

$$f(q) = 0 \text{ or } f(a) = 1 \text{ for every } a \in J. \tag{5.6}$$

In (5.4) we have that $\gcd(a, dn) = 1$. Hence $f(adn) = f(a) \cdot f(dn)$. Since $f(1) = 1$, we see that (5.4) is equivalent to the single equation

$$1 + \sum_{d\mid q, 1 < d < q} \prod_{p \in P(d)} \left(1 - \frac{1}{p^{\phi(q)}}\right)^{-1} \sum_{n \in S(d)} \frac{f(dn)}{dn} + \frac{f(q)}{\phi(q)} = 0.$$

Since

$$\frac{f(q)}{\phi(q)} = \sideset{}{^*}\sum_{n} \frac{f(qn)}{qn},$$

where the summation is extended over all positive integers n which are composed of prime divisors of q, the left-hand side equals

$$\sideset{}{^*}\sum_{n} \frac{f(n)}{n} = \prod_{p \mid q} \left(1 + \frac{f(p)}{p} + \frac{f(p^2)}{p^2} + \cdots\right).$$

Hence, for multiplicative functions f with period q, condition (5.4) is equivalent to

$$1 + \frac{f(p)}{p} + \frac{f(p^2)}{p^2} + \cdots = 0 \text{ for some prime divisor } p \text{ of } q. \tag{5.7}$$

It follows immediately from (5.7) and Theorem 8 with (5.7) in place of (5.4) that the only periodic multiplicative functions $f \colon \mathbb{Z} \to \mathbb{Q}$ with period 4 are given by (4.1); those with period 6 are given by $f(n) = 0, 1, -3, 0, 3, -1$ and $f(n) = 2, 1, -1, -2, -1, 1$ for $n \equiv 0, 1, 2, 3, 4, 5 \bmod 6$, respectively; and those with period 8 by $f(n) = 0, 1, -2, -1, 0, 1, 2, -1$ and $f(n) = 4 + t, 1, t, 1, -3t - 8, 1, t, 1$ for $n \equiv 0, 1, 2, 3, 4, 5, 6, 7 \bmod 8$, respectively, where $t \in \mathbb{Q}$ can be chosen arbitrarily. On the other hand, we now derive the following result.

Theorem 9. *Let $f \colon \mathbb{Z} \to \mathbb{Q}$ be multiplicative and have period q. If $|f(p^k)| < p - 1$ for every prime divisor p of q and every positive integer k, then $\sum_{n=1}^{\infty} f(n)/n \neq 0$.*

Proof. It follows from (5.7) that there is a prime divisor p of the period q such that $f(p^k) = -(p - 1)$ for $k = 1, 2, \ldots$. This yields a contradiction to (5.5), applied to this prime p. $\qquad\square$

It follows from Theorem 9 that Erdős' statement is true for multiplicative functions f.

If f is completely multiplicative and periodic mod q, then (5.7) implies that q is even and $f(p) = -p$ for some prime divisor p of q. Hence $f(p^k) = (-p)^k$ for every positive integer k, which is impossible for a periodic function. Thus we have:

Theorem 10. *Let $f: \mathbb{Z} \to \mathbb{Q}$ be completely multiplicative and periodic. Then $\sum_{n=1}^{\infty} f(n)/n \neq 0$.*

By combining this result with Theorem 5 we obtain Corollary 5.2.

Acknowledgements. I thank F. Beukers, Yu. F. Bilu, J.-H. Evertse, K. Győry, Yu. V. Nesterenko and T. N. Shorey for useful discussions and comments on early versions of the paper.

References

[1] S. D. Adhikari, N. Saradha, T. N. Shorey, and R. Tijdeman, *Transcendental infinite sums*, Indag. Math. **12** (2001), 1–14.

[2] A. Baker, *Transcendental number theory*, Cambridge University Press, 1975.

[3] ———, *The theory of linear forms in logarithms*, Transcendence Theory: Advances and Applications (A. Baker and D. W. Masser, eds.), Academic Press, 1977, pp. 1–27.

[4] A. Baker, B. J. Birch, and E. A. Wirsing, *On a problem of Chowla*, J. Number Theory **5** (1973), 224–236.

[5] A. Baker and J. Coates, *Integral points on curves of genus 1*, Proc. Cambridge Philos. Soc. **67** (1970), 595–602.

[6] A. Baker and G. Wüstholz, *Logarithmic forms and group varieties*, J. Reine Angew. Math. **442** (1993), 19–62.

[7] C. D. Bennett, J. Blass, A. M. W. Glass, D. B. Meronk, and R. P. Steiner, *Linear forms in the logarithms of three positive rational integers*, J. Théor. Nombres Bordeaux **9** (1997), 97–136.

[8] M. Bennett, *Rational approximation to algebraic numbers of small height: The Diophantine equation $|ax^n - by^n| = 1$*, J. Reine Angew. Math. **535** (2001), 1–49.

[9] F. Beukers, *The diophantine equation $Ax^p + By^q = Cz^r$*, Duke Math. J. **91** (1998), 61–88.

[10] F. Beukers and H. P. Schlickewei, *The equation $x + y = 1$ in finitely generated groups*, Acta Arith. **78** (1996), 189–199.

[11] Y. F. Bilu, G. Hanrot, and P. M. Voutier, *Existence of primitive divisors of Lucas and Lehmer numbers (with an appendix by M. Mignotte)*, J. Reine Angew. Math. **539** (2001), 75–122.

[12] C. Breuil, B. Conrad, F. Diamond, and R. Taylor, *On the modularity of elliptic curves over $\mathbb{Q}$*, J. Amer. Math. Soc. **14** (2001), 843–939.

[13] N. Bruin, *Chabauty methods and covering techniques applied to generalised Fermat equations*, Ph.D. thesis, Leiden University, 1999.

[14] _____, *The diophantine equations $x^2 \pm y^4 = \pm z^6$ and $x^2 + y^8 = z^3$*, Compositio Math. **118** (1999), 305–321.

[15] _____, *On powers as sums of two cubes*, Algorithmic Number Theory (W. Bosma, ed.), LNCS 1838, Springer, 2000, pp. 169–184.

[16] Y. Bugeaud and M. Laurent, *Minoration effective de la distance p-adique entre puissances de nombres algébriques*, J. Number Theory **61** (1996), 311–342.

[17] Y. Bugeaud and M. Mignotte, *On integers with identical digits*, Mathematika **46** (1999), 411–417.

[18] Y. Bugeaud, M. Mignotte, and Y. Roy, *On the diophantine equation $\frac{x^n-1}{x-1} = y^q$*, Pacific J. Math. **193** (2000), 257–268.

[19] Y. Bugeaud and T. N. Shorey, *On an equation of Goormaghtigh, II*, to appear.

[20] S. Chowla, *The Riemann zeta and allied functions*, Bull. Amer. Math. Soc. **58** (1952), 287–305.

[21] _____, *The nonexistence of nontrivial linear relations between the roots of a certain irreducible equation*, J. Number Theory **2** (1970), 120–123.

[22] H. Darmon, *The equations $x^n + y^n = z^2$ and $x^n + y^n = z^3$*, Int. Math. Res. Notices **10** (1993), 263–274.

[23] H. Darmon and A. Granville, *On the equations $z^m = F(x,y)$ and $Ax^p + By^q = Cz^r$*, Bull. London Math. Soc. **27** (1995), 513–543.

[24] H. Darmon and L. Merel, *Winding quotients and some variants of Fermat's last theorem*, J. Reine Angew. Math. **490** (1997), 81–100.

[25] P. Erdős, *On a diophantine equation*, J. London Math. Soc. **26** (1951), 176–178.

[26] P. Erdős, C. L. Stewart, and R. Tijdeman, *Some diophantine equations with many solutions*, Compositio Math. **66** (1988), 37–56.

[27] J.-H. Evertse, *On sums of S-units and linear recurrences*, Compositio Math. **53** (1984), 225–244.

[28] ———, *The number of solutions of decomposable form equations*, Invent. Math. **122** (1995), 559–601.

[29] J.-H. Evertse, H. P. Schlickewei, and W. M. Schmidt, *Linear equations in variables which lie in a multiplicative group*, to appear.

[30] J.-H. Evertse, C. L. Stewart, and R. Tijdeman, *Multivariate diophantine equations with many solutions*, to appear.

[31] K. Győry, *On the diophantine equation $\binom{n}{k} = x^l$*, Acta Arith. **80** (1997), 289–295.

[32] A. Kraus, *Sur l'équation $a^3 + b^3 = c^p$*, Experiment. Math. **7** (1998), 1–13.

[33] M. Laurent, M. Mignotte, and Y. V. Nesterenko, *Formes linéaires en deux logarithmes et déterminants d'interpolation*, J. Number Theory **55** (1995), 285–321.

[34] D. H. Lehmer, *Euler constants for arithmetical progressions*, Acta Arith. **27** (1975), 125–142.

[35] A. E. Livingston, *The series $\sum_1^\infty f(n)/n$ for periodic f*, Canad. Math. Bull. **8** (1965), 413–432.

[36] E.M. Matveev, *An explicit lower bound for a homogeneous rational linear form in logarithms of algebraic numbers. II*, Izv. Math. **64** (2000), 1217 – 1269.

[37] R. D. Mauldin, *A generalization of Fermat's Last Theorem: the Beal conjecture and prize problem*, Notices Amer. Math. Soc. **44** (1997), 1436–1437.

[38] M. Mignotte, *Lower bounds for Catalan's equation*, The Ramanujan J. **1** (1997), 351–356.

[39] _____, *A corollary to the Laurent-Mignotte-Nesterenko theorem*, Acta Arith. **86** (1998), 101–111.

[40] P. Mihailescu, *A class number free criterion for Catalan's conjecture*, to appear.

[41] T. Okada, *On a certain infinite series for a periodic arithmetical function*, Acta Arith. **40** (1982), 143–153.

[42] H. P. Schlickewei, *The p-adic Thue-Siegel-Roth-Schmidt theorem*, Arch. Math. **29** (1977), 267–270.

[43] _____, *The quantitative Subspace Theorem for number fields*, Compositio Math. **82** (1992), 245–274.

[44] _____, *Multiplicities of recurrence sequences*, Acta Math. **176** (1996), 171–243.

[45] W. M. Schmidt, *Norm form equations*, Ann. Math. **96** (1972), 526–551.

[46] _____, *The subspace theorem in diophantine approximation*, Compositio Math. **69** (1989), 121–173.

[47] _____, *The zero multiplicity of linear recurrences*, Acta Math. **182** (1999), 243–282.

[48] T. N. Shorey, *Some conjectures in the theory of exponential diophantine equations*, Publ. Math. Debrecen **56** (2000), 631–641.

[49] T. N. Shorey and R. Tijdeman, *Exponential diophantine equations*, Cambridge University Press, 1986.

[50] C. L. Stewart, *On divisors of Fermat, Fibonacci, Lucas and Lehmer numbers*, Proc. Lond. Math. Soc. **35** (1977), 425–447.

[51] C. L. Stewart and R. Tijdeman, *On the Oesterlé-Masser conjecture*, Monatsh. Math. **102** (1986), 251–257.

[52] C. L. Stewart and K. R. Yu, *On the abc-conjecture*, Math. Ann. **291** (1991), 225–230.

[53] _____, *On the abc-conjecture II*, Duke Math. J. **108** (2001), 169–181.

[54] R. L. Taylor and A. Wiles, *Ring-theoretic properties of certain Hecke algebras*, Ann. Math. **141** (1995), 553–572.

[55] R. Tijdeman, *On integers with many small prime factors*, Compositio Math. **26** (1973), 319–330.

[56] ———, *On the maximal distance between numbers composed of fixed primes*, Compositio Math. **28** (1974), 159–162.

[57] ———, *On the equation of Catalan*, Acta Arith. **29** (1976), 197–209.

[58] ———, *Diophantine equations and diophantine approximations*, Number Theory and Applications (R. Mollin, ed.), Kluwer, Dordrecht etc., 1989, pp. 215–243.

[59] A. J. van der Poorten and H. P. Schlickewei, *Additive relations in fields*, J. Austral. Math. Soc. Ser. A **51** (1991), 154–170.

[60] M. van Frankenhuysen, *A lower bound in the abc-conjecture*, J. Number Theory **82** (2000), 91–95.

[61] P. M. Voutier, *Primitive divisors of Lucas and Lehmer numbers III*, Math. Proc. Cambridge Philos. Soc. **123** (1998), 407–419.

[62] M. Waldschmidt, *Minorations de combinaisons linéaires de logarithmes de nombres algébriques*, Canadian J. Math. **45** (1993), 176–224.

[63] A. Wiles, *Modular elliptic curves and Fermat's Last Theorem*, Ann. Math. **141** (1995), 443–551.

[64] K. Yu, *p-adic logarithmic forms and group varieties I*, J. Reine Angew. Math. **502** (1998), 29–92.

[65] ———, *p-adic logarithmic forms and group varieties II*, Acta Arith. **89** (1999), 337–378.

Waring's Problem: A Survey

R. C. Vaughan[1] and T. D. Wooley[2]

1 The Classical Waring Problem

"Omnis integer numerus vel est cubus, vel e duobus, tribus, 4, 5, 6, 7, 8, vel novem cubis compositus, est etiam quadrato-quadratus vel e duobus, tribus, &c. usque ad novemdecim compositus, & sic deinceps."

Waring [150, pp. 204-5].

"Every integer is a cube or the sum of two, three, ... nine cubes; every integer is also the square of a square, or the sum of up to nineteen such; and so forth."

Waring [152, p. 336].

It is presumed that by this, in modern notation, Waring meant that for every $k \geq 3$ there are numbers s such that every natural number is the sum of at most s k-th powers of natural numbers and that the smallest such number $g(k)$ satisfies $g(3) = 9$, $g(4) = 19$.

By the end of the nineteenth century, the existence of $g(k)$ was known for only a finite number of values of k. There is an account of this work in Dickson [48], and as far as we have been able to ascertain, by 1909 its existence was known for $k = 3, 4, 5, 6, 7, 8, 10$, but not for any larger k (of course, with the natural extension of the definition of $g(k)$, Lagrange proved in 1770 that $g(2) = 4$). However, starting with Hilbert [69], who showed that $g(k)$ does indeed exist for every k, the twentieth century has seen an almost complete solution of this problem. Let $[x]$ denote the greatest integer not exceeding x and write $\{x\}$ for $x - [x]$. As the result of the work of many mathematicians we now know that

$$g(k) = 2^k + [(3/2)^k] - 2,$$

provided that

$$2^k\{(3/2)^k\} + [(3/2)^k] \leq 2^k. \tag{1.1}$$

If this fails, then

$$g(k) = 2^k + [(3/2)^k] + [(4/3)^k] - \theta$$

[1] Research supported by NSF grant DMS-9970632.
[2] Packard Fellow, and supported in part by NSF grant DMS-9970440.

where θ is 2 or 3 according as

$$[(4/3)^k][(3/2)^k] + [(4/3)^k] + [(3/2)^k]$$

equals or exceeds 2^k.

The condition (1.1) is known to hold (Kubina & Wunderlich [85]) whenever $k \leq 471,600,000$, and Mahler [91] has shown that there are at most a finite number of exceptions. To complete the proof for all k it would suffice to know that $\{(3/2)^k\} \leq 1-(3/4)^{k-1}$. Beukers [3] has shown that whenever $k > 5,000$ one has $\{(3/2)^k\} \leq 1 - a^k$, where $a = 2^{-0.9} = 0.5358...$, and this has been improved slightly by Dubitskas [49] to $a = 0.5769...$, so long as k is sufficiently large (see also Bennett [1] for associated estimates). A problem related to the evaluation of $g(k)$ now has an almost definitive answer. Let $g_r(k)$ denote the smallest integer s with the property that every natural number is the sum of at most s elements from the set $\{1^k, r^k, (r+1)^k, ...\}$. Then Bennett [2] has shown that for $4 \leq r \leq (k+1)^{1-1/k} - 1$, one has $g_r(k) = r^k + [(1+1/r)^k] - 2$.

By the way, before turning to the modern form of Waring's problem, it is perhaps worth observing that in the 1782 edition of *Meditationes Algebraicæ*, Waring makes an addition:

"confimilia etiam affirmari possunt (exceptis excipiendis) de eodem numero quantitatum earundem dimensionum."

Waring [151, p. 349].

"similar laws may be affirmed (exceptis excipiendis) for the correspondingly defined numbers of quantities of any like degree."

Waring [152, p. 336].

It would be interesting to know exactly what Waring had in mind. This, taken with some of the observations which immediately follow the remark, suggest that for more general polynomials than the k-th powers he was aware that some kind of local conditions can play a rôle in determining when representations occur.

2   The Modern Problem

The value of $g(k)$ is determined by the peculiar behaviour of the first three or four k-th powers. A much more challenging question is the value, for $k \geq 2$, of the function $G(k)$, the smallest number t such that every *sufficiently large* number is the sum of at most t k-th powers of positive integers. The function $G(k)$ has only been determined for two values of k, namely $G(2) = 4$, by Lagrange in 1770, and $G(4) = 16$, by Davenport [30]. The bulk of what is known about $G(k)$ has been obtained through the medium of the Hardy-Littlewood method. This has its genesis in a celebrated paper of Hardy and Ramanujan [64] devoted to the partition function. In this paper (section 7.2) there is also a brief discussion about the representation of a natural number as the sum of a fixed number of squares of integers, and there seems little doubt that it is the methods described therein which inspired the later work of Hardy and Littlewood.

Our knowledge concerning the function $G(k)$ currently leaves much to be desired. If, instead of insisting that all sufficiently large numbers be represented in a prescribed form, one rather asks that *almost all* numbers (in the sense of natural density) be thus represented, then the situation is somewhat improved. Let $G_1(k)$ denote the smallest number u such that almost every number n is the sum of at most u k-th powers of positive integers. The function $G_1(k)$ has been determined in five non-trivial instances as follows:

$$\text{Davenport [29],} \quad G_1(3) = 4,$$
$$\text{Hardy and Littlewood [62],} \quad G_1(4) = 15,$$
$$\text{Vaughan [121],} \quad G_1(8) = 32,$$
$$\text{Wooley [155],} \quad G_1(16) = 64,$$
$$\text{Wooley [155],} \quad G_1(32) = 128$$

(of course, the conclusion $G_1(2) = 4$ is classical).

3 General Upper Bounds for G(k)

The first explicit general upper bound for $G(k)$, namely

$$G(k) \le (k-2)2^{k-1} + 5,$$

was obtained by Hardy and Littlewood [61] (in [58] and [59], only the existence of $G(k)$ is stated, although it is already clear that in principle their method gave an explicit upper bound). In Hardy and Littlewood [62] this is improved to

$$G(k) \le (k-2)2^{k-2} + k + 5 + [\zeta_k],$$

where

$$\zeta_k = \frac{(k-2)\log 2 - \log k + \log(k-2)}{\log k - \log(k-1)}.$$

There has been considerable activity reducing this upper bound over the years, and Table 1 below presents upper bounds for $G(k)$ that were probably the best that were known at the time they appeared, at least for

Vinogradov [136], $32(k \log k)^2$,

Vinogradov [137][139], $k^2 \log 4 + (2 - \log 16)k$ $(k \geq 3)$,

Vinogradov [135] [138] [140] [143], $6k \log k + 3k \log 6 + 4k$ $(k \geq 14)$,

Vinogradov [147], $k(3 \log k + 11)$,

Tong [114], $k(3 \log k + 9)$,

Jing-Run Chen [24], $k(3 \log k + 5.2)$,

Vinogradov [148], $2k(\log k + 2 \log \log k + O(\log \log \log k))$,

Vaughan [124], $2k(\log k + \log \log k + O(1))$,

Wooley [155], $k(\log k + \log \log k + O(1))$.

Table 1. General upper bounds for $G(k)$

k sufficiently large. This list is not exhaustive. In particular, there is a long sequence of papers by Vinogradov between 1934 and 1947, and for further details we refer the reader to the Royal Society obituary of I. M. Vinogradov (see Cassels and Vaughan [23]).

The last entry on this list has been refined further by Wooley [159], and this provides the estimate

$$G(k) \leq k(\log k + \log \log k + 2 + O(\log \log k / \log k))$$

that remains the sharpest available for larger exponents k.

4   Cubes

For small values of k there are many special variants of the Hardy-Littlewood method that have been developed. However, in the case of cubes, until recently the best upper bounds were obtained by rather different methods that related cubes to quadratic forms, especially sums of squares. Thus Landau [86] had shown that $G(3) \leq 8$, and this bound was reduced by Linnik [87][88] to $G(3) \leq 7$, with an alternative and simpler proof given by Watson [153]. Only with the advent of refinements to the circle method utilising efficient differencing did it become feasible (Vaughan [119]) to give a proof of the bound $G(3) \leq 7$ via the Hardy-Littlewood method. Subsequent developments involving the use of smooth numbers (see Vaughan [124][125] and Wooley [158]) have provided a more powerful approach to this problem that, from a practical point of view, is more direct than earlier treatments. Complicated nonetheless, these latter proofs yield much more information concerning Waring's problem for cubes. We can illustrate the latter observation with two examples which, in the absence of foreseeable progress on the upper bound for $G(3)$, provide the problems central to current activity surrounding Waring's problem for cubes.

When X is a large real number, denote by $E(X)$ the number of positive integers not exceeding X that *cannot* be written as a sum of four positive integral cubes. Then the conclusion $G_1(3) = 4$, attributed above to Davenport [29], is an immediate consequence of the estimate $E(X) \ll X^{29/30+\epsilon}$ established in the latter paper. Following subsequent work of Vaughan [119], Brüdern [10][12], and Wooley [158], the sharpest conclusion currently available (see Wooley [162]) shows that $E(X) \ll X^{1-\beta}$ for any positive number β smaller than

$$(422 - 6\sqrt{2833})/861 = 0.119215\ldots.$$

It is conjectured that $G(3) = 4$ (see §10 below), and this would imply that $E(X) \ll 1$.

Consider next the density of integers represented as a sum of three positive integral cubes. When X is a large real number, let $N(X)$ denote the number of positive integers of the latter type not exceeding X. It is conjectured that $N(X) \gg X$, and following work of Davenport [29][33], Vaughan [118][119], Ringrose [100], Vaughan [124] and Wooley [158], the sharpest currently available conclusion due to Wooley [162] establishes that $N(X) \gg X^{1-\alpha}$ for any real number α exceeding

$$(\sqrt{2833} - 43)/123 = 0.083137\ldots.$$

We remark that, subject to the truth of an unproved Riemann Hypothesis concerning certain Hasse-Weil L-functions, one has the conditional estimate $N(X) \gg X^{1-\epsilon}$ due to Hooley [73][74], and Heath-Brown [67]. Unfortunately, the underlying L-functions are not yet known even to have an analytic continuation inside the critical strip.

We finish our discussion of Waring's problem for cubes by noting that, while Dickson [47] was able to show that 23 and 239 are the only positive integers not represented as the sum of eight cubes of natural numbers, no such conclusion is yet available for sums of seven or fewer cubes (but see McCurley [92] for sums of seven cubes, and Deshouillers, Hennecart and Landreau [44] for sums of four cubes).

5 Biquadrates

Davenport's definitive statement that $G(4) = 16$ is not the end of the story for sums of fourth powers (otherwise known as biquadrates). Let

$G^{\#}(4)$ denote the least integer s_0 such that whenever $s \geq s_0$, and $n \equiv r$ (mod 16) for some integer r with $1 \leq r \leq s$, then n is the sum of at most s biquadrates. Then Davenport [30] showed that $G^{\#}(4) \leq 14$, and this has been successively reduced by Vaughan [121][124] to $G^{\#}(4) \leq 12$. In an ironic twist of fate, the polynomial identity

$$x^4 + y^4 + (x+y)^4 = 2(x^2 + xy + y^2)^2,$$

reminiscent of identities employed in the nineteenth century, has recently been utilised to make yet further progress. Thus, when $n \equiv r$ (mod 16) for some integer r with $1 \leq r \leq 10$, Kawada and Wooley [82] have shown that n is the sum of at most 11 biquadrates. This and allied identities have also permitted the proof of an effective version of Davenport's celebrated theorem. Thus, as a consequence of work of Deshouillers, Hennecart and Landreau [45] and Deshouillers, Kawada and Wooley [46], it is now known that all integers exceeding $13,792$ may be written as the sum of at most sixteen biquadrates. A detailed history of Waring's problem for biquadrates is provided in Deshouillers, Hennecart, Kawada, Landreau and Wooley [43].

6   Upper Bounds for G(k) when $5 \leq k \leq 20$

Although we have insufficient space to permit a comprehensive account of the historical evolution of available upper bounds for $G(k)$ for smaller values of k, in Table 2 we have recorded many of the key developments, concentrating on the past twenty-five years. Each row in this table presents the best upper bound known for $G(k)$, for the indicated values of k, at the time of publication of the cited work. We note that the claimed bound $G(7) \leq 52$ of Sambasiva Rao [99] is based on an arithmetical error, and hence we have attributed the bound $G(7) \leq 53$ parenthetically to Davenport's methods [28] [32]. Also, it is worth remarking that the work of Vaughan and Wooley [130] [131] and [132] appeared in print in an order reversed from its chronological development (indeed, this work was first announced in 1991). The bounds parenthetically attributed to Vaughan and Wooley [132] follow directly from the methods therein, and were announced on that occasion, though details (with additional refinements) appeared only in Vaughan and Wooley [134]. Meanwhile, the bounds recorded in Wooley [159] were an immediate consequence of the methods of Wooley [155] combined with the new estimates for smooth Weyl sums obtained in the former work (no attempt was made therein to exploit the methods of Vaughan and Wooley [132]).

	k	5	6	7	8	9	10	11	12	13	14	15	16	17	18	19	20
Hardy & Littlewood [62]		41	87	192	425	949	2113										
James [81]		35		164		824											
Heilbronn [68]						190	217	244	272	300	329	359	388	418	449	480	511
Estermann [52]		29	42	59	78	101	125	153	184	217	253	292	333	377	424	474	
Hua [76]		28															
Davenport [28][32]		23	36														
Narasimhamurti [96]				(53)	73	99	122										
Chen [24]						96											
Cook [26]							121										
Vaughan [115]						91	107	122	137	153	168	184	200	216			
Thanigasalam [107]						90	106	121	136	152	167	183	199	215	231	248	264
Thanigasalam [108]						88	104	119	134	150	165	181	197	213	229	245	262
Thanigasalam [109]				50	68	87	103										
Vaughan [120][121]		21	31	45	62	82											
Vaughan [124][126]		19	29	41	57	75	93	109	125	141	156	171	187	202	217	232	248
Bründern [11]		18															
Vaughan & Wooley [129]		18	28				92	108	124	139	153	168	184	198	213	228	243
Wooley [155]			27	36	47	55	63	70	79	87	95	103	112	120	129	138	146
Wooley [159]								62	78	86	94	102	110	118	127	135	144
Vaughan & Wooley [132]		17	25	33	43	51	(59)	(67)	(76)	(84)	(92)	(100)	110	118	127	135	144
Vaughan & Wooley [130]					42												
Vaughan & Wooley [131]			24														
Meng [94]																	143
Vaughan & Wooley [134]						50	59	67	76	84	92	100	109	117	125	134	142
Conjectured		6	9	8	32	13	12	12	16	14	15	16	64	18	27	20	25

Table 2. Upper bounds for $G(k)$ when $5 \leq k \leq 20$

7 The Hardy-Littlewood Method

Practically all of the above conclusions have been obtained via the Hardy-Littlewood method. Here is a quick introduction. Let n be a large natural number, and write $P = n^{1/k}$ and

$$f(\alpha) = \sum_{x \leq P} e(\alpha x^k)$$

(here we follow the standard convention of writing $e(z)$ for $e^{2\pi i z}$). Then on writing $R(n)$ for the number of representations of n as the sum of s kth powers of natural numbers, it follows from orthogonality that

$$R(n) = \int_0^1 f(\alpha)^s e(-\alpha n) d\alpha.$$

When α is "close" to a rational number a/q with $(a, q) = 1$ and q "small", we expect that

$$f(\alpha) \sim q^{-1} S(q, a) v(\alpha - a/q),$$

where

$$S(q, a) = \sum_{r=1}^q e(ar^k/q) \quad \text{and} \quad v(\beta) = \int_0^P e(\beta \gamma^k) d\gamma.$$

This relation is straightforward to establish in an interval about a/q, so long as "close" and "small" are interpreted suitably. Now put

$$R_A(n) = \int_A f(\alpha)^s e(-\alpha n) d\alpha.$$

For a suitable union $\mathfrak{M}$ of such intervals centred on a/q (the *major arcs*), and for s sufficiently large in terms of k, one can establish that as $n \to \infty$, the asymptotic relation

$$R_{\mathfrak{M}}(n) \sim \frac{\Gamma(1 + 1/k)^s}{\Gamma(s/k)} n^{s/k-1} \mathfrak{S}(n)$$

holds, where $\mathfrak{S}(n)$ is the *singular series*

$$\mathfrak{S}(n) = \sum_{q=1}^{\infty} T(q; n)$$

and

$$T(q; n) = \sum_{\substack{a=1 \\ (a,q)=1}}^{q} q^{-s} S(q, a)^s e(-an/q).$$

8   The Necessary Congruence Condition

For each prime p, let

$$U(p;n) = \sum_{k=0}^{\infty} T(p^k;n).$$

The function $T(q;n)$ is multiplicative. Thus, when the singular series converges absolutely, one has

$$\mathfrak{S}(n) = \prod_{p} U(p;n).$$

It is helpful to the success of the circle method that the singular series should satisfy $\mathfrak{S}(n) \gg 1$. With this observation in mind, Hardy and Littlewood [61] defined $\Gamma(k)$ to be the least integer s with the property that, for every prime number p, there is a positive number $C(p)$ such that $U(p;n) \geq C(p)$ uniformly in n. Subsequently, Hardy and Littlewood [62] showed that indeed $\mathfrak{S}(n) \gg 1$ whenever $s \geq \max\{\Gamma(k), 4\}$. Next let $\Gamma_0(k)$ be the least number s with the property that the equation

$$x_1^k + \cdots + x_s^k = n \tag{8.1}$$

has a non-singular solution in $\mathbb{Q}_p$ (or rather, that the corresponding congruence modulo q always has a solution with $(x_1, q) = 1$). Hardy and Littlewood [63] were able to show that $\Gamma_0(k) = \Gamma(k)$ (see Theorem 1 of the aforementioned paper). Thus one sees that the singular series reflects the local properties of sums of k-th powers. In particular, the singular series is zero whenever the equation (8.1) fails to have a p-adic solution, for some prime p, and this reflects the trivial observation that the equation can be soluble over $\mathbb{Z}$ only if it is soluble everywhere locally.

Hardy and Littlewood [63] conjecture that $\Gamma(k) \to \infty$ as $k \to \infty$, but it is not even known whether or not one has

$$\liminf_{k \to \infty} \Gamma(k) \geq 4.$$

When $k > 2$, they showed that $\Gamma(k) = 4k$ when k is a power of 2 and that $\Gamma(k) \leq 2k$ otherwise. They also computed $\Gamma(k)$ exactly when $3 \leq k \leq 36$, and established that $\Gamma(k) \geq 4$ when $3 \leq k \leq 3000$. Here they showed that equality occurs only when $k = 3, 7, 19$, and possibly (but improbably) when $k = 1163, 1637, 1861, 1997, 2053$. These values of k can probably be settled by modern computing methods, and doubtless the calculations could be carried a good deal further. As far as we are aware, nothing has been done in this direction.

For a more detailed exposition of the Hardy-Littlewood method and the analysis of the major arcs and the singular series, see Vaughan [128] (especially Chapters 2 and 4).

9   The Minor Arcs

In order to establish an asymptotic formula for $R(n)$ it suffices to show that $R_{\mathfrak{m}}(n) = o(n^{s/k-1})$, where $\mathfrak{m} = [0, 1) \setminus \mathfrak{M}$ (the *minor arcs*). One needs to show that the minor arc contribution $R_{\mathfrak{m}}(n)$ is smaller by a factor $o(n^{-1})$, or equivalently $o(P^{-k})$, than the trivial estimate P^s. Routinely this is established via an inequality of the kind

$$\int_{\mathfrak{m}} |f(\alpha)|^s \, d\alpha \leq \left(\sup_{\mathfrak{m}} |f(\alpha)| \right)^{s-2t} \int_0^1 |f(\alpha)|^{2t} \, d\alpha.$$

The integral on the right hand side of this inequality may be interpreted as the number of solutions of an underlying diophantine equation, and it is from here that most of the savings usually come. On the other hand, non-trivial estimates for $|f(\alpha)|$, when $\alpha \in \mathfrak{m}$, may be obtained from estimates stemming from work of Weyl [154] and Vinogradov [144] (see Vaughan [128] for more modern estimates). When successful, this leads to the relation

$$R(n) \sim \frac{\Gamma(1+1/k)^s}{\Gamma(s/k)} n^{s/k-1} \mathfrak{S}(n). \tag{9.1}$$

It is in finding ways of dealing with the minor arcs, or in modifying the method so as to make the minor arcs more amenable, that most of the research has concentrated in the eighty years that have elapsed since the pioneering investigations of Hardy and Littlewood.

10   The Asymptotic Formula

Define $\widetilde{G}(k)$ to be the smallest natural number s_0 such that whenever $s \geq s_0$, the asymptotic relation (9.1) holds. Work of Hardy and Littlewood [62], described already in §3, established the general upper bound

$$\widetilde{G}(k) \leq (k-2)2^{k-1} + 5.$$

Progress on upper bounds for $\widetilde{G}(k)$ has since been achieved on two fronts. In Table 3 we present upper bounds for $\widetilde{G}(k)$ relevant for small values of k.

When k is large, bounds stemming from Vinogradov's mean value theorem provide dramatic improvements over the estimates recorded in Table 3.

$$\text{Hua [75]}, \quad 2^k + 1,$$
$$\text{Vaughan [119][122]}, \quad 2^k \quad (k \geq 3),$$
$$\text{Heath-Brown [65][66]}, \quad 7 \cdot 2^{k-3} + 1 \quad (k \geq 6),$$
$$\text{Boklan [5]}, \quad 7 \cdot 2^{k-3} \quad (k \geq 6).$$

Table 3. Upper bounds for $\widetilde{G}(k)$: smaller k

$$\text{Vinogradov [142]}, \quad 183k^9 (\log k + 1)^2 + 1,$$
$$\text{Vinogradov [141]}, \quad 91k^8 (\log k + 1)^9 + 1 \quad (k \geq 20),$$
$$\text{Vinogradov [147]}, \quad 10k^2 \log k,$$
$$\text{Hua [77]}, \quad 4k^2 (\log k + \tfrac{1}{2} \log \log k + 8),$$
$$\text{Wooley [156]}, \quad 2k^2 (\log k + O(\log \log k)),$$
$$\text{Ford [53]}, \quad k^2 (\log k + \log \log k + O(1)).$$

Table 4. Upper bounds for $\widetilde{G}(k)$: larger k

In Table 4 we present upper bounds for $\widetilde{G}(k)$ of use primarily when k is large.

We note that the methods underlying the last two bounds can be adapted to give explicit bounds for $\widetilde{G}(k)$ when k is of moderate size. Thus the method of Ford yields a bound for $\widetilde{G}(k)$ that is superior to the best recorded in Table 3 as soon as $k \geq 9$, and indeed unpublished work of Boklan and Wooley pushes this transition further to $k \geq 8$.

The bounds recorded in Tables 3 and 4 are likely to be a long way from the truth. One might expect that $G(k) = \max\{k+1, \Gamma_0(k)\}$, and, with an appropriate interpretation of the asymptotic formula when $\mathfrak{S}(n) = 0$, that $\widetilde{G}(k) = k + 1$.

One curiosity is that when $k = 3$ and $s = 7$, it can be shown that $R(n) \gg n^{4/3}$ (see Vaughan [125]), yet we are unable to show that $R(n) \ll n^{4/3}$. Indeed, it is currently the case that, quite generally, when s lies in the range between the known upper bounds for $G(k)$ and $\widetilde{G}(k)$, we can show that $R(n) \gg n^{s/k-1}$, but not $R(n) \ll n^{s/k-1}$.

11   Diminishing Ranges

In the Hardy-Littlewood method as outlined above the main problem is that of obtaining a suitable estimate for the mean value

$$\int_0^1 |f(\alpha)|^{2t} d\alpha,$$

that is, the number of integral solutions of the equation

$$x_1^k + \cdots + x_t^k = y_1^k + \cdots + y_t^k, \tag{11.1}$$

with $1 \le x_j, y_j \le P$ $(1 \le j \le t)$. Available estimates can be improved significantly if the variables are restricted, an idea already present in Hardy and Littlewood [62]. Define

$$P_1 = \tfrac{1}{2}P, \quad P_j = \tfrac{1}{2}P_{j-1}^{1-1/k} \quad (2 \le j \le t),$$

and consider the equation (11.1) subject to the constraints $P_j < x_j, y_j \le 2P_j$ $(1 \le j \le t)$. Inspecting the expression $|x_j^k - y_j^k|$ successively for $j = 1, 2, \ldots$ when $x_j \ne y_j$, we find that only the diagonal solutions in which $x_j = y_j$ $(1 \le j \le t)$ occur. Thus we find that the number of solutions of this type is at most $O(P_1 \ldots P_t)$. This saves $P_1 \ldots P_t$ over the trivial bound, which is of order $(P_1 \ldots P_t)^2$. Now $P_1 \ldots P_t \approx P^\lambda$, where

$$\lambda = 1 + (1 - 1/k) + \cdots + (1 - 1/k)^{t-1} = k - k(1 - 1/k)^t.$$

Already when $t \sim Ck \log k$, for a suitable positive constant C, this exponent is close to k. Vinogradov and Davenport have exploited and developed this idea in a number of ways (see, for example, Davenport [27][34], and Vinogradov [147][149]; see also Davenport and Erdős [35]).

There is a "p-adic" analogue of this idea, first exploited by Davenport [31], in which one considers expressions of the kind

$$x_1^k + p_2^k(x_2^k + p_3^k(x_3^k + \cdots))$$

on each side of the equation. Here the p_i denote suitably chosen prime numbers. The analysis rests on congruences of the type $x_1 \equiv y_1 \pmod{p^k}$. When $p > P^{1/k}$ and $1 \le x_1, y_1 \le P$, this congruence implies that $x_1 = y_1$, and so on, just as in the diminishing ranges device. This idea has the merit of returning the various k-th powers $p_2^k \ldots p_j^k x_j^k$ to being in comparable size ranges. However in each of these methods the variables in (11.1) have varying natures and the homogeneity is essentially lost.

12 Smooth Numbers and Efficient Differences

In modern variants of the circle method as applied to Waring's problem, starting with the work of Vaughan [124], homogeneity is restored by considering the number of solutions $S_t(P, R)$ of the equation (11.1) with $x_j, y_j \in \mathcal{A}(P, R)$, where $\mathcal{A}(P, R)$ denotes the set of R-smooth numbers up to P, namely

$$\mathcal{A}(P, R) = \{n \in [1, P] \cap \mathbb{Z} : p|n \implies p \leq R\}.$$

In applications, one takes R to be a suitably small, but positive power of P. The set $\mathcal{A}(P, R)$ has the extremely convenient property that, given any positive number M with $M \leq P$, and an element $x \in \mathcal{A}(P, R)$ with $x > M$, there is always an integer m with $m \in [M, MR]$ for which $m|x$. Moreover, this integer m can be coaxed into playing the rôle of the prime p in the p-adic argument mentioned above. Finally, and of great importance in what follows, the set $\mathcal{A}(P, R)$ has positive density whenever R is no smaller than a positive power of P.

The objective now is to find good exponents λ_s with the property that whenever $\varepsilon > 0$, there exists a positive number $\eta_0 = \eta_0(s, k, \varepsilon)$ such that whenever $R = P^\eta$ with $0 < \eta \leq \eta_0$, one has

$$S_s(P, R) \ll P^{\lambda_s + \varepsilon}.$$

Such exponents are established via an iterative process in which a sequence of sets of exponents $\boldsymbol{\lambda}^{(n)} = (\lambda_1^{(n)}, \lambda_2^{(n)}, \ldots)$ is constructed by finding an expression for each $\lambda_s^{(n+1)}$ in terms of the elements of $\boldsymbol{\lambda}^{(n)}$. Boundedness is trivial, so there is always a convergent subsequence. In fact, our arguments produce monotonicity, and the convergence is fairly rapid. For a more detailed introduction and motivation for the underlying ideas in using smooth numbers in Waring's problem, see the survey article Vaughan [127].

Beginning with the work of Wooley [155], a key element in the iterations is the repeated use of *efficient differencing*, and this procedure is fully exploited in subsequent work of Vaughan and Wooley [130] [131] [132] [134]. For each $s \in \mathbb{N}$, we take $\phi_i = \phi_{i,s}$ $(i = 1, \ldots, k)$ to be real numbers with $0 \leq \phi_i \leq 1/k$. For $0 \leq j \leq k$, we then define

$$P_j = 2^j P, \quad M_j = P^{\phi_j}, \quad H_j = P_j M_j^{-k}, \quad Q_j = P_j(M_1 \ldots M_j)^{-1},$$

$$\widetilde{H}_j = \prod_{i=1}^{j} H_i \quad \text{and} \quad \widetilde{M}_j = \prod_{i=1}^{j} M_i R.$$

Define the modified forward difference operator, Δ_j^*, recursively by taking

$$\Delta_1^*(f(x); h; m) = m^{-k}(f(x + hm^k) - f(x)),$$

and when $j \geq 1$, by inductively defining

$$\Delta_{j+1}^*(f(x); h_1, \ldots, h_{j+1}; m_1, \ldots, m_{j+1})$$
$$= \Delta_1^*(\Delta_j^*(f(x); h_1, \ldots, h_j; m_1, \ldots, m_j); h_{j+1}; m_{j+1}).$$

For $0 \leq j \leq k$, let

$$f(z) = (z - h_1 m_1^k - \cdots - h_j m_j^k)^k,$$

and define the difference polynomial

$$\Psi_j = \Psi_j(z; h_1, \ldots, h_j; m_1, \ldots, m_j)$$

by taking

$$\Psi_j = \Delta_j^*(f(z); 2h_1, \ldots, 2h_j; m_1, \ldots, m_j).$$

Here we adopt the convention that $\Psi_0(z; \mathbf{h}; \mathbf{m}) = z^k$. We write

$$f_j(\alpha) = \sum_{x \in A(Q_j, R)} e(\alpha x^k),$$

and

$$F_j(\alpha) = \sum_{z, \mathbf{h}, \mathbf{m}} e(\alpha \Psi_j(z; \mathbf{h}; \mathbf{m})),$$

where the summation is over z, $\mathbf{h}$, $\mathbf{m}$ with

$$1 \leq z \leq P_j, \quad 1 \leq h_i \leq 2^{j-i} H_i,$$

$$M_i < m_i \leq M_i R, \quad m_i \in A(P, R),$$

for $1 \leq i \leq j$. Finally, we define

$$T(j, s) = \int_0^1 |F_j(\alpha) f_j(\alpha)^{2s}| d\alpha.$$

Now, on considering the underlying diophantine equation, we have

$$S_{s+1}(P, R) \leq \int_0^1 |F_0(\alpha)^2 f_0(\alpha)^{2s}| d\alpha.$$

The starting point in the iterative process is to bound the latter expression in terms of $S_s(Q_1, R)$ and $T(1, s)$. This corresponds to taking the first difference in the classical Weyl differencing argument, and extracting the contribution arising from those terms with $x_1 = y_1$. Thus one obtains

$$S_{s+1}(P, R) \ll P^\varepsilon M_1^{2s-1}(PM_1 S_s(Q_1, R) + T(1, s)),$$

and this inequality we write symbolically as

$$F_0^2 f_0^{2s} \longmapsto F_1 f_1^{2s}.$$

One way to proceed is by means of a repeated efficient differencing step. In principle this is based on the Cauchy-Schwarz inequality, applied in the form

$$\int_0^1 |F_j f_j^{2s}| \, d\alpha \le \left(\int_0^1 |f_j|^{2t} d\alpha \right)^{1/2} \left(\int_0^1 |F_j^2 f_j^{4s-2t}| \, d\alpha \right)^{1/2},$$

where for the sake of concision we have written f_j for $f_j(\alpha)$ and likewise F_j for $F_j(\alpha)$. Thus, for $j = 1, 2, \ldots$, the mean value $T(j, s)$ can be related to $S_t(Q_j, R)$ and $T(j+1, 2s-t)$, where $t < 2s$ is a parameter at our disposal, via inequalities of the shape

$$T(j, s) \ll P^\varepsilon (S_t(Q_j, R))^{1/2} (\tilde{H}_j \tilde{M}_j M_{j+1}^{4s-2t-1} \Xi_{j+1})^{1/2},$$

where we write

$$\Xi_{j+1} = P \tilde{H}_j \tilde{M}_{j+1} S_{2s-t}(Q_{j+1}, R) + T(j+1, 2s-t).$$

This is the $(j+1)$-th step in the differencing process and can be portrayed by

$$F_j f_j^{2s} \quad \longrightarrow \quad F_{j+1} f_{j+1}^{4s-2t}$$
$$\downarrow$$
$$f_j^{2t}$$

There are more sophisticated variants of this procedure wherein it may be useful to restrict some of the variables to a range $(\frac{1}{2} Q_j R^{-j}, Q_j]$, or to replace the set $\mathcal{A}(Q_j, R)$ by $\mathbb{N} \cap [1, Q_j]$ (see §2 of Vaughan and Wooley [134] for details, and a more complete discussion).

Another option is to use Hölder's inequality to bound $T(j, s)$. Thus we obtain an inequality of the type

$$T(j, s) = \int_0^1 |F_j(\alpha) f_j(\alpha)^{2s}| \, d\alpha \ll I_l^a I_{l+1}^b U_v^c U_w^d,$$

where

$$I_m = \int_0^1 |F_j(\alpha)|^{2^m} d\alpha \quad (m = l, l+1),$$

$$U_u = \int_0^1 |f_j(\alpha)|^{2u} d\alpha \quad (u = v, w),$$

and l, v, w, a, b, c, d are non-negative numbers satisfying the equations

$$a + b + c + d = 1, \quad 2^l a + 2^{l+1} b = 1, \quad vc + wd = s.$$

There is clearly great flexibility in the possible choices of the parameters here. We can summarise this process by

$$F_j f_j^{2s} \implies (F_j^{2^l})^a (F_j^{2^{l+1}})^b (f_j^{2v})^c (f_j^{2w})^d.$$

Yet another option is to apply the Hardy-Littlewood method to $T(j, s)$. In practice we expect that the minor arc contribution dominates, although this is not guaranteed. But if it does, then

$$T(j, s) \ll \left(\sup_{\alpha \in m} |F_j(\alpha)| \right) S_s(Q_j, R), \tag{12.1}$$

and this we abbreviate to

$$F_j f_j^{2s} \implies (F_j)(f_j^{2s}).$$

By optimising choices for the parameters in order to obtain the sharpest estimates at each stage of the iteration process, one ultimately obtains relations describing $\lambda^{(n+1)}$ in terms of $\lambda^{(n)}$. The sharpest permissible exponents λ attainable by these methods are in general not easy to describe, and require substantial computations to establish (see, for example, Vaughan and Wooley [134]). However, one can describe in general terms the salient features of the permissible exponents λ_s. When s is rather small compared to k, it transpires that permissible exponents $\lambda_s = s + \delta_s$ can be derived with δ_s positive but small (see the next section for a consequence of this fact). Further, the simplest versions of the repeated efficient differencing method (see Wooley [155][157]) establish that the exponent $\lambda_s = 2s - k + ke^{1-2s/k}$ is permissible for every natural number s. Roughly speaking, therefore, one may compare the respective strengths of the diminishing ranges argument, and the repeated efficient differencing method, by comparing how rapidly the respective functions $k(1 - 1/k)^s \sim ke^{-s/k}$ and $ke^{1-2s/k}$ tend to zero as s increases.

The improvements in the most recent work (Vaughan and Wooley [134]) come about mostly through the following technical improvements:

• Better use of the Hardy-Littlewood method to estimate

$$T(j, s) = \int_0^1 |F_j(\alpha) f_j(\alpha)^{2s}| \, d\alpha.$$

In particular, tighter control is exercised in mean on the behaviour of the exponential sum $F_j(\alpha)$ on the major arcs, and this permits the assumption of (12.1) for a larger range of parameters ϕ than previously available.

- Better estimates for

$$I_m = \int_0^1 |F_j(\alpha)|^{2^m} \, d\alpha,$$

established largely via estimates for the number of integral points on certain affine plane curves.

- Better estimates (see Wooley [159]) for

$$\sup_{\alpha \in m} |f_0(\alpha)|.$$

Such estimates might be described as playing a significant rôle in the estimation of $G(k)$ in the final stages of the analysis.

13 Breaking Classical Convexity

All of the methods thus far described depend, in a fundamental manner, on the natural interpretation of even moments of exponential sums in terms of the number of solutions of certain underlying diophantine equations. In §12, for example, one is limited to permissible exponents λ_s corresponding to integral values of s, and in this setting the most effective method for bounding odd and fractional moments of smooth Weyl sums is to apply Hölder's inequality to interpolate between even moments. With the natural extension of the notion of a permissible exponent λ_s from integral values of s to arbitrary positive numbers s, the resulting exponents form the convex hull of the set of permissible exponents $\{\lambda_s : s \in \mathbb{N}\}$. A perusal of §12 reveals that extra flexibility in choice of parameters, and therefore the potential for further improvements, will be achieved by the removal of this "classical convexity" barrier, and such has recently become available.

In Wooley [158], a method is established which, loosely speaking, enables one to replace the inequality

$$S_{s+1}(P, R) \ll P^\varepsilon M_1^{2s-1}(P M_1 S_s(Q_1, R) + T(1, s))$$

that occurred in §12 with s restricted to be a natural number, by the new inequality

$$S_{s+t}(P, R) \ll P^\varepsilon M_1^{2s-t}(P^t M_1^t S_s(Q_1, R) + T^*),$$

where

$$T^* = \int_0^1 |F(\alpha)^t f_1(\alpha)^{2s}| \, d\alpha$$

and

$$F(\alpha) = \sum_{\substack{u \in \mathcal{A}(M_1 R, R) \\ u > M_1}} \sum_{\substack{z_1, z_2 \in \mathcal{A}(P,R) \\ z_1 \equiv z_2 \pmod{u^k} \\ z_1 \neq z_2}} e(\alpha u^{-k}(z_1^k - z_2^k)).$$

In this latter estimate, the parameter s is no longer restricted to be integral, and the parameter t may be chosen freely with $0 < t \leq 1$. Moreover, the mean value T^* is very much reminiscent of $T(1, s)$, with $F(\alpha)$ substituted for $F_1(\alpha)$ and exhibiting similar properties. Thus, in addition to removing the integrality constraint on s, one may also iterate with a fractional number $2t$ of variables.

As might be expected, the additional flexibility gained in this way leads to improved permissible exponents λ_s even for integral s, since our methods are so highly iterative. The overall improvements are usually quite small and are largest for smaller values of k. Such progress has not yet delivered sharper bounds for $G(k)$, but this work provides the sharpest results available concerning sums of cubes (see §4, and also Brüdern and Wooley [18]), and has also permitted new conclusions to be derived in certain problems involving sums of mixed powers (see Brüdern and Wooley [16] [19] [17], and also §15 below). Also, this "breaking convexity" device provides the best available lower bounds for the number $\mathcal{N}_{k,s}(X)$ of natural numbers not exceeding X that are the sum of s kth powers of positive integers, at least when s is small compared to k. Thus, when $2 < s \leq 2e^{-1}\sqrt{k}$, one has

$$\mathcal{N}_{k,s}(X) \gg X^{s/k - e^{-\gamma_s k}},$$

where $\gamma_s = 16/(es)^2$ (see Theorems 1.3 and 1.4 of Wooley [158], and the associated discussion). For comparison, one conjecturally has the lower bound $\mathcal{N}_{k,s}(X) \gg X^{s/k}$ whenever $s < k$.

Much remains to be investigated for fractional moments, in part owing to the substantial increase in complexity of the underlying computations (see Wooley [161] for more on this). However, such developments presently appear unlikely to have a large impact on the central problem of bounding $G(k)$ in the classical version of Waring's problem.

14 Variants of Waring's Problem: Primes

Much work has been devoted to various generalisations of the classical version of Waring's problem, and it seems appropriate to discuss some of the more mainstream variants.

We begin with the Waring-Goldbach problem, in which one seeks to represent integers as sums of kth powers of prime numbers. In order to describe the associated local conditions, suppose that k is a natural number and p is a prime number. We denote by $\theta = \theta(k, p)$ the integer with $p^\theta | k$ and $p^{\theta+1} \nmid k$, and then define $\gamma = \gamma(k, p)$ by

$$\gamma(k, p) = \begin{cases} \theta + 2, & \text{when } p = 2 \text{ and } \theta > 0, \\ \theta + 1, & \text{otherwise.} \end{cases}$$

Finally, we put

$$K(k) = \prod_{(p-1)|k} p^\gamma.$$

Denote by $H(k)$ the least integer s such that every sufficiently large positive integer congruent to s modulo $K(k)$ may be written as a sum of s kth powers of prime numbers. Note that when $(p-1)|k$, one has $p^\theta(p-1)|k$, whence $a^k \equiv 1 \pmod{p^\gamma}$ whenever $(p, a) = 1$. This explains our seemingly awkward definition of $H(k)$, since whenever n is the sum of s kth powers of primes exceeding $k + 1$, then necessarily $n \equiv s \pmod{K(k)}$. Naturally, further congruence conditions may arise from primes p with $(p-1) \nmid k$.

Following the pioneering investigations of Vinogradov [145][146] (see also Vinogradov [147]), Hua comprehensively investigated additive problems involving prime numbers in his influential book (see Hua [79], but also Hua [78]). Thus, it is known that for every natural number k one has

$$H(k) \leq 2^k + 1,$$

and, when k is large, that

$$H(k) \leq 4k(\log k + \tfrac{1}{2}\log\log k + O(1)).$$

In the conventional plan of attack on the Waring-Goldbach problem, one applies the Hardy-Littlewood method in a manner similar to that outlined above, but in interpreting the number of solutions of an analogue of the equation (11.1) over prime numbers, one obtains an upper bound by discarding the primality condition. With sufficiently many variables employed to save a factor of n via such an approach, one additional variable suffices to save the extra power of $\log n$ required by primality considerations. Although this strategy evidently prohibits the use of smooth numbers, the diminishing ranges technology perfected by Davenport, and refined by Vaughan [121] and Thanigasalam [109]–[113] plays a prominent rôle in establishing the best available upper bounds for $H(k)$ when k is small. We should also mention that recent progress depends on good estimates of Weyl-type for exponential sums over primes, and allied sums, available from the use of Vaughan's identity (see Vaughan [117]), combined with the linear sieve equipped with a switching principle (see Kawada and Wooley [83]). Thus, for $4 \leq k \leq 10$, the best known upper bounds for $H(k)$ are as follows:

$$\text{Kawada and Wooley [83]}, \quad H(4) \leq 14, \; H(5) \leq 21,$$
$$\text{Thanigasalam [111]}, \quad H(6) \leq 33, \; H(7) \leq 47, \; H(8) \leq 63,$$
$$H(9) \leq 83, \; H(10) \leq 107.$$

Despite much effort on the Waring-Goldbach problem for exponents 1, 2 and 3, further progress remains elusive. Improvements are feasible, however, if one is prepared to accept almost-primes in place of prime numbers (see, in particular, Chen [25], Brüdern [13][14], and Brüdern and Fouvry [15]). Difficulties related to those associated with the Waring-Goldbach problem are encountered when other sequences are substituted for prime numbers. For Waring's problem with smooth variables, see Harcos [57] and Brüdern and Wooley [18]. Also, see Nechaev [97] for work on Waring's problem with polynomial summands (Wooley [160] and Ford [55] have restricted improvements employing smooth numbers).

15 Variants of Waring's Problem: Sums of Mixed Powers

Suppose that $k_1, k_2, \ldots, k_s$ are natural numbers with $2 \leq k_1 \leq k_2 \leq \cdots \leq k_t$. Then an optimistic counting argument suggests that whenever the equation

$$x_1^{k_1} + x_2^{k_2} + \cdots + x_s^{k_s} = n \tag{15.1}$$

has p-adic solutions for each prime p, and

$$k_1^{-1} + k_2^{-1} + \cdots + k_s^{-1} > 1, \tag{15.2}$$

then n should be represented as the sum of mixed powers of positive integers (15.1) whenever it is sufficiently large in terms of $\mathbf{k}$. When $s = 3$ such an assertion may fail in certain circumstances (see Jagy and Kaplansky [80], or Exercise 5 of Chapter 8 of Vaughan [128]), but a heuristic application of the Hardy-Littlewood method suggests, at least, that the condition (15.2) should ensure that *almost all* integers in the expected congruence classes are thus represented. Moreover, subject instead to the condition

$$k_1^{-1} + k_2^{-1} + \cdots + k_s^{-1} > 2, \tag{15.3}$$

a formal application of the circle method suggests that *all* integers in the expected congruence classes should be represented in the form (15.1). Meanwhile, a simple counting argument shows that in circumstances in which the condition (15.2) does not hold, then arbitrarily large integers are not represented in the desired form.

The investigation of such analogues of Waring's problem for mixed powers has, since the early days of the Hardy-Littlewood method, stimulated progress in technology of use even in the classical version of Waring's problem. Additive problems in which the summands are restricted to be squares, cubes or biquadrates are perhaps of greater interest than those with higher powers, and here the current situation is remarkably satisfactory. We summarise below the current state of knowledge in the simpler problems of this nature. In Tables 5 and 6 we list constellations of powers whose sum represents, respectively, almost all, and all, integers subject to the expected congruence conditions. The tables are arranged, roughly speaking, starting with predominantly smaller exponents, and ending with predominantly larger exponents, and therefore not in chronological order of the results.

We have been unable to trace the origin in the literature of the conclusion on a square, two biquadrates and a kth power, but refer the reader to Exercise 6 of §2.8 of Vaughan [128] for related ideas (see also Roth

Davenport & Heilbronn [36],	two squares, one kth power,
Davenport & Heilbronn [37],	one square, two cubes,
Roth [101],	one square, one cube, one biquadrate,
Vaughan [116],	one square, one cube, one fifth power,
Folklore (?)	one square, two biquadrates, one kth power,
Hooley [70],	one square, one cube, one sixth power, one kth power,
Davenport [29],	four cubes,
Brüdern [9][8],	three cubes, one biquadrate,
Brüdern [8], Lu [89],	three cubes, one fifth power,
Brüdern & Wooley [19],	three cubes, one sixth power ,
Kawada & Wooley [82],	one cube, four biquadrates,
Vaughan [124],	six biquadrates,
Kawada & Wooley [82],	five biquadrates, one kth power (k odd).

Table 5. Representation of almost all integers

Gauss [56],	three squares,
Hooley [71],	two squares, three cubes,
Hooley [70],	two squares, assorted powers,
Vaughan [123],	one square, five cubes,
Brüdern & Wooley [16],	one square, four cubes, one biquadrate,
Ford [54],	one square, one cube, one biquadrate,...,
	one fifteenth power,
Linnik [87][88],	seven cubes,
Brüdern [9],	six cubes, two biquadrates,
Brüdern [9],	five cubes, three biquadrates,
Kawada & Wooley [82],	three cubes, six biquadrates,
Brüdern & Wooley [17],	two cubes, seven biquadrates,
Kawada & Wooley [82],	one cube, nine biquadrates
Vaughan [124],	twelve biquadrates,
Kawada & Wooley [82],	ten biquadrates, one kth power
	(k odd).

Table 6. Representation of all integers

[101]). We remark that all ternary problems of interest have been solved, since for non-trivial triples (k_1, k_2, k_3) not accounted for in Table 5, one has $k_1^{-1} + k_2^{-1} + k_3^{-1} \leq 1$. Also, energetic readers may be interested in tackling a problem which presently defies resolution only by the narrowest of margins, namely the problem of showing that almost all integers are represented as the sum of two cubes and two biquadrates of positive integers.

Note that although the three square theorem is commonly ascribed to Legendre, his "proof" depended on an unsubstantiated assumption only later established by Dirichlet, and the first complete proof is due to Gauss. We finish by noting that in problems involving sums of two squares, methods more effective than the circle method can be brought into play (see especially Hooley [70][71] and Brüdern [7]).

16   Variants of Waring's Problem: Beyond ℤ

Given the considerable energy expended on the investigation of Waring's problem over the rational integers, it seems natural to extend this work

to algebraic number fields. Here one encounters the immediate difficulty
of deciding what precisely Waring's problem should mean in this broader
context. It is possible, for example, that an algebraic integer in a number
field K may not be a sum of any finite number of kth powers of algebraic
integers of that field (consider, say, the parity of the imaginary part of $\mathfrak{a}^2$
when $\mathfrak{a} \in \mathbb{Z}[i]$). With this in mind, when K is a number field and $\mathfrak{O}_K$
is its ring of integers, we define J_k to be the subring of $\mathfrak{O}_K$ generated by
the kth powers of integers of K. We must also provide an analogue of the
positivity of the kth powers inherent in the classical version of Waring's
problem. Thus we define $G_K(k)$ to be the smallest positive integer s with
the property that, for some positive number $c = c(k, K)$, and for all totally
positive integers $\nu \in J_k$ of sufficiently large norm, the equation

$$\nu = \lambda_1^k + \lambda_2^k + \cdots + \lambda_s^k \tag{16.1}$$

is always soluble in totally non-negative integers λ_j of K with $N(\lambda_j) \leq$
$cN(\nu)^{1/k}$ $(1 \leq j \leq s)$.

Following early work of Meissner [93] and Mordell [95] for a restricted
class of number fields, Siegel [102][103] was the first to obtain quite general
conclusions for sums of squares, and hence, via the method of Hilbert [69],
for sums of kth powers. Siegel later developed a proper generalisation of
the Hardy-Littlewood method to number fields, and here the dissection
into major and minor arcs is a particular source of difficulty. In this way,
Siegel [104][105] obtained the upper bound

$$G_K(k) \leq dk(2^{k-1} + d) + 1,$$

where d denotes the degree of the field K. If one were to break the equation (16.1) into components with respect to an integral basis for $\mathfrak{O}_K$, then
one would obtain d equations of degree k in ds variables, and so one might
optimistically expect the analytic part of the circle method to apply with a
number of variables roughly the same in both K and $\mathbb{Q}$. Perhaps motivated
by such considerations, Siegel asked for reasonable bounds on $G_K(k)$ independent of d. This question was ultimately addressed through the work of
Birch [4] and Ramanujam [98], who provided the upper bound

$$G_K(k) \leq \max\{8k^5, 2^k + 1\}. \tag{16.2}$$

It is evident that the uniform bound (16.2) is far from the above cited
bound $G(k) \leq (1 + o(1))k \log k$ of Wooley [155], and the only slightly
weaker precursors of Vinogradov. An important desideratum, therefore, is
the reduction of the bound (16.2) to one of similar order to that presently
available for $G(k)$, or at least a reduction to a bound polynomial in k.

Failing this, effort has been expended in pursuit of bounds of order $k \log k$, but with a modest dependence on the degree d of the field K. Progress towards this objective has mirrored developments in the classical version of Waring's problem. Thus, building on work generalising that of Vinogradov to the number field setting by Körner [84] and Eda [50], methods employing smooth numbers and repeated efficient differences have recently been applied by Davidson [39] to establish the bound

$$G_K(k) \leq (3 + o(1))k \log k + c_d k,$$

where

$$c_d = 4d + 3 \log(\tfrac{1}{2}(d^2 + 1)) + 7,$$

and the term $o(1)$ is independent of d (apparently, the term $3 + o(1)$ here can be replaced by $2 + o(1)$ with only modest effort). Moreover, when K has class number one, Davidson [40][42] has obtained the bound

$$G_K(k) \leq k(\log k + \log \log k + c_d),$$

where c_d is approximately $4d$. Finally, Davidson [41] has improved on earlier work of Tatuzawa [106] to establish the strikingly simple conclusion that, for every number field K, one has

$$G_K(k) \leq 2d(\widehat{G}(k) + 2k),$$

where $\widehat{G}(k)$ denotes the least number s satisfying the property that the set of rational integers that can be expressed as the sum of s kth powers of natural numbers has positive density (in particular, of course, $\widehat{G}(k) \leq G_1(k)$).

More exotic still than the variants of Waring's problem over number fields are those in which one works over the polynomial rings $\mathbb{F}_q[t]$. Here one must again impose restrictions on the size of polynomials employed in the representation (and in this situation, the degree of a polynomial provides a measure of its size). Analogues of the Hardy-Littlewood method have been devised in this polynomial ring setting (see, for example, Effinger and Hayes [51]). Unfortunately, however, Weyl differencing proves ineffective whenever kth powers are considered over $\mathbb{F}_q[t]$ with $k \geq \text{char}(\mathbb{F}_q)$, for in such circumstances the factor $k!$, introduced into the argument of the exponential sum over kth powers via the differencing argument, is equal to zero in $\mathbb{F}_q[t]$. Consequently, one frequently restricts attention to kth powers with k smaller than the characteristic (but see Car and Cherly [22] for results on sums of 11 cubes in $\mathbb{F}_{2^h}[t]$). With this restriction, Car [20][21] has shown that every polynomial M, with $M \in \mathbb{F}_q[t]$, of sufficiently large degree, can be written in the form

$$M = M_1^k + \cdots + M_s^k,$$

with $M_i \in \mathbb{F}_q[t]$ of degree smaller than $1 + (\deg(M))/k$ $(1 \leq s \leq k)$, provided that

$$s \geq \min\{2^k + 1, \, 2k(k-1)\log 2 + 2k + 3\}.$$

17 Open Problems and Conjectures

Returning temporarily to the methods of §12, there are a number of problems connected with the mean value

$$I_m = \int_0^1 |F_j(\alpha)|^{2^m} \, d\alpha$$

which suggest some interesting questions. We will concentrate on the situation with $m = 1$, and remark only that the available results become less satisfactory as m increases.

A simple combinatorial argument reveals that the difference polynomial $\Psi_j = \Psi_j(z; \mathbf{h}; \mathbf{m})$ is given explicitly by the formula

$$\Psi_j = k! 2^j h_1 \ldots h_j \sum_{u \geq 0} \sum_{v_1 \geq 0} \cdots \sum_{v_j \geq 0} \frac{z^u (h_1 m_1^k)^{2v_1} \ldots (h_j m_j^k)^{2v_j}}{u!(2v_1 + 1)! \ldots (2v_j + 1)!},$$

where the summation is subject to the condition $u + 2v_1 + \cdots + 2v_j = k - j$. Consequently, one has

$$\Psi_j = h_1 \ldots h_j z^d \sum_{r=0}^{\frac{1}{2}(k-j-d)} c_r(h_1 m_1^k, \ldots, h_j m_j^k) z^{2r},$$

where

$$d = \begin{cases} 0, & \text{when } k - j \text{ is even,} \\ 1, & \text{when } k - j \text{ is odd,} \end{cases}$$

and for $0 \leq r \leq (k - j - d)/2$, the coefficients $c_r(\boldsymbol{\xi})$ are polynomials with positive integral coefficients that are symmetric in $\xi_1^2, \ldots, \xi_j^2$, and have total degree $k - j - 2r - d$. Now the mean value

$$I_1 = \int_0^1 |F_j(\alpha)|^2 \, d\alpha$$

is equal to the number of solutions of the diophantine equation

$$\Psi_j(z; \mathbf{h}; \mathbf{m}) = \Psi_j(z'; \mathbf{h}'; \mathbf{m}'), \tag{17.1}$$

with the variables in ranges discernible from the definition of $F_j(\alpha)$ in §12, and one might hope that the total number of solutions is close to the number of diagonal solutions, which is to say that

$$I_1 \ll P^{1+\varepsilon}\widetilde{M}_j\tilde{H}_j.$$

When $k-j$ is odd (so $d=1$), the presence of the term z^d makes it especially easy to deal with equation (17.1) by exploiting the inherent multiplicative structure, and indeed one can achieve the desired bound provided also that $j \le (k-d)/3$. The cases $j=1$ and $k-j=2$ or 4 are also doable. However, when $k-j$ is even and $1 < j \le k-6$, the situation is not so easy. By the way, this difficulty already occurs in Davenport's work [31]. To illustrate this situation, the simplest special case that we cannot handle directly corresponds to $k=8$ and $j=2$, and here one has

$$\begin{aligned}
\Psi_j = h_1 h_2 \big(224z^6 &+ 1120z^4(h_1^2 m_1^{16} + h_2^2 m_2^{16}) \\
&+ z^2(672h_1^4 m_1^{32} + 2240h_1^2 h_2^2 m_1^{16} m_2^{16} + 672h_2^4 m_2^{32}) \\
&+ 32h_1^6 m_1^{48} + 224h_1^4 h_2^2 m_1^{32} m_2^{16} + 224h_1^2 h_2^4 m_1^{16} m_2^{32} + 32h_2^6 m_2^{48}\big).
\end{aligned}$$

This suggests various general questions.

• Suppose that $f, g \in \mathbb{Z}[x]$. Are there simple conditions on f, g such that the number N of integral points $(x, y) \in [-P, P]^2$ for which $f(x) = g(y)$ satisfies

$$N \ll \big(P\mathcal{H}(f)\mathcal{H}(g)\big)^\varepsilon?$$

Here $\mathcal{H}(h)$ denotes the height of h. A qualitative version of this has already been considered by Davenport, Lewis and Schinzel [38], and if $f(x) - g(y)$ is irreducible over $\mathbb{C}$, then a celebrated theorem of Siegel shows that the number of solutions is finite unless there is a rational parametric solution of special form.

By the way, in view of the above, it is perhaps not surprising that in our treatment of I_m the bound of Bombieri and Pila [6] plays a rôle.

• Suppose that $A \subset \mathbb{Z}^k \cap [-X, X]^k$. Let $R(n; A)$ denote the number of solutions of the equation

$$a_1 x + a_2 x^2 + \cdots + a_k x^k = n$$

with $x \in \mathbb{Z} \cap [-P, P]$ and $\mathbf{a} \in A$. Are there any simple conditions under which it is true that

$$\sum_n R(n; A)^2 \ll \operatorname{card}(A)P(XP)^\varepsilon?$$

• A well-known conjecture in connection with Waring's problem is Hypothesis K (Hardy and Littlewood [62]). Let

$$R_{k,s}(n) = \text{card}\{\mathbf{x} \in \mathbb{N}^s \, : \, x_1^k + \cdots + x_s^k = n\}.$$

Then Hypothesis K asserts that for each natural number k, one has

$$R_{k,k}(n) \ll n^\varepsilon. \tag{17.2}$$

From this it would follow that

$$G(k) \le \max\{2k + 1, \Gamma_0(k)\} \tag{17.3}$$

and

$$G_1(k) = \max\{k + 1, \Gamma_0(k)\}. \tag{17.4}$$

The conjecture (17.2) was later shown by Mahler [90] to be false for $k = 3$, and indeed his counter-example shows that, infinitely often, one has $R_{3,3}(n) > 9^{-1/3} n^{1/12}$. However, the conjecture is still open when $k \ge 4$, and for (17.3) and (17.4), it suffices to know that

$$\sum_{n \le N} R_{k,k}(n)^2 \ll N^{1+\varepsilon}.$$

Hooley [74] established this when $k = 3$, under the assumption of the Riemann Hypothesis for a certain Hasse-Weil L-function. As far as we know, no simple conjecture of this kind is known from which it would follow that $G(k) = \max\{k + 1, \Gamma_0(k)\}$.

• It may well be true that, when $k \ge 3$, one has

$$\sum_{n \le N} R_{k,k}(n)^2 \sim CN.$$

However Hooley [72] has shown, at least when $k = 3$, that the constant C here is larger than what would arise simply from the major arcs. This leads to some interesting speculations. The number of solutions of the equation

$$x_1^k + \cdots + x_s^k = y_1^k + \cdots + y_s^k \le N,$$

in which the variables on the right hand side are a permutation of those on the left hand side, is asymptotic to $C_1 N^{s/k}$, for a certain positive number $C_1 = C_1(k, s)$, and when $s < k$ the contribution arising from the major arcs is smaller. Maybe one should think of these solutions as being "trivial", "parametric", or as arising from some "degenerate" property of the geometry of the surface. Anyway, their contribution is mostly concentrated on the minor arcs. It seems rather likely that this phenomenon persists for $s \geq k$ and explains Hooley's discovery. This leads to the philosophy that the major arcs correspond to non-trivial solutions, and the minor arcs to trivial solutions. There is an example of this phenomenon in Vaughan and Wooley [133].

• Recall the definitions of $f(\alpha)$, $S(q,a)$ and $v(\beta)$ from §7. One can conjecture that, whenever $(a,q) = 1$, one has

$$f(\alpha) - q^{-1}S(q,a)v(\alpha - a/q) \ll \left(q + P^k|q\alpha - a|\right)^{1/k}. \tag{17.5}$$

Possibly the exponent has to be weakened to $1/k + \varepsilon$, but any counterexamples would be interesting.

From (17.5) it would follow that

$$\sum_{n \leq N} R_{k,k}(n)^2 \ll N.$$

Also it is just conceivable that (17.5), in combination with a variant of the Hardy-Littlewood-Kloosterman method, would achieve the bound $G(k) \leq \max\{\mathfrak{G}(k), \Gamma_0(k)\}$, where $\mathfrak{G}(k) < 2k + 1$.

• The inequality (17.5) is a special case of conjectures that can be made about the exponential sum

$$f(\alpha) = \sum_{x \leq P} e(\alpha_1 x + \cdots + \alpha_k x^k)$$

that would have many consequences in analytic number theory. For example one can ask if something like

$$f(\alpha) - q^{-1}S(q,\mathbf{a})v(\alpha - \mathbf{a}/q) \ll \sum_{j=2}^{k} \left(\frac{q + P^j|\alpha_j q - a_j|}{(q, a_j)}\right)^{1/j} \tag{17.6}$$

is true. Here

$$S(q,\mathbf{a}) = \sum_{r=1}^{q} e((a_1 r + \cdots + a_k r^k)/q),$$

$$v(\beta) = \int_0^P e(\beta_1 \gamma + \cdots + \beta_k \gamma^k)d\gamma,$$

and $a_1 = c(q, a_2, \ldots, a_k)$, where $c = c(q, \mathbf{a})$ is the unique integer with

$$-\frac{1}{2} < c - \alpha_1 q/(q, a_2, \ldots, a_k) \leq \frac{1}{2}.$$

The inequality (17.6) is known to hold for $k = 2$ (Vaughan unpublished) with the right hand side weakened slightly to

$$\left(\frac{q}{(q, a_2)} \right)^{1/2} \left(\log \frac{2q}{(q, a_2)} + P|\alpha_2 - a_2/q_2|^{1/2} \right).$$

• One way of viewing the Hardy-Littlewood method is that we begin by considering the Fourier transform with respect to Lebesgue measure on $[0, 1)$ for an appropriate generating function defined in terms of the additive characters, and then approximate to it by a product of discrete measures. Since part of the problem when one has relatively few variables is that the geometry genuinely intrudes, one can ask whether we are using the best measure for the problem at hand. In this situation something more closely related to the underlying geometry might be more useful.

In conclusion, it is clear that although we have come a long way in the twentieth century, there remains plenty still to be done!

References

[1]   M. A. Bennett, *Fractional parts of powers of rational numbers*, Math. Proc. Cambridge Philos. Soc. **114** (1993), 191–201.

[2]   _____ , *An ideal Waring problem with restricted summands*, Acta Arith. **66** (1994), 125–132.

[3]   F. Beukers, *Fractional parts of powers of rationals*, Math. Proc. Cambridge Philos. Soc. **90** (1981), 13–20.

[4]   B. J. Birch, *Waring's problem in algebraic number fields*, Proc. Cambridge Philos. Soc. **57** (1961), 449–459.

[5]   K. D. Boklan, *The asymptotic formula in Waring's problem*, Mathematika **41** (1994), 147–161.

[6]   E. Bombieri and J. Pila, *The number of integral points on arcs and ovals*, Duke Math. J. **59** (1989), 337–357.

[7]   J. Brüdern, *Sums of squares and higher powers*, J. London Math. Soc. (2) **35** (1987), 233–243.

[8]   _____ , *Iterationsmethoden in der additiven Zahlentheorie*, Dissertation, Göttingen, 1988.

[9]   ———, *On Waring's problem for cubes and biquadrates*, J. London Math. Soc. (2) **37** (1988), 25–42.

[10]   ———, *Sums of four cubes*, Monatsh. Math. **107** (1989), 179–188.

[11]   ———, *On Waring's problem for fifth powers and some related topics*, Proc. London Math. Soc. (3) **61** (1990), 457–479.

[12]   ———, *On Waring's problem for cubes*, Math. Proc. Cambridge Philos. Soc. **109** (1991), 229–256.

[13]   ———, *Sieves, the circle method, and Waring's problem for cubes*, Habilitationschrift, Göttingen, 1991, Mathematica Göttingensis **51**.

[14]   ———, *A sieve approach to the Waring-Goldbach problem, II. On the seven cube theorem*, Acta Arith. **72** (1995), 211–227.

[15]   J. Brüdern and É. Fouvry, *Lagrange's four squares theorem with almost prime variables*, J. Reine Angew. Math. **454** (1994), 59–96.

[16]   J. Brüdern and T. D. Wooley, *On Waring's problem: a square, four cubes and a biquadrate*, Math. Proc. Cambridge Philos. Soc. **127** (1999), 193–200.

[17]   ———, *On Waring's problem: two cubes and seven biquadrates*, Tsukuba J. Math. **24** (2000), 387–417.

[18]   ———, *On Waring's problem for cubes and smooth Weyl sums*, Proc. London Math. Soc. (3) **82** (2001), 89–109.

[19]   ———, *On Waring's problem: three cubes and a sixth power*, Nagoya Math. J. **163** (2001), 13–53.

[20]   M. Car, *Le problème de Waring pour l'anneau des polynômes sur un corps fini*, Séminaire de Théorie des Nombres 1972–1973 (Univ. Bordeaux I, Talence), Lab. Théorie des Nombres, CNRS, Talence, 1973, Exp. No. 6, 13pp.

[21]   ———, *Waring's problem in function fields*, Proc. London Math. Soc. (3) **68** (1994), 1–30.

[22]   M. Car and J. Cherly, *Sommes de cubes dans l'anneau $\mathbb{F}_{2^h}[X]$*, Acta Arith. **65** (1993), 227–241.

[23]   J. W. S. Cassels and R. C. Vaughan, *Ivan Matveevich Vinogradov*, Biographical Memoirs of Fellows of the Royal Society **31** (1985), 613–631, reprinted in: Bull. London Math. Soc. **17** (1985), 584–600.

[24]   J.-R. Chen, *On Waring's problem for n-th powers*, Acta Math. Sinica **8** (1958), 253–257.

[25]   _____, *On the representation of a large even integer as the sum of a prime and the product of at most two primes*, Sci. Sinica **16** (1973), 157–176.

[26]   R. J. Cook, *A note on Waring's problem*, Bull. London Math. Soc. **5** (1973), 11–12.

[27]   H. Davenport, *Sur les sommes de puissances entières*, C. R. Acad. Sci. Paris, Sér. A **207** (1938), 1366–1368.

[28]   _____, *On sums of positive integral kth powers*, Proc. Roy. Soc. London, Ser. A **170** (1939), 293–299.

[29]   _____, *On Waring's problem for cubes*, Acta Math. **71** (1939), 123–143.

[30]   _____, *On Waring's problem for fourth powers*, Ann. of Math. **40** (1939), 731–747.

[31]   _____, *On sums of positive integral kth powers*, Amer. J. Math. **64** (1942), 189–198.

[32]   _____, *On Waring's problem for fifth and sixth powers*, Amer. J. Math. **64** (1942), 199–207.

[33]   _____, *Sums of three positive cubes*, J. London Math. Soc. **25** (1950), 339–343.

[34]   _____, *The collected works of Harold Davenport, Volume III (B. J. Birch, H. Halberstam and C. A. Rogers, eds.)*, Academic Press, London, 1977.

[35]   H. Davenport and P. Erdős, *On sums of positive integral kth powers*, Ann. of Math. **40** (1939), 533–536.

[36]   H. Davenport and H. Heilbronn, *Note on a result in the additive theory of numbers*, Proc. London Math. Soc. (2) **43** (1937), 142–151.

[37]   _____, *On Waring's problem: two cubes and one square*, Proc. London Math. Soc. (2) **43** (1937), 73–104.

[38]   H. Davenport, D. J. Lewis, and A. Schinzel, *Equations of the form $f(x) = g(y)$*, Quart. J. Math. Oxford (2) **12** (1961), 304–312.

[39] M. Davidson, *Sums of k-th powers in number fields*, Mathematika **45** (1998), 359–370.

[40] _____, *A new minor-arcs estimate for number fields*, Topics in number theory (University Park, PA, 1997), Kluwer Acad. Publ., Dordrecht, 1999, pp. 151–161.

[41] _____, *On Waring's problem in number fields*, J. London Math. Soc. (2) **59** (1999), 435–447.

[42] _____, *On Siegel's conjecture in Waring's problem*, to appear, 2001.

[43] J.-M. Deshouillers, F. Hennecart, K. Kawada, B. Landreau, and T. D. Wooley, *Sums of biquadrates: a survey*, in preparation, 2001.

[44] J.-M. Deshouillers, F. Hennecart, and B. Landreau, $7, 373, 170, 279, 850$, Math. Comp. **69** (2000), 421–439.

[45] _____, *Waring's problem for sixteen biquadrates – Numerical results*, J. Théorie Nombres Bordeaux **12** (2000), 411–422, Colloque International de Théorie des Nombres (Talence, 1999).

[46] J.-M. Deshouillers, K. Kawada, and T. D. Wooley, *On sums of sixteen biquadrates*, in preparation, 2001.

[47] L. E. Dickson, *All integers except 23 and 239 are sums of eight cubes*, Bull. Amer. Math. Soc. **45** (1939), 588–591.

[48] _____, *History of the theory of numbers, Vol. II: Diophantine analysis*, Chelsea, New York, 1966.

[49] A. K. Dubitskas, *A lower bound on the value $\|(3/2)^k\|$*, Uspekhi Mat. Nauk **45** (1990), 153–154, translation in: Russian Math. Surveys **45** (1990), 163–164.

[50] Y. Eda, *On Waring's problem in an algebraic number field*, Rev. Columbiana Math. **9** (1975), 29–73.

[51] G. Effinger and D. Hayes, *Additive number theory of polynomials over a finite field*, Oxford University Press, Oxford, 1991.

[52] T. Estermann, *On Waring's problem for fourth and higher powers*, Acta Arith. **2** (1937), 197–211.

[53] K. B. Ford, *New estimates for mean values of Weyl sums*, Internat. Math. Res. Notices (1995), 155–171.

[54] _____, *The representation of numbers as sums of unlike powers. II,* J. Amer. Math. Soc. **9** (1996), 919–940.

[55] _____, *Waring's problem with polynomial summands,* J. London Math. Soc. (2) **61** (2000), 671–680.

[56] C. F. Gauss, *Disquisitiones arithmeticæ,* Leipzig, 1801.

[57] G. Harcos, *Waring's problem with small prime factors,* Acta Arith. **80** (1997), 165–185.

[58] G. H. Hardy and J. E. Littlewood, *A new solution of Waring's problem,* Quart. J. Math. Oxford **48** (1920), 272–293.

[59] _____, *Some problems of "Partitio Numerorum": I. A new solution of Waring's problem,* Göttingen Nachrichten (1920), 33–54.

[60] _____, *Some problems of "Partitio Numerorum": II. Proof that every large number is the sum of at most 21 biquadrates,* Math. Z. **9** (1921), 14–27.

[61] _____, *Some problems of "Partitio Numerorum": IV. The singular series in Waring's Problem and the value of the number $G(k)$,* Math. Z. **12** (1922), 161–188.

[62] _____, *Some problems of "Partitio Numerorum" (VI): Further researches in Waring's problem,* Math. Z. **23** (1925), 1–37.

[63] _____, *Some problems of "Partitio Numerorum" (VIII): The number $\Gamma(k)$ in Waring's problem,* Proc. London Math. Soc. (2) **28** (1928), 518–542.

[64] G. H. Hardy and S. Ramanujan, *Asymptotic formulae in combinatory analysis,* Proc. London Math. Soc. (2) **17** (1918), 75–115.

[65] D. R. Heath-Brown, *Weyl's inequality, Hua's inequality, and Waring's problem,* J. London Math. Soc. (2) **38** (1988), 396–414.

[66] _____, *Weyl's inequality and Hua's inequality,* Number Theory (Ulm, 1987), Lecture Notes in Math., vol. 1380, Springer-Verlag, Berlin, 1989, pp. 87–92.

[67] _____, *The circle method and diagonal cubic forms,* Philos. Trans. Roy. Soc. London Ser. A **356** (1998), 673–699.

[68] H. Heilbronn, *Über das Waringsche Problem,* Acta Arith. **1** (1936), 212–221.

[69]   D. Hilbert, *Beweis für Darstellbarkeit der ganzen Zahlen durch eine feste Anzahl nter Potenzen Waringsche Problem*, Nach. Königl. Ges. Wiss. Göttingen math.-phys. Kl. (1909), 17–36, Math. Ann. **67** (1909), 281–300.

[70]   C. Hooley, *On a new approach to various problems of Waring's type*, Recent progress in analytic number theory, Vol. 1 (Durham, 1979), Academic Press, London, 1981, pp. 127–191.

[71]   _____, *On Waring's problem for two squares and three cubes*, J. Reine Angew. Math. **328** (1981), 161–207.

[72]   _____, *On some topics connected with Waring's problem*, J. Reine Angew. Math. **369** (1986), 110–153.

[73]   _____, *On Waring's problem*, Acta Math. **157** (1986), 49–97.

[74]   _____, *On Hypothesis K* in Waring's problem*, Sieve methods, exponential sums, and their applications in number theory (Cardiff, 1995), Cambridge Univ. Press, Cambridge, 1997, pp. 175–185.

[75]   L.-K. Hua, *On Waring's problem*, Quart. J. Math. Oxford **9** (1938), 199–202.

[76]   _____, *Some results on Waring's problem for smaller powers*, C. R. Acad. Sci. URSS (2) **18** (1938), 527–528.

[77]   _____, *An improvement of Vinogradov's mean-value theorem and several applications*, Quart. J. Math. Oxford **20** (1949), 48–61.

[78]   _____, *Die Abschätzungen von Exponentialsummen und ihre Anwendung in der Zahlentheorie*, B.G. Teubner, Leipzig, 1959.

[79]   _____, *Additive theory of prime numbers*, American Math. Soc., Providence, Rhode Island, 1965.

[80]   W. C. Jagy and I. Kaplansky, *Sums of squares, cubes, and higher powers*, Experiment. Math. **4** (1995), 169–173.

[81]   R. D. James, *On Waring's problem for odd powers*, Proc. London Math. Soc. **37** (1934), 257–291.

[82]   K. Kawada and T. D. Wooley, *Sums of fourth powers and related topics*, J. Reine Angew. Math. **512** (1999), 173–223.

[83]   _____, *On the Waring-Goldbach problem for fourth and fifth powers*, Proc. London Math. Soc. (3) **83** (2001), 1–50.

[84] O. Körner, *Über das Waringsche Problem in algebraischen Zahlkörpern*, Math. Ann. **144** (1961), 224–238.

[85] J. M. Kubina and M. C. Wunderlich, *Extending Waring's conjecture to 471,600,000*, Math. Comp. **55** (1990), 815–820.

[86] E. Landau, *Über eine Anwendung der Primzahlen auf das Waringsche Problem in der elementaren Zahlentheorie*, Math. Ann. **66** (1909), 102–105.

[87] Ju. V. Linnik, *On the representation of large numbers as sums of seven cubes*, Dokl. Akad. Nauk SSSR **35** (1942), 162.

[88] ———, *On the representation of large numbers as sums of seven cubes*, Mat. Sb. **12** (1943), 218–224.

[89] M. G. Lu, *On Waring's problem for cubes and fifth power*, Sci. China Ser. A **36** (1993), 641–662.

[90] K. Mahler, *Note on hypothesis K of Hardy and Littlewood*, J. London Math. Soc. **11** (1936), 136–138.

[91] ———, *On the fractional parts of the powers of a rational number (II)*, Mathematika **4** (1957), 122–124.

[92] K. S. McCurley, *An effective seven cube theorem*, J. Number Theory **19** (1984), 176–183.

[93] O. E. Meissner, *Über die Darstellung der Zahlen einiger algebraischen Zahlkörper als Summen von Quadratzahlen des Körpers*, Arch. Math. Phys. (3) **7** (1904), 266–268.

[94] Z. Z. Meng, *Some new results on Waring's problem*, J. China Univ. Sci. Tech. **27** (1997), 1–5.

[95] L. J. Mordell, *On the representation of algebraic numbers as a sum of four squares*, Proc. Cambridge Philos. Soc. **20** (1921), 250–256.

[96] V. Narasimhamurti, *On Waring's problem for 8th, 9th, and 10th powers*, J. Indian Math. Soc. (N.S.) **5** (1941), 122.

[97] V. I. Nechaev, *Waring's problem for polynomials*, Trudy Mat. Inst. Steklov, Izdat. Akad. Nauk SSSR, Moscow **38** (1951), 190–243.

[98] C. P. Ramanujam, *Sums of m-th powers in p-adic rings*, Mathematika **10** (1963), 137–146.

[99]   K. Sambasiva Rao, *On Waring's problem for smaller powers*, J. Indian Math. Soc. **5** (1941), 117–121.

[100] C. J. Ringrose, *Sums of three cubes*, J. London Math. Soc. (2) **33** (1986), 407–413.

[101] K. F. Roth, *Proof that almost all positive integers are sums of a square, a positive cube and a fourth power*, J. London Math. Soc. **24** (1949), 4–13.

[102] C. L. Siegel, *Darstellung total positiver Zahlen durch Quadrate*, Math. Z. **11** (1921), 246–275.

[103] _____, *Additive Theorie der Zahlkörper II*, Math. Ann. **88** (1923), 184–210.

[104] _____, *Generalisation of Waring's problem to algebraic number fields*, Amer. J. Math. **66** (1944), 122–136.

[105] _____, *Sums of m-th powers of algebraic integers*, Ann. of Math. **46** (1945), 313–339.

[106] T. Tatuzawa, *On Waring's problem in algebraic number fields*, Acta Arith. **24** (1973), 37–60.

[107] K. Thanigasalam, *On Waring's problem*, Acta Arith. **38** (1980), 141–155.

[108] _____, *Some new estimates for $G(k)$ in Waring's problem*, Acta Arith. **42** (1982), 73–78.

[109] _____, *Improvement on Davenport's iterative method and new results in additive number theory, I*, Acta Arith. **46** (1985), 1–31.

[110] _____, *Improvement on Davenport's iterative method and new results in additive number theory, II. Proof that $G(5) \leq 22$*, Acta Arith. **46** (1986), 91–112.

[111] _____, *Improvement on Davenport's iterative method and new results in additive number theory, III*, Acta Arith. **48** (1987), 97–116.

[112] _____, *On sums of positive integral powers and simple proof of $G(6) \leq 31$*, Bull. Calcutta Math. Soc. **81** (1989), 279–294.

[113] _____, *On admissible exponents for kth powers*, Bull. Calcutta Math. Soc. **86** (1994), 175–178.

[114] K.-C. Tong, *On Waring's problem*, Advancement in Math. **3** (1957), 602–607.

[115] R. C. Vaughan, *Homogeneous additive equations and Waring's problem*, Acta Arith. **33** (1977), 231–253.

[116] _____, *A ternary additive problem*, Proc. London Math. Soc. (3) **41** (1980), 516–532.

[117] _____, *Recent work in additive prime number theory*, Proceedings of the International Congress of Mathematicians (Helsinki, 1978), Acad. Sci. Fennica, 1980, pp. 389–394.

[118] _____, *Sums of three cubes*, Bull. London Math. Soc. **17** (1985), 17–20.

[119] _____, *On Waring's problem for cubes*, J. Reine Angew. Math. **365** (1986), 122–170.

[120] _____, *On Waring's problem for sixth powers*, J. London Math. Soc. (2) **33** (1986), 227–236.

[121] _____, *On Waring's problem for smaller exponents*, Proc. London Math. Soc. (3) **52** (1986), 445–463.

[122] _____, *On Waring's problem for smaller exponents. II*, Mathematika **33** (1986), 6–22.

[123] _____, *On Waring's problem: one square and five cubes*, Quart. J. Math. Oxford (2) **37** (1986), 117–127.

[124] _____, *A new iterative method in Waring's problem*, Acta Math. **162** (1989), 1–71.

[125] _____, *On Waring's problem for cubes II*, J. London Math. Soc. (2) **39** (1989), 205–218.

[126] _____, *A new iterative method in Waring's problem II*, J. London Math. Soc. (2) **39** (1989), 219–230.

[127] _____, *The use in additive number theory of numbers without large prime factors*, Philos. Trans. Roy. Soc. London Ser. A **345** (1993), 363–376.

[128] _____, *The Hardy-Littlewood Method, 2nd edition*, Cambridge University Press, Cambridge, 1997.

[129] R. C. Vaughan and T. D. Wooley, *On Waring's problem: some refinements*, Proc. London Math. Soc. (3) **63** (1991), 35–68.

[130] _____, *Further improvements in Waring's problem, III: Eighth powers*, Philos. Trans. Roy. Soc. London Ser. A **345** (1993), 385–396.

[131] _____, *Further improvements in Waring's problem, II: Sixth powers*, Duke Math. J. **76** (1994), 683–710.

[132] _____, *Further improvements in Waring's problem*, Acta Math. **174** (1995), 147–240.

[133] _____, *On a certain nonary cubic form and related equations*, Duke Math. J. **80** (1995), 669–735.

[134] _____, *Further improvements in Waring's problem, IV: Higher powers*, Acta Arith. **94** (2000), 203–285.

[135] I. M. Vinogradov, *A new estimate for $G(n)$ in Waring's problem*, Dokl. Akad. Nauk SSSR **5** (1934), 249–253.

[136] _____, *A new solution of Waring's problem*, Dokl. Akad. Nauk SSSR **2** (1934), 337–341.

[137] _____, *On some new results in the analytic theory of numbers*, C. R. Acad. Sci. Paris, Sér. A **199** (1934), 174–175.

[138] _____, *On the upper bound for $G(n)$ in Waring's problem*, Izv. Akad. Nauk SSSR Ser. Phiz.-Matem. **10** (1934), 1455–1469.

[139] _____, *A new variant of the proof of Waring's theorem*, Trud. Matem. instituta it. V. A. Steklov **9** (1935), 5–15.

[140] _____, *A new variant of the proof of Waring's theorem*, C. R. Acad. Sci. Paris, Sér. A **200** (1935), 182–184.

[141] _____, *An asymptotic formula for the number of representations in Waring's problem*, Mat. Sb. **42** (1935), 531–534.

[142] _____, *New estimates for Weyl sums*, Dokl. Akad. Nauk SSSR **8** (1935), 195–198.

[143] _____, *On Waring's problem*, Ann. of Math. **36** (1935), 395–405.

[144] _____, *On Weyl's sums*, Mat. Sb. **42** (1935), 521–530.

[145] _____, *Representation of an odd number as a sum of three primes*, Dokl. Akad. Nauk SSSR **15** (1937), 6–7.

[146] ———, *Some theorems concerning the theory of primes*, Mat. Sb. **44** (1937), 179–195.

[147] ———, *The method of trigonometrical sums in the theory of numbers*, Trav. Inst. Math. Stekloff **23** (1947), 109 pp.

[148] ———, *On an upper bound for G(n)*, Izv. Akad. Nauk SSSR Ser. Mat. **23** (1959), 637–642.

[149] ———, *Selected Works (translated from the Russian by N. Psv)*, Springer-Verlag, Berlin, 1985.

[150] E. Waring, *Meditationes Algebraicæ, second edition*, Archdeacon, Cambridge, 1770.

[151] ———, *Meditationes Algebraicæ, third edition*, Archdeacon, Cambridge, 1782.

[152] ———, *Meditationes Algebraicæ*, American Math. Soc., Providence, 1991, translation by Dennis Weeks of the 1782 edition.

[153] G. L. Watson, *A proof of the seven cube theorem*, J. London Math. Soc. **26** (1951), 153–156.

[154] H. Weyl, *Über die Gleichverteilung von Zahlen mod Eins*, Math. Ann. **77** (1916), 313–352.

[155] T. D. Wooley, *Large improvements in Waring's problem*, Ann. of Math. **135** (1992), 131–164.

[156] ———, *On Vinogradov's mean value theorem*, Mathematika **39** (1992), 379–399.

[157] ———, *The application of a new mean value theorem to the fractional parts of polynomials*, Acta Arith. **65** (1993), 163–179.

[158] ———, *Breaking classical convexity in Waring's problem: sums of cubes and quasi-diagonal behaviour*, Invent. Math. **122** (1995), 421–451.

[159] ———, *New estimates for smooth Weyl sums*, J. London Math. Soc. (2) **51** (1995), 1–13.

[160] ———, *On exponential sums over smooth numbers*, J. Reine Angew. Math. **488** (1997), 79–140.

[161] ———, *Quasi-diagonal behaviour and smooth Weyl sums*, Monatsh. Math. **130** (2000), 161–170.

[162] ———— , *Sums of three cubes*, Mathematika, to appear, 2001.

Solving the Pell Equation

H. C. Williams

1 Introduction

Let D be a positive nonsquare integer. The misnamed Pell equation is an expression of the form

$$T^2 - DU^2 = 1, \tag{1.1}$$

where we constrain the values of T and U to be integers. For example, the Pell equation

$$T^2 - 7U^2 = 1 \tag{1.2}$$

has the solutions $(\pm 1, 0), (8, 3), (-8, 3), (8, -3), (-8, -3), (127, 48)$, etc. In fact, Lagrange, completing earlier work of Euler, had established by 1768 (see Weil [95, pp. 314–315]) that (1.1) will always have a non-trivial solution (a solution with $U \neq 0$). Furthermore, there exists an infinitude of such solutions (T, U) given by

$$T + \sqrt{D}U = \pm \left(t + \sqrt{D}u \right)^n, \tag{1.3}$$

where n is any integer and (t, u) is the *fundamental solution* (with $t + u\sqrt{D} > 1$) of (1.1), and no other solutions of (1.1) exist except those given by (1.3). The fundamental solution of (1.1) is the least positive integral solution; for example, the fundamental solution of (1.2) is $(8, 3)$. In view of these observations, we will say that the problem of solving the Pell equation (1.1) is that of determining t and u.

This very simple Diophantine equation has been the object of study by mathematicians for over 2000 years. It is named after John Pell because of an error in attribution made by Euler [20] to a method of solving it in "Wallis's works." This was most likely the result of a cursory reading by him of Wallis's *Algebra*. As noted by many authorities, most recently Weil [95, p. 174], Pell's name occurs frequently in *Algebra*, but never in connection with the Pell equation. In fact, it seems most likely that the method referred to by Euler for solving (1.1) is a technique that Wallis credits to Brouncker. In spite of ample evidence attesting to Euler's carelessness (see also footnote 3 on p. 59 of [97]), there has been an effort made to connect Pell with (1.1) (see, for example, Scott [73, p. 208]). This seems to have begun with a misunderstanding of a remark of Hankel [27, p. 203], who actually stated that "Pell has done it no other service than to set it forth again in a much read work." The "much read work" was the English translation [70] of the *Teutschen Algebra* of Rahn. However, a careful examination of this work by Konen [42, pp. 33–34, footnote 1], Wertheim [96] and Eneström [19] did not result in the discovery of any mention of this equation. There can be little doubt that much of this book, particularly pages 100–192, were due to Pell (see Scriba [74]), yet the only mention of anything resembling the Pell equation in it is the equation.

$$x = 12y^2 - z^2 \tag{1.4}$$

on p. 143. This caused Whitford [97, p. 2] to believe that Pell had some acquaintance with (1.1) and seems also to have served as the reason that Pell's biography [3, pp. 1973–1975] suggests that this might have been the case. A thorough inspection of the context in which (1.4) arises in [70] reveals that it is to be used to find x after values for y and z have been selected. This can scarcely be regarded as the Pell equation. Thus, there is no evidence whatsoever linking Pell with (1.1). Nevertheless, as Weil [95, p. 174] asserts, the "traditional designation [of (1.1)] as 'Pell's equation' is unambiguous and convenient." Consequently, it is the term used here for (1.1), even though it is both historically wrong and unjust to those early individuals who did make important contributions to its study.

Much of the very early work on (1.1) was episodic, but since the rediscovery of the equation by Fermat in 1657, research involving it has been continual, and a considerable literature has accumulated, including two books: *Geschichte der Gleichung* $t^2 - Du^2 = 1$ by H. Konen [42] and *The Pell Equation* by E. E. Whitford [97]. Indeed, research on the Pell equation continues to be very active today; at least one hundred articles dealing with it in various contexts have appeared within the last decade. This is simply because it is fundamental to the problem of solving the general second degree Diophantine equation in two unknowns:

$$g(x, y) = ax^2 + bxy + cy^2 + dx + ey + f = 0, \tag{1.5}$$

which means that it finds many applications in number theory and elsewhere. Furthermore, the problem of solving (1.1) is very much connected to the problem of solving the discrete logarithm problem in a real quadratic field (see Jacobson [33], [34]), a problem of much interest to cryptographers [36].

The purpose of this paper is to provide a survey of results concerning the problem of solving (1.1) as we have characterized that problem above. As such, it will be largely influenced by the history of this problem.

Before beginning this, however, it will be convenient to make one further simplifying assumption. We will assume that D has no square integer factor, i.e., D is square-free. We can do this because Lehmer [48], [49] has shown that if p is any prime and we define

$$\psi_D\left(p^k\right) = \begin{cases} 2^k & \text{when } p = 2 \\ p^k & \text{when } p \text{ is odd and } p|D \\ p^{k-1}\left(p - (D/p)\right)/2 & \text{when } p \text{ is odd and } p \nmid D, \end{cases}$$

then $p^k|U_s$, where $s = \psi_D\left(p^k\right)$,

$$T_n + \sqrt{D}U_n = \left(t + u\sqrt{D}\right)^n \qquad (n \in \mathbb{Z})$$

and (D/p) is the Legendre symbol. If

$$m = \prod_{i=1}^{k} p_i^{\alpha_i},$$

where p_i $(i = 1, 2, \ldots, k)$ are distinct primes and we define

$$\psi_D(m) = LCM\left[\psi_D\left(p_i^{\alpha_i}\right), i = 1, 2, \ldots, k\right],$$

then $m|U_s$, where $s = \psi_D(m)$. Furthermore, if $\omega_D(m)$ is the least positive value of s such that $m|U_s$, then

$$\omega_D(m)|\psi_D(m). \qquad (1.6)$$

Now let $D = EF^2$, where E has no square factor. If (t_1, u_1) is the fundamental solution of $T^2 - EU^2 = 1$ and

$$T_n + \sqrt{E}U_n = \left(t_1 + \sqrt{E}u_1\right)^n,$$

then the fundamental solution (t, u) of (1.1) is given by

$$t = T_n, u = U_n/F,$$

where $n = \omega_E(F)$. Since by (1.5) all the possibilities for $\omega_E(F)$ must be divisors of an easily computed integer, it is not difficult to find (t, u) once (t_1, u_1) is known, provided we have the prime factorization of F.

2   Contributions of the Greeks and Indians

There is no direct evidence extant that the ancient Greeks knew how to solve the Pell equation. In fact, there is little mention of the equation at all in their surviving work. Theon of Smyrna (c. 130 AD) knew how to find successive solutions to

$$T^2 - 2U^2 = \pm 1$$

by putting $T_1 = U_1 = 1$ and finding solution (T_{n+1}, U_{n+1}) from (T_n, U_n) by

$$T_{n+1} = 2T_n + U_n, \qquad U_{n+1} = T_n + U_n.$$

Tannery [87] showed that this idea could be extended to any equation of the form

$$T^2 - DU^2 = r, \tag{2.1}$$

but it is unclear whether the Greeks knew this, and even if they did, it would provide little evidence that they knew how to solve (1.1) for t and u.

Diophantus (c. 250 AD) in his *Arithmetica* [28] did solve

$$T^2 - 26U^2 = 1$$

and

$$T^2 - 30U^2 = 1$$

for integral values of T and U (see §§9 and 11 in Book V of [28]), but the method given would, in general, only find rational values of T and U for other values of D. Also, if r is a square, he showed how to find a second rational solution of (2.1) from a given rational solution (see the lemma in §14 of Book VI). This concentration on techniques that produce rational solutions suggests the possibility that the Greeks had not developed an interest in computing integral solutions of (1.1), and in spite of Tannery's [86, p. 101ff] suggestion that Diophantus might have considered such problems in the lost seven books of the *Arithmetica*, we might feel justified in thinking that the Greeks made little contribution to the study of (1.1). However, there is an important piece of evidence that we have not yet considered: the Cattle Problem of Archimedes.

This problem (see Vardi [91] for an English translation of the problem, further discussion and references) consists of two parts. The first of these asks us to determine 8 unknowns (numbers of bulls and cows in each of four herds of cattle), given seven relatively simple linear relationships among these numbers. From this it is not difficult to deduce that the solution of this part of the problem can be given by

$$(10366482, 7460514, 4149387, 7358060, 7206360, 4893246, 543913, 3515820)n,$$

where n is any positive integer. The second, far more difficult part constrains two of the numbers to add to a perfect square and two others to add to a triangular number. Symbolically, this means that $n = 4456749U^2$ where

$$T^2 - 41028642327842U^2 = 1 \qquad (2.2)$$

and T, U are integers.[1]

This problem was attributed to Archimedes (c. 250 BC) in ancient times, and Krumbiegel [44] has argued that, while the text of the problem which we currently possess did not likely originate with Archimedes, the problem itself is likely due to him. However, Fraser [23, p. 402, pp. 407–409] has provided good reasons for accepting Archimedes, not only as the source of the problem, but also as the likely composer of the poetic form in which we now have it. Certainly, Weil [95, p. 17] is correct when he asserts that "there is indeed every reason to accept the attribution to Archimedes, and none for putting it in doubt."

There has been some dispute about the exact wording of the problem (see, for example, Schreiber [72] and Waterhouse [93]) and some concerning the first condition of the second part (see, however, Dijksterhuis [17, pp. 399–400, footnote 3]), but by and large the bulk of expert opinion on the problem is that it reduces to the form stated above. It was first solved by Amthor in 1880, but he was unable to write down the numbers because the value of u for the fundamental solution of (2.2) is approximately 1.86×10^{103265}. The total number of cattle was first computed (using a computer) in 1965 [101] and published later by Nelson [63] in 12 pages of fine print. Vardi has given a very elegant representation of the total number of cattle as

$$\left\lceil \tfrac{25194541}{184119142} \left(109931986732829734979866232821433543901088049 \right.\right.$$
$$\left.\left. + 50549485234315033074477819735540408986340\sqrt{4729494} \right)^{4658} \right\rceil.$$

This number by almost any measure is very large $\left(\approx 7.76 \times 10^{206544} \right)$, but compared to some of the values for t and u which we will find later, it will appear to be rather small.

While it is extremely unlikely that Archimedes was able to solve the cattle problem, we are left with the puzzle of why it was posed and what Archimedes knew when he posed it. The lightly satirical tone of the text of the problem prompted Hultsch to suggest that the problem was Archimedes' response to Apollonius' improvement to his measurement of

---

[1] This is a case of (1.1) in which D is not squarefree. Amthor [4] noted that

$$41028642327842 = 4729494(9314)^2,$$

a fact he was able to exploit in deriving his solution of (2.2).

the circle and his work on naming large numbers. On the other hand, Knorr has speculated that Eratosthenes composed the first part of the problem and that the second part is Archimedes' response. Whatever the reason, the truly interesting problem is determining the state of Archimedes' knowledge of how to solve the Pell equation.

In his work on the measurement of the circle, he gives without any explanation the inequality

$$265/153 < \sqrt{3} < 1351/780.$$

This suggests that he knew of some method of finding good rational approximations to the square roots of integers, but as there are many possible methods he might have used (see Knorr [41, pp. 136–137]) to do this, it is unclear as to how much he might have known concerning discovering integer solutions of (1.1). Nevertheless, as knowledge of how to solve problems of this kind is important for solving the Pell equation, it is certainly reasonable to infer that Archimedes might have been interested in problems like solving (1.1) and likely had devised some technique for accomplishing this for small values of D, at least. Possibly these investigations prompted him to believe that the Pell equation is always solvable, but that when D is large it is a difficult problem.

Knorr [41] is certainly correct when he states that "Greek mathematics was founded on a strong tradition of practical arithmetic competence." Further, Fowler [22] has developed a convincing reconstruction of some of the arithmetic techniques that might have been known to the Greeks. In particular, he has provided strong evidence that continued fractions (see §3) were fundamental to the way that ratios were understood to them. Thus, it is certainly possible that the Greeks possessed much more information about the Pell equation than that suggested by the pitifully few fragments that have survived. Unfortunately, we simply cannot be sure of what it was they did know.

The situation is much different when we consider the achievements of the Indian mathematicians of the Middle Ages. As early as the fifth century, Aryabhata I had developed a method for solving the linear Diophantine equation

$$ax - by = c \tag{2.3}$$

for integers x, y, given integers a, b, c. This technique was subsequently refined and extended by later workers until the Indian mathematicians had essentially produced what is known today as the extended Euclidean algorithm for solving (2.3). It is often assumed by number theorists that the Greeks had found a method of solving (2.3); this may well be true, but very little direct evidence has come down to us to justify this assumption.

In 628 AD Brahmagupta discovered that if

$$A^2 - DB^2 = Q \tag{2.4}$$

and

$$P^2 - DR^2 = S, \tag{2.5}$$

then

$$(AP + DBR)^2 - D(AR + BP)^2 = QS. \tag{2.6}$$

He made use of this observation to develop an *ad hoc* method of solving (1.1) for integers. However, the crowning achievement of Indian mathematics with respect to the Pell equation was the development of the cyclic method for solving it. The technique, described by Bhaskara II in 1150 AD, and its history are well described by Selenius [75], [76], and the interested reader should consult this work for further details and references. We will only sketch, with additional information, one variant (there were several) of the algorithm here.

We will assume that $Q, A, B \in \mathbb{Z}$ and that $(A, B) = 1$ in (2.4); this means that $(B, Q) = 1$. As the technique for solving (2.3) was known, the step of finding an integer P such that $Q | BP + A$ could be easily achieved. It follows that since $(B, Q) = 1$, we must have $Q | P^2 - D$ and $Q | AP + DB$. By putting $R = 1$ in (2.5) we see from (2.6) that

$$\left(\frac{AP + DB}{Q}\right)^2 - D\left(\frac{A + BP}{Q}\right)^2 = \frac{P^2 - D}{Q}. \tag{2.7}$$

From this simple observation we can develop the cyclic method for solving the Pell equation.

Given integers n, A_{n-1}, B_{n-1}, Q_n where $(A_{n-1}, B_{n-1}) = 1$ such that

$$\left|A_{n-1}^2 - DB_{n-1}^2\right| = Q_n,$$

find an integer P_{n+1} such that $Q_n | P_{n+1} B_{n-1} + A_{n-1}$ and $\left|P_{n+1}^2 - D\right|$ is minimal. Put $Q_{n+1} = \left|P_{n+1}^2 - D\right| / Q_n$,

$$A_n = (A_{n-1} P_{n+1} + DB_{n-1}) / Q_n, \ B_n = (B_{n-1} P_{n+1} + A_{n-1}) / Q_n. \tag{2.8}$$

By (2.7) we get

$$\left|A_n^2 - DB_n^2\right| = Q_{n+1}, \tag{2.9}$$

and $(A_n, B_n) = 1$. The latter result follows easily by observing that $|A_n B_{n-1} - B_n A_{n-1}| = 1$. The method terminates when, for some n, $Q_{n+1} = 1, 2, 4$ because Brahmagupta had previously developed methods of solving (1.1) for integers from an integer solution of

$$T^2 - DU^2 = -1, \pm 2, \pm 4.$$

Concerning this technique Hankel [27, p. 202], stated, "It is beyond all praise; it is certainly the finest thing that was achieved in the theory of numbers before Lagrange." Unfortunately, the Indians did not provide a proof that the cyclic method would always work. They were content, it seems, in the empirical knowledge that it always seemed to do so. They used it to solve (1.1) for $D = 61, 67, 97, 103$.

A number of misconceptions continue to circulate concerning the cyclic method. One of these is that it was rediscovered by Lagrange. This, as Selenius has vigorously pointed out, is not strictly speaking the case, but Lagrange did make use of a method similar to it. Often the algorithm is attributed to Bhaskara II, but as mentioned by Shankar Shukla [84, p. 1, p. 20], Bhaskara made no claim to being the originator of the method, and as Jayadeva, who worked in the 10th century or earlier, had discovered a variant of the technique, it seems that it must have been developed much earlier than the time of Bhaskara. Finally, there is the belief, apparently due to Tannery [87], that the cyclic method derives from Greek influences. There seems, apart from possible wishful thinking on the part of Tannery, to be little solid evidence in support of this. The simple fact is that, as mentioned earlier, we don't really know what the Greeks knew about the Pell equation. What we do know, however, is that the Indian methods display a history of steady development and refinement up to and including the discovery of the cyclic method, and this very strongly suggests that Hankel's [27, pp. 203–204] position that the Indians evolved the technique by themselves is the correct one.

3   Continued Fractions

The story of the Pell equation resumes with the challenge [21, pp. 333–335], issued in 1657 to Frénicle in particular and mathematicians in general by Fermat. Fermat had most likely, through his research, come to recognize the fundamental nature of the Pell equation. An English translation of his general challenge can be found in [28, pp. 285–286]. It asks for a proof of the following statement:

> Given any [positive] number [D] whatever that is not a square,
> there are also given an infinite number of squares such that,
> if the square is multiplied into the given number and unity is
> added to the product, the result is a square.

It next requests a general rule by which solutions of this problem could be determined, and as examples asks for solutions when $D = 109, 149, 433$.

The story of how the second part of this challenge was answered by Brouncker and Wallis has been very well told by Weil [95] and Mahoney [57] and needs no elaboration here. Suffice it to say that Brouncker developed a technique for solving (1.1) in integers, but neither he nor Wallis nor Frénicle was able to provide a proof that the Pell equation could always be solved (non-trivially) for any positive non-square value of D. Fermat [21, p. 433], took notice of this and stated that he had such a proof "by means of *descente* duly and appropriately applied." Unfortunately, Fermat provided no further information concerning his proof than this. Hofmann [29] and, with greater success, Weil [94], [95, §XIII] have attempted to reconstruct what Fermat's method might have been. While we may never really know what this was it is nevertheless very likely that Fermat did have a proof. The fact that he selected 109, 149, 433 for values of D as challenge examples is particularly suggestive because the corresponding Pell equations have large values of t and u.

The method of Brouncker was modified and extended by Euler, who realized that continued fractions could be used to provide an efficient algorithm for solving the Pell equation. However, even though he had devised all of the important tools, he just fell short of proving that his method would work for any non-square D. As mentioned earlier, the development of such a technique was first done by Lagrange in a rather clumsy work, which he later improved. For further information on this particularly interesting part of mathematical history, the reader is referred to [95].

While the technique of the Indians and those of Brouncker, Euler and Lagrange for solving the Pell equation are different to some degree, they can all be unified by considering the theory of semiregular continued fractions. For a given real number ϕ, put $\phi_0 = \phi$ and define

$$\phi_{i+1} = \frac{a_{i+1}}{\phi_i - b_i},$$

where $a_{i+1} = \pm 1$ and $b_i \in \mathbb{Z}$ such that $\phi_{i+1} > 1$ $(i = 0, 1, 2, \ldots)$. Then

$$\phi = b_0 + \cfrac{a_1}{b_1 + \cfrac{a_2}{b_2 + \cfrac{a_3}{b_3 + \cdots + \cfrac{a_n}{b_n + \cfrac{a_{n+1}}{\phi_{n+1}}}}}} \tag{3.1}$$

If $b_i \geq 1$, $b_i + a_{i+1} \geq 1$ $(i \geq 1)$, the above expression is called a *semiregular continued fraction expansion of* ϕ. (See, for example, Perron [67, ch. 5].) For example, in the case of $b_i = \lfloor \phi_i \rfloor$ and $a_{i+1} = 1$ $(i \geq 0)$, we say that (3.1) is the *regular* continued fraction (RCF) expansion of ϕ; in the case of $b_i =$

$\lfloor \phi_i + 1/2 \rfloor$, the nearest integer to ϕ_i, and $sign\,(a_{i+1}) = sign\,(\phi_i - b_i)$ $(i \geq 0)$, we have what is called the *nearest integer* continued fraction (NICF) expansion of ϕ.

If

$$C_n = b_0 + \cfrac{a_1}{b_1 + \cfrac{a_2}{b_2 + \cfrac{a_3}{b_3 + \cdots + \cfrac{a_n}{b_n}}}},$$

we call C_n a *convergent* of (3.1), and it can be shown that $C_n = A_n/B_n$, where $A_{-2} = 0, A_{-1} = 1, B_{-2} = 1, B_{-1} = 0, a_0 = 1$, and

$$A_{i+1} = b_{i+1}A_i + a_{i+1}A_{i-1}, B_{i+1} = b_{i+1}B_i + a_{i+1}B_{i-1} \quad (i \geq 1).$$

Also,

$$A_n B_{n-1} - B_n A_{n-1} = (-1)^{n-1}a_1 a_2 \ldots a_n \quad (n \geq 1).$$

If $\phi = \left(P_0 + \sqrt{D}\right)/Q_0$, where $P_0, Q_0 \in \mathbb{Z}$ and $Q_0 | D - P_0^2$, then

$$\phi_{n+1} = \left(P_{n+1} + \sqrt{D}\right)/Q_{n+1},$$

where,

$$P_{n+1} = b_n Q_n - P_n, Q_{n+1} = a_{n+1}\left(D - P_{n+1}^2\right)/Q_n > 0.$$

Let $\alpha \in \mathbb{Q}\left(\sqrt{D}\right)$. We denote by $\bar{\alpha}$, the conjugate of α in $\mathbb{Q}\left(\sqrt{D}\right)$ and by $N(\alpha)$ (the norm of α) the value of $\alpha\bar{\alpha} \in \mathbb{Q}$. It is easy to deduce that

$$N\,(A_n - \phi B_n) = (-1)^{n+1}\,(a_1 a_2 \ldots a_{n+1})\,Q_{n+1}/Q_0$$

and

$$A_n - \phi B_n = \prod_{i=1}^{n+1} \frac{P_i - \sqrt{D}}{Q_{i-1}}.$$

If we define $G_n = Q_0 A_n - P_0 B_n \,(= A_n$ when $\phi = \sqrt{D})$, then

$$Q_n G_n = P_{n+1}G_{n-1} + DB_{n-1}, Q_n B_n = G_{n-1} + P_{n+1}B_{n-1}, \quad (3.2)$$

$$G_n^2 - DB_n^2 = (-1)^{n+1}\,(a_1 a_2 \ldots a_{n+1})\,Q_{n+1}Q_0, \quad (3.3)$$

and

$$\left(G_n + \sqrt{D}B_n\right)/Q_0 = A_n - \bar{\phi}B_n = \prod_{i=1}^{n+1} \frac{P_i + \sqrt{D}}{Q_{i-1}}. \quad (3.4)$$

Notice that the formulas (3.2) are generalizations of (2.7) and (3.3) is a generalization of (2.8).

Now put $c_n = \lfloor \phi_n \rfloor$, $P'_{n+1} = c_n Q_n - P_n$, $Q'_{n+1} = \left(D - P'^2_{n+1}\right)/Q_n$, $P''_{n+1} = P'_{n+1} + Q_n$, $Q''_{n+1} = \left(P''^2_{n+1} - D\right)/Q_n$. We next define $b_n = c_n + 1$, $a_{n+1} = -1$ when $Q''_{n+1} \leq Q'_{n+1}$ and $b_n = c_n$, $a_{n+1} = 1$ otherwise. Note that in the former case we get $P_{n+1} = P''_{n+1}$, $Q_{n+1} = Q''_{n+1}$ and in the latter case we get $P_{n+1} = P'_{n+1}$, $Q_{n+1} = Q'_{n+1}$. If b_i and a_{i+1} are defined in this way for $i = 0, 1, 2, \ldots$, we get the semiregular continued fraction that is implicitly produced by the cyclic method when $\phi = \phi_0 = \sqrt{D}$.

Indeed, when $\phi = \sqrt{D}$ it can be shown in the case of the cyclic method, the RCF (used by Euler and Lagrange) and the NICF that

$$t + u\sqrt{D} = A_n + B_n\sqrt{D},$$

where n is the least non-negative integer such that

$$(-1)^{n+1}(a_1 a_2 \ldots a_{n+1}) Q_{n+1} = 1. \tag{3.5}$$

Furthermore, for any nonsquare positive value of D, (3.5) must always occur for some $n \geq 0$.

Subsequent to the development of the RCF algorithm for solving the Pell equation, several authors, starting with Legendre in 1798, produced tables of values of t, u for certain ranges of values of D. These tables are described in Lehmer [50, pp. 54–59]. This work culminated in the unpublished table of Lehmer [47] in 1926 which dealt with all nonsquare D in the range $1700 < D \leq 2000$. At this point (1.1) had been solved for all positive $D \leq 2000$. Considering that these tables were all produced by hand calculation, it is easy to see why this was about as much as could be done for several years. In 1941 and more extensively in 1955 Patz [65], [66] produced tables of the RCF of $\sqrt{D}$ for all nonsquare values of D such that $1 < D < 10000$. Finally, in 1961 Kortum and McNeil [43], [77] used a computer to produce a very large table, giving, among other things, the RCF of $\sqrt{D}$ and the least solution of $T^2 - DU^2 = \pm 1$ for all nonsquare D such that $1 < D < 10000$. For reasons that will be made clear in the next section, no further tables have ever been published. It should, however, be mentioned that before the advent of computers, several authors attempted to solve the Pell equation for values of D in excess of 2000. One of the more impressive of these is the solution for $D = 9817$ by Martin [58] in 1877. Not only is his solution correct, but the value of t is a number of 97 digits.

4   The Regulator

One of the main reasons why it is not practicable to produce large tables of solutions of the Pell equation is that the values of t and u tend to become very large. For example, the values of t and u when $D = 95419$ have 376 and 374 decimal digits, respectively. Of course, another important reason why tables are no longer produced is the easy availability of special purpose computer software which will very quickly and accurately produce a solution of the Pell equation on input of the value of D. Nevertheless, large values of D still present a problem, as we shall see.

Let $\mathcal{K} = \mathbb{Q}\left(\sqrt{D}\right)$, where D is a square-free positive integer and put

$$r = \begin{cases} 2 & \text{when } D \equiv 1 \pmod 4, \\ 1 & \text{otherwise.} \end{cases}$$

We define $\omega = \left(r - 1 + \sqrt{D}\right)/r$ and d, the *discriminant* of $\mathcal{K}$, as $(\omega - \overline{\omega})^2 = 4D/r^2$. The ring of algebraic integers $\mathcal{O}$ of $\mathcal{K}$ is given by $\mathbb{Z} + \omega\mathbb{Z}$. If $\eta \in \mathcal{O}$ and $N(\eta) = \pm 1$, then η is a *unit* of $\mathcal{K}$ (or $\mathcal{O}$). It is well known that the set $\mathcal{O}^*$ of all units of $\mathcal{O}$ is an abelian group under multiplication with generators -1 and $\varepsilon_d = \left(x + y\sqrt{d}\right)/2$, where $x, y \in \mathbb{Z}$ and $\varepsilon_d(> 1)$ is called the *fundamental unit* of $\mathcal{K}$ (or $\mathcal{O}$). If we define $\varepsilon(D) = t + u\sqrt{D}$, then

$$\varepsilon(D) = \varepsilon_d^\nu,$$

where $\nu \in \{1, 2, 3, 6\}$.

It is a relatively straight-forward exercise to show that the value of ν can be determined from the table below. Thus, we can always easily determine ν when we know the value of $d \pmod{16}$ and the values of $x, y \pmod 8$. It follows that, given ε_d, we can find $\varepsilon(D)$, and as $t = \lceil \varepsilon(D)/2 \rceil$ and $u = \lfloor \varepsilon(D)/\left(2\sqrt{D}\right) \rfloor$, we can compute t and u. However, the problem with computing ε_d is that the values of x and y also tend to become very large as D increases. So instead we work with the *regulator*[2] R_d of $\mathcal{K}$, which is defined by $R_d = \log \varepsilon_d$. This is much easier to deal with from the point of view of the storage of information; for example, the regulator[3] when $D = 95419$ is 865.5675.

---

[2] The term "regulator" was first used (in a much wider context than this) by Dedekind [51, p. 597], in 1893. However, as pointed out in a footnote, he borrowed the term from an 1844 paper of Eisenstein [18, p. 313], who used it in a somewhat different context.

[3] The regulator is an irrational number; therefore, we cannot provide its exact value. The values given in this paper are approximations rounded to the number of figures written.

$d \pmod{16}$	y	x	ν
12	—	—	1
8	$2\mid y$	—	1
8	$2 \nmid y$	—	2
$1, 5, 9, 13$	$4\mid y$	—	1
$1, 5, 9, 13$	$2 \parallel y$	—	2
5	$2 \nmid y$	$x \equiv \pm 3y \pmod 8$	3
		$x \equiv \pm y \pmod 8$	6
13	$2 \nmid y$	$x \equiv \pm y \pmod 8$	3
		$x \equiv \pm 3y \pmod 8$	6

One of the interesting aspects of the RCF expansion of $\phi = \omega$ or $\sqrt{D}$ is its symmetry property. We find that if $n + 1$ is the least positive integer such that $Q_{n+1} = Q_0$ (=1 or 2), then

$$Q_i = Q_{n+1-i}, \quad P_{i+1} = P_{n+1-i} \quad (i = 0, 1, 2, \ldots, n),$$

$$b_i = b_{n+1-i} \quad (i = 1, 2, \ldots, n).$$

These results mean that we can rewrite (3.4) as

$$\frac{\left(G_n + \sqrt{D}B_n\right)}{Q_0} = \begin{cases} \frac{P_{k+1}+\sqrt{D}}{Q_0} \prod_{i=1}^{k} \left(\frac{P_i+\sqrt{D}}{Q_i}\right)^2 & \text{when } n+1 = 2k+1, \\ \frac{Q_k}{Q_0} \prod_{i=1}^{k} \left(\frac{P_i+\sqrt{D}}{Q_i}\right)^2 & \text{when } n+1 = 2k. \end{cases}$$

Furthermore, we know that $n + 1 = 2k + 1$ whenever we find some minimal $k(\geq 0)$ such that $Q_k = Q_{k+1}$; on the other hand we get $n + 1 = 2k$ whenever we find some minimal $k(> 0)$ such that $P_k = P_{k+1}$. In the case where $\phi = \omega$, we get

$$\varepsilon_d = \left(G_n + \sqrt{D}B_n\right)/Q_0;$$

hence,

$$R_d = \begin{cases} \log\left(\left(P_{k+1} + \sqrt{D}\right)/Q_0\right) + 2S & \text{when } Q_k = Q_{k+1}, \\ \log\left(Q_n/Q_0\right) + 2S & \text{when } P_k = P_{k+1}, \end{cases} \qquad (4.1)$$

where

$$S = \sum_{i=1}^{k} \log \left(\left(P_i + \sqrt{D} \right) / Q_i \right).$$

Formulas like (4.1) were used in 1976 by Williams and Broere [99] to determine all the values of R_d for $D < 1.5 \times 10^5$.

The NICF expansion of ω also has symmetry properties which make it possible to produce a formula like (4.1) for R_d. However, instead of only 2 mid-period criteria, there are 6. Williams and Buhr [100] made use of the NICF to compute R_d for all square-free values of $D < 10^6$. The use of the NICF is recommended over that of the RCF because, as noted by Adams [2], the value for n for the NICF tends to be about 70% of that for the RCF expansion of ϕ. However, the extra work needed to determine the proper mid-period criterion tends to cause the run-time of the NICF algorithm to increase to 75% of the RCF algorithm. For larger values of D of a certain type (the type that tends to produce large values of n), the results of Williams [98] could be used to improve this speed, but as we shall see below, there are much better methods for evaluating R_d for large values of D. Before beginning to deal with these, it should be noted that all the continued fraction methods for computing R_d described to this point execute in time which is $O(R_d) = O\left(d^{1/2+\varepsilon}\right)$ for any $\varepsilon > 0$.

If we let $h(d)$ denote the class number of $\mathcal{K}$, we know from the analytic class number formula that

$$2h(d)R_d = \sqrt{d}L(1, \chi_d),$$

where $\chi_d(j)$ is the Kronecker symbol (d/j) and

$$L(1, \chi_d) = \sum_{j=1}^{\infty} \frac{\chi_d(j)}{j} = \prod_q \left(\frac{q}{q - (d/q)} \right). \tag{4.2}$$

Here the (Euler) product is taken over all the primes q. Hua [30] has shown that $L(1, \chi_d) < (\log d)/2 + 1$, a result that has recently been improved by Louboutin [56], but under a generalized Riemann Hypothesis (GRH), we know from a result of Littlewood [55] and a small improvement by Shanks [81] that

$$\{1 + o(1)\} (c_1 \log\log d)^{-1} < L(1, \chi_d) < \{1 + o(1)\} c_2 \log\log d, \tag{4.3}$$

where

$$c_2 = re^\gamma, c_1 = 4(r+1)e^\gamma/\pi^2, e^\gamma = 1.781072418.$$

This result has been tested numerically by Shanks [81], Jacobson, Lukes and Williams [35] and Jacobson [31], [32], and it has always been found to hold.

We now note that the Cohen-Lenstra heuristics [15], [16] on the distribution of the values of $h^*(d)$ (the odd part of the class number; $h(d) = 2^s h^*(d)$, $2 \nmid h^*(d)$) strongly suggest that $h^*(d)$ tends to be small. Indeed, about 75% of the values of $h^*(d)$ are 1. Also, under these heuristics we find that

$$P_r\left(h^*(d) > x\right) = \frac{1}{2x} + O\left(\frac{\log x}{x^2}\right)$$

(see [31]), where $P_r\left(h^*(d) > x\right)$ is the probability that $h^*(d) > x$. Since there are many values of d for which $h(d) = h^*(d)$ (for example, prime values of d), we would expect that the value of R_d can be large very often, and this was certainly found to be the case in the numerical study [35].

To gain some understanding of just how large R_d can get, we define the *regulator index* RI for $\mathcal{K}$ by

$$RI = 2R_d/(c_2\sqrt{d}\log\log d) = L\left(1, \chi_d\right)/\left(c_2 h(d)\log\log d\right).$$

From our previous observations we would expect that RI should only infrequently be very small and that $RI < 1 + o(1)$. For example, when $D = 95419$, the value of RI is .61612. In fact, the largest value of RI for all $d < 10^9$ is $RI = .68698$ [32] for $d = 513215704$ and the largest value of RI currently known is $RI = .78354$ for the 18 digit $D = 5749111115184562766$ (see [31, p. 93]). Although it has been shown that as d runs through prime values ($\equiv 1 \pmod 4$)

$$\limsup L\left(1, \chi_d\right)/\log\log d \geq e^\gamma$$

(see, for example, Joshi [40]), it's usually rather difficult to produce a very large value of D such that $RI > 1/2$. Indeed, very recent investigations by Granville and Soundararajan [24] have led them to believe that there exists a constant C such that if

$$f(d) = 2c_2^{-1}L\left(1, \chi_d\right) - \log\log d - \log\log\log d,$$

then

$$\limsup f(d) = C.$$

This, of course, suggests that, as d becomes very large, we would not expect the value of RI to be much larger than $1/2$.

What is particularly important to observe here is that if we are getting values of R_d that are roughly of the same order of magnitude as $\sqrt{d}$, then for such values of d, particularly as d becomes large, the problem of simply writing t and u down is of exponential complexity. In other words we cannot record the values of t and u in the conventional way in polynomial time;

indeed, we may very often need exponential time in order to do this. Thus, we seem to be forced in such cases to record the value of R_d only. Now R_d is certainly a fundamental invariant of $\mathcal{K}$, but it's a transcendental number (and therefore can only be written to a certain precision) and, on the face of it, does not seem to reveal much about t and u. However, as we will point out in §7 we can compute an approximation to R_d which will allow us to develop an expression for $\varepsilon(D)$ that can be written in polynomial time. Before doing that, we must examine the problem of computing R_d more efficiently than we have done so far.

5   Ideals and Ideal Classes

In this section we will review several results that will be important for the discussion of the infrastructure of the class group of $\mathcal{K}$, which will follow. Most of this material can be found in any algebraic number theory textbook. There is also a computationally oriented discussion in Mollin and Williams [61]. Thus, our treatment of this information will be brief. We first note that $\mathcal{O}$ (the maximal order of $\mathcal{K}$) is the set of algebraic integers in $\mathcal{K}$ and serves in $\mathcal{K}$ in much the same way as $\mathbb{Z}$ does in the rationals $\mathbb{Q}$. Since $\mathcal{O}$ is a ring, it has ideals. If $\mathfrak{a}$ is any (integral) ideal of $\mathcal{O}$, then $\mathfrak{a}$ can be written

$$\mathfrak{a} = a\mathbb{Z} + \beta\mathbb{Z},$$

where $\beta = b + c\omega$; $a, b, c \in \mathbb{Z}$; $c|a$, $c|b$; $a, c > 0$; $|b| < a$. We call $\{a, \beta\}$ a *normal* $\mathbb{Z}$-basis of $\mathfrak{a}$. The quantity a is unique for $\mathfrak{a}$; it is the least positive rational integer in $\mathfrak{a}$. Thus, if m is any other rational integer in $\mathfrak{a}$, then $a|m$; this means that if $\gamma \in \mathfrak{a}$, then $a|N(\gamma)$. The norm of $\mathfrak{a}$, written $N(\mathfrak{a})$, is equal to ac. If $c = 1$, we say that $\mathfrak{a}$ is a *primitive* ideal; that is, $\mathfrak{a}$ has no rational integer divisors. It is well known that if $\mathfrak{a}$ is primitive, then

$$\left(N\left(\mathfrak{a}\right), \beta + \overline{\beta}, N\left(\beta\right)/N\left(\mathfrak{a}\right)\right) = 1. \tag{5.1}$$

Let $\theta_1, \theta_2, \ldots, \theta_k \in \mathcal{O}$ and define $\mathfrak{a} = (\theta_1, \theta_2, \theta_3, \ldots, \theta_k)$ to be $\{\theta_1\gamma_1 + \theta_2\gamma_2 + \ldots + \theta_k\gamma_k : \gamma_1, \gamma_2, \ldots, \gamma_k \in \mathcal{O}\}$. This is clearly an ideal of $\mathcal{O}$, and we call it the ideal *generated* by $\theta_1, \theta_2, \ldots, \theta_k$. It can be shown that if $\mathfrak{a}$ is any ideal of $\mathcal{O}$, then there exists $\theta_1, \theta_2 \in \mathcal{O}$ such that $\mathfrak{a} = (\theta_1, \theta_2)$. If $\mathfrak{a} = (\theta)$, we say that $\mathfrak{a}$ is the *principal ideal generated* by θ. For such an ideal we get $N(\mathfrak{a}) = |N(\theta)|$. If $\mathfrak{a} = (\theta_1, \theta_2, \ldots, \theta_k)$ and $\mathfrak{b} = (\phi_1, \phi_2, \ldots, \phi_n)$, then we define the product ideal $\mathfrak{a}\mathfrak{b}$ to be the ideal generated by the kn generators $\theta_i\phi_j$ $(i = 1, 2, \ldots, k; j = 1, 2\ldots, n)$. Given normal $\mathbb{Z}$-bases for the ideals $\mathfrak{a}$ and $\mathfrak{b}$, we can compute a normal $\mathbb{Z}$-basis for the ideal $\mathfrak{a}\mathfrak{b}$ by a deterministic algorithm that executes in $O(\log N(\mathfrak{a}) + \log N(\mathfrak{b}) + \log d)$

multiplication/division operations.[4] This algorithm is essentially the algorithm that Gauss discovered for the composition of quadratic forms and can be found, for example, on pp. 284–285 of [61]. Also, if $\mathfrak{a} = a\mathbb{Z} + \beta\mathbb{Z}$, then we define the ideal $\bar{\mathfrak{a}}$ conjugate to $\mathfrak{a}$ by $\bar{\mathfrak{a}} = a\mathbb{Z} + \bar{\beta}\mathbb{Z}$. It is easy to show that $\mathfrak{a}\bar{\mathfrak{a}} = (N(\mathfrak{a}))$, a principal ideal. If $\mathfrak{a} = \mathfrak{bc}$, we say that the ideal $\mathfrak{b}$ divides $\mathfrak{a}$; a necessary and sufficient condition that $\mathfrak{b}|\mathfrak{a}$ is that $\mathfrak{b} \supseteq \mathfrak{a}$. A property of the norm is that $N(\mathfrak{ab}) = N(\mathfrak{a})N(\mathfrak{b})$.

If $\mathfrak{a}$ and $\mathfrak{b}$ are non-zero ideals of $\mathcal{O}$ and there exist $\alpha, \beta \in \mathcal{O}$ such that $\alpha, \beta \neq 0$ and

$$(\alpha)\mathfrak{a} = (\beta)\mathfrak{b},$$

we say that $\mathfrak{a}$ and $\mathfrak{b}$ are *equivalent* and denote this by $\mathfrak{a} \sim \mathfrak{b}$. This equivalence relationship partitions all the ideals in $\mathcal{O}$ into $h(d)$ disjoint equivalence classes. If we denote by $[\mathfrak{a}]$ the class of all ideals of $\mathcal{O}$ equivalent to $\mathfrak{a}$, we can define an operation of multiplication of these classes by $[\mathfrak{a}][\mathfrak{b}] = [\mathfrak{ab}]$. Under this operation, it is easy to see that the set of all these ideal classes forms a group Cl with identity $[(1)]$, the class of all principal ideals.

We say that a primitive ideal $\mathfrak{a}$ of $\mathcal{O}$ is *reduced* if there does not exist any non-zero $\alpha \in \mathfrak{a}$ such that

$$|\alpha| < N(\mathfrak{a}), |\bar{\alpha}| < N(\mathfrak{a}).$$

An interesting and important consequence of this definition is that if $\mathfrak{a}$ is a reduced ideal, then $N(\mathfrak{a}) < \sqrt{d}$. It follows that there can only be a finite number of reduced ideals in $\mathcal{O}$. Also, if $\mathfrak{a}$ is any primitive ideal and $N(\mathfrak{a}) < \sqrt{d}/2$, then $\mathfrak{a}$ must be reduced.

If $\{a, \beta\}$ is a normal basis of $\mathfrak{a}$ and we produce the RCF expansion of $\beta/a = \left(P_0 + \sqrt{D}\right)/Q_0$, then we can produce a sequence of primitive ideals

$$\mathfrak{a}_1, \mathfrak{a}_2, \mathfrak{a}_3 \ldots \tag{5.2}$$

such that

$$\mathfrak{a}_i = (Q_{i-1}/r)\,\mathbb{Z} + \left(\left(P_{i-1} + \sqrt{D}\right)/r\right)\mathbb{Z} \tag{5.3}$$

and $\mathfrak{a}_i \sim \mathfrak{a}\,(=\mathfrak{a}_1)$ $(i = 1, 2, \ldots)$. Here, the values of the Q_{i-1} and P_{i-1} are those defined in §3. It can be shown that for a value $i = O(\log N(\mathfrak{a}))$, we must have $N(\mathfrak{a}_i) < \sqrt{d}/2$. Thus, the RCF algorithm provides us with a technique that will produce a reduced ideal $\mathfrak{b}$ such that $\mathfrak{b} \sim \mathfrak{a}$ and this algorithm will execute in $O(\log N(\mathfrak{a}))$ operations.

Let $\mathfrak{a}_1 = (1) = \mathcal{O}$. We can use the RCF algorithm on ω to produce a sequence of reduced ideals (5.2) ((1) is evidently a reduced ideal) such that

---

[4]We will use the term "operation" in the sequel to refer to an operation of multiplication or division (or addition or subtraction).

$\mathfrak{a}_i = (\theta_i)$, where $\theta_i = \left(G_{i-2} + \sqrt{D}B_{i-2}\right)/r$ (see §3). We can also show that

$$1 + 1/\sqrt{d} < \theta_{i+1}/\theta_i < \sqrt{d}$$
$$\theta_{i+2}/\theta_i > 2 \quad (i = 1, 2, \ldots). \tag{5.4}$$

Hence $\theta_1 = 1$ and

$$\theta_1 < \theta_2 < \theta_3 < \ldots < \theta_n < \theta_{n+1} < \ldots.$$

Indeed, we must have $\mathfrak{a}_{n+1} = \mathfrak{a}_1$ for some minimal $n > 0$. It follows, as mentioned above, that $\varepsilon_d = \theta_{n+1}$. If we put $\mathcal{R} = \{\mathfrak{a}_1, \mathfrak{a}_2, \ldots, \mathfrak{a}_n\}$, then it can be shown that $\mathcal{R}$ is the set of all reduced principal ideals of $\mathcal{O}$. That is, the RCF algorithm produces all the reduced principal ideals of $\mathcal{O}$, something that other semiregular continued fractions, like the NICF, do not in general do.

6   The Infrastructure of Cl

As mentioned earlier, in the early 1970s Shanks was testing the truth of (4.3) by doing some numerical work. He had defined what he called the upper Littlewood index (ULI) as

$$\text{ULI} = L(1, \chi_d) / (c_2 \log \log d)$$

and was attempting to find values of it that were large. An examination of (4.2) suggests that $L(1, \chi_d)$ will be large when $(d/q) = 1$ for as many of the small (in particular) primes as possible. In October of 1971, Lehmer had found the 14 digit $d = D = 26437680473689$ on his delay line sieve DLS-157; this number is a prime such that $(d/q) = 1$ for all primes $q \leq 149$. Shanks knew through the empirical work of Kloss et al. (see Shanks [78]) that it was likely that $h(d) = 1$ when d is a prime congruent to 1 modulo 4; thus, by estimating $L(1, \chi_d)$ and using the analytic class number formula, assuming $h(d) = 1$, he concluded that $R_d \approx 2.17 \times 10^7$. With the algorithms for evaluating R_d at his disposal, however, he realized that he would not be able to compute an accurate value of R_d, which he needed to improve his estimate of $L(1, \chi_d)$. It was this problem which caused him to examine the inner structure of the various ideal classes, structures that he collectively referred to as the infrastructure of Cl (or $\mathcal{K}$).

Although the ideas described below can be applied to any ideal class, we will confine (as did Shanks in [80]) our attention here to the principal class.

Let $c \in \mathcal{R}$; then $c = a_i$ for some i such that $1 \leq i \leq n$. Define $\delta(c) = \log \theta_i$. By (5.4) we see that $\delta(a_{i+2}) > \delta(a_i) + \log 2$; hence $j = O(\delta(a_j))$. Now let $a, b \in \mathcal{R}$ and consider the ideal ab. This ideal is the product of two principal ideals and must therefore be principal, but it is not necessarily reduced. Suppose we apply our reduction algorithm to ab until we get c, the first ideal such that $N(c) < \sqrt{d}/2$. We will denote this c by red(ab) and observe that c is reduced.[5] Thus, $c \in \mathcal{R}$. Furthermore

$$\delta(c) \equiv \delta(a) + \delta(b) + \kappa \pmod{R_d} \tag{6.1}$$

and

$$-\log d < \kappa < \log 2.$$

What is important to note here is that the values of $\delta(a), \delta(b), \delta(c)$ can be quite large (near R_d, say), but κ is quite small. Also, there is a simple formula for κ which allows it to be computed quickly and accurately. Thus, if we start with some ideal $b \in \mathcal{R}$ such that $\delta(b)$ is large by comparison to $\log d$, but smaller than R_d, and define

$$b_{i+1} = \text{red}(bb_i), \tag{6.2}$$

where $b_1 = b$, then by (6.1) we get

$$\delta(b_j) \geq j\delta(b) - (j-1)\log d.$$

Having arrived at this deduction, Shanks was next able to apply his baby-step giant-step [79] technique[6] to find R_d.

First, use the RCF to compute the set of ideals $\mathcal{L} = \{a_1 = (1), a_2, a_3, \ldots, a_t, a_{t+1}, a_{t+2}\} \subseteq \mathcal{R}$. These are the $t + 2$ baby-steps. Here we need t to be large enough that $\delta(a_t) > \sqrt[4]{d}$. Next, put $b_1 = a_t$ and compute $b_2, b_3, \ldots$ as in (6.2). These are the giant steps. Now notice that

$$\delta(b_{i+1}) - \delta(b_i) < \delta(b_1) + \log 2 = \delta(a_t) + \log 2 < \delta(a_{t+2})$$

---

[5]It should be noted that Shanks [83] also developed a method of computing a reduced ideal $\sim ab$ which did not first compute ab and then perform a reduction technique on ab. He called this algorithm NUCOMP. At first it was thought that this algorithm was not much more efficient than the technique described above, but a new analysis by van der Poorten [89] and unpublished computational results of Jacobson suggest that it is at least 20% faster when $d > 0$.

[6]Curiously, it was also numbers produced from Lehmer's sieve (the DLS-127, an earlier version of the DLS-157) which inspired Shanks' discovery of the baby-step giant-step technique. In 1968 Lehmer had produced 3 values of d for which he wanted the value of $h(-d)$, the class number of $\mathbb{Q}(\sqrt{-d})$. These values were selected in order to minimize the value of $h(-d)/\sqrt{d}$ (see Teske and Williams [88]), but they were too large for computing $h(-d)$ by the means currently available. The first time the baby-step giant-step idea was used was in the determination of $h(-229565917267) = 29351$ in August of 1968.

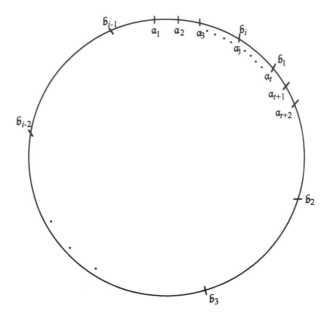

by (6.1) and (5.4). Consider the diagram above where each ideal is placed on the circumference of the circle according to the value of $\delta(\mathfrak{c})$. Since $\mathfrak{a}_{n+1} = \mathfrak{a}_1$, the length of the circumference of the circle is R_d. If we examine this diagram we see that we must get some $\mathfrak{b}_i \in \mathcal{L}$. This means that $\mathfrak{b}_i = \mathfrak{a}_j$ and we can compute

$$R_d = \delta(\mathfrak{b}_i) - \delta(\mathfrak{a}_j).$$

The complexity of computing R_d by this procedure can be easily estimated. We know that

$$d^{1/2+\varepsilon} > R_d > \delta(\mathfrak{b}_{i-1})$$
$$> (i-1)\delta(\mathfrak{a}_t) - (i-2)\log d > (i-1)\sqrt[4]{d} - (i-2)\log d.$$

Hence $i = O\left(d^{1/4+\varepsilon}\right)$. Furthermore, if $\delta(\mathfrak{a}_{t-1}) < \sqrt[4]{d}$ and $\delta(\mathfrak{a}_t) > \sqrt[4]{d}$, then $t = O\left(d^{1/4+\varepsilon}\right)$. It follows that the total complexity of determining R_d is $O\left(d^{1/4+\varepsilon}\right)$, a considerable saving over the previous methods of using continued fractions, particularly when d is large.

By using this technique, Shanks was able to compute R_d for Lehmer's value of d as $R_d = 21737796.43$ and $L(1, \chi_d) = 8.45539$. Also $RI = .69180$. In fact, as computed later [82], $n = 18334815$. This is much too large to compute R_d by the continued fraction method with the computer power available to Shanks at the time. It should be stressed here that although we have used an ideal theoretic setting in which to describe Shanks' ideas and methods, he used the language and techniques of the theory of quadratic forms.

Shanks' discovery led to a number of results of both theoretic and computational interest. For example, in 1979 Lagarias [45] used the infrastructure in his sketch of a proof of the following interesting result.

Theorem 6.1. *Let $L(m)$ denote the length of the binary encoding of the integer m and let $L(g)$ be the length of the binary encoding (the total number of bits in $a, b, \ldots, f$) of the quadratic Diophantine equation (1.5). Suppose (1.5) has a non-negative integer solution (x, y) satisfying*

$$x \equiv x_1 \ (mod \ m)$$
$$y \equiv y_1 \ (mod \ m).$$

Then there exists a certificate (proof) showing that $f(x, y) = 0$ has such a solution which requires only $O\left(L^5 \log L \log \log L\right)$ elementary (bit) operations to verify, where $L = L(g) + 3L(m)$.

This result, in view of what we have said about the solution of the Pell equation in §4, is surprising because we know that these are equations of the form (1.5) whose smallest non-negative integer solution is so large that it takes time which is exponential in $L(g)$ just to write the solution down in the usual binary or base 10 representation. Unfortunately, the importance of this result was not recognized at the time and a full proof of it never appeared in any refereed journal. It was published as a technical report [46], but did not circulate sufficiently for the result to receive the attention it merited.

Later, in 1982, Lenstra [54] showed how to use the infrastructure, together with estimates on the value of $L(1, \chi_d)$, to improve the speed of computing R_d to $O\left(d^{1/5+\varepsilon}\right)$. The value of R_d that Lenstra's algorithm produces is correct (to the number of figures written); but the complexity of his algorithm is conditional on the truth of an appropriate generalization of the Riemann hypothesis (GRH). A year later Schoof [71] implemented Lenstra's algorithm on a computer in order to determine the class groups of certain real quadratic fields. He also showed that under the GRH any method that can compute R_d can be converted to a method which will obtain the integer factorization of D and that this only requires a polynomial (in $\log D$) multiple of the time needed to find R_d to execute. Thus, computing R_d is at least (from a computational complexity point of view) as difficult as factoring D under the GRH.

In 1988 Buchmann and Williams [9] showed that R_d could be computed unconditionally with complexity $O\left(R_d^{1/2}d^\varepsilon\right)$. This is still the best unconditional complexity measure known for the computation of R_d. More recently, Lenstra's technique, incorporating a method of Bach [6] for obtaining more accurate estimates for $L(1, \chi_d)$, was implemented on a computer by Jacobson, Lukes and Williams [35] and Jacobson [32] to compute all R_d for $d < 10^9$. Indeed, to illustrate the power of this technique, we computed R_d for the 30 digit

$$D = 990676090995853870156271607886.$$

Here $R_d = 4770372955851343.43$ and $RI = .63242$. This is the largest value of D with a regulator index $> .5$ for which a non-conditional regulator has ever been computed.

Finally, we point out that Srinivasan [85] has recently produced a version of Lenstra's algorithm in which she obtains an estimate of $L(1, \chi_d)$ without assuming any Riemann hypotheses, but by using a new method called the random summation technique. As a result she is able to compute the regulator deterministically in expected time $O(d^{1/5+\varepsilon})$. The regulator that the algorithm produces is correct — only the running time is probabilistic.

7 Compact Representation

In view of the result of Lagarias mentioned above, it should be possible to write $\varepsilon_d = (x + y\sqrt{d})/2$ in a more compact way than by using the standard binary or base 10 representation of x and y. Indeed, Buchmann, Thiel and Williams [8] have shown that there always exists a *compact representation* of ε_d given by

$$\varepsilon_d = \prod_{j=1}^{k} (\alpha_j/d_j)^{2^{k-j}}, \tag{7.1}$$

where

1) $k < \log_2 R_d + 3$,

2) $d_j \in \mathbb{Z}; \alpha_j = (a_j + b_j\sqrt{d})/2 \in \mathcal{O}; a_j, b_j \in \mathbb{Z}$ $(j = 1, 2, \ldots k)$,

3) $0 < d_j < d^{1/2}, |a_j| \le 2d^{5/2}, |b_j| < 2d^2$ $(j = 1, 2, \ldots, k)$.

Thus, to represent ε_d we require only $O((\log d)^2)$ bits instead of the possibly exponential number of bits needed by the standard representation. Also, a refinement of the methods of [8] will yield better bounds of $|a_j| < 30d$ and $|b_j| < 30\sqrt{d}$ $(j = 1, 2, \ldots, k)$. We also have the following important theorem.

Theorem 7.1. *There is a deterministic polynomial (in log d) time algorithm which, given $\hat{R}_d$ such that $\left| \hat{R}_d - R_d \right| < 1$, will compute a compact representation of ε_d.*

In fact, this algorithm, which is easily implemented on a computer, executes in $O\left((\log d)^{3+\varepsilon}\right)$ operations and requires only $O\left((\log d)^{2+\varepsilon}\right)$ space.

Thus, we have solved the two problems mentioned in §4 concerning the use of R_d instead of $\varepsilon(D)$. We need only compute R_d to within 1 of its actual value in order to compute ε_d. On the face of it, this seems an utterly amazing result because we can compute the enormous ε_d from very little information, but the constraint that $N(\varepsilon_d) = \pm 1$ is so restrictive that even the relatively tiny amount of data represented by $\hat{R}_d$ is sufficient to allow for the complete determination of ε_d.

To give some idea of how effective the notion of a compact representation is, we point out that for the 30 digit value of D mentioned in the last section, we get $\varepsilon(D) = \varepsilon_d$ and, from the value of R_d, we can easily deduce that $t, u > 10^{2 \cdot 10^{15}}$. This means that it would require over 6,000,000 books, each of 1000 pages, of the same format used by Nelson to record the solution of the Cattle Problem, to record t and u. However, it requires less than one page to write $\varepsilon(D)$ as a compact representation.

We also have a companion theorem to Theorem 7.1

Theorem 7.2. *Given any positive integer q and a compact representation of ε_d, there exists a polynomial (in log d and q) time algorithm which will compute R_d such that*

$$\left| \hat{R}_d - R_d \right| < 2^{-q}.$$

This means that we can easily obtain a very good approximation to R_d, once we have a compact representation of ε_d.

It should be emphasized that there is nothing unique about an expression of the form (7.1). We can easily find others, although we may not necessarily have the inequalities given in (3). In the process of determining a compact representation by the algorithm given in [8], we produce a set of reduced ideals of $\mathcal{O}$:

$$\mathfrak{g}_0, \mathfrak{g}_1, \mathfrak{g}_2, \ldots \mathfrak{g}_k$$

such that $\mathfrak{g}_0 = (1)$, $\mathfrak{g}_j = (\gamma_j)$ has a normal basis $\{d_j, \mu_j\}$ ($\mathfrak{g}_i = d_i \mathfrak{A}_i$ in [8]) $(j = 1, 2, \ldots, k)$, where

$$\gamma_j = \alpha_j \prod_{i=1}^{j-1} (\alpha_i/d_i)^{2^{j-i}} \in \mathcal{O}.$$

We also have

$$(d_{j-1})^2 \, \mathfrak{g}_j = (\alpha_j) \, \mathfrak{g}_{j-1}^2 \quad (j = 1, 2, \ldots, k). \tag{7.2}$$

Now for any i $(0 \le i \le k)$, let $\lambda_i \in \mathfrak{g}_i$; there must exist some ideal $\mathfrak{h}_i$ of $\mathcal{O}$ such that

$$(N(\mathfrak{h}_i)) \, \mathfrak{g}_i = (\lambda_i) \, \mathfrak{h}_i; \tag{7.3}$$

hence

$$|N(\lambda_i)| = N(\mathfrak{h}_i) \, N(\mathfrak{g}_i)$$

by the multiplicative property of the norm. Also,

$$\mathfrak{h}_i = (\kappa_i),$$

where

$$\kappa_i = \overline{\lambda}_i \gamma_i / d_i \in O \tag{7.4}$$

Put $\lambda_0 = 1$. Then $\mathfrak{h}_0 = \mathfrak{g}_0$, $\kappa_0 = 1$ and since

$$\gamma_i = \alpha_i \, (\gamma_{i-1}/d_{i-1})^2 \quad (i = 1, 2, \ldots k),$$

we get

$$\kappa_i = \nu_i \, (\kappa_{i-1}/N(\mathfrak{h}_{i-1}))^2 \quad (i = 1, 2, \ldots, k), \tag{7.5}$$

where

$$\nu_i = \frac{\overline{\lambda}_i \alpha_i N(\mathfrak{h}_{i-1})^2}{d_i \overline{\lambda}_{i-1}^2} = \frac{\overline{\lambda}_i \alpha_i \lambda_{i-1}^2}{d_{i-1}^2} \pm \frac{N(\mathfrak{h}_i) \, \alpha_i \lambda_{i-1}^2}{d_{i-1}^2 \lambda_i} \, .$$

By (7.2) and (7.3) we get

$$\left(N(\mathfrak{h}_{i-1})^2 \, d_{i-1}^2 \lambda_i\right) \mathfrak{h}_i = \left(\alpha_i \lambda_{i-1}^2 N(\mathfrak{h}_i)\right) \mathfrak{h}_{i-1}^2;$$

hence $\nu_i \in \mathfrak{h}_i \subseteq \mathcal{O}$. If we put $\lambda_k = N(\mathfrak{g}_k) = 1$, then $\mathfrak{h}_k = \mathfrak{g}_k = (1)$ and from (7.4) we get

$$\varepsilon_d = \kappa_k = \prod_{i=1}^{k} (\nu_i / N(\mathfrak{h}_i))^{2^{k-i}} \, . \tag{7.6}$$

Since α_k and ν_k must have the same sign, we may assume that

$$\nu_i = \frac{N(\mathfrak{h}_i) \, \alpha_i \lambda_{i-1}^2}{d_{i-1}^2 \lambda_i} \quad (i = 1, 2, \ldots, k) \tag{7.7}$$

in (7.6).

It may be thought that a representation of the form (7.1) for ε_d might be somewhat unwieldy for computing, but as shown in [8] there are many operations that can be performed quite expeditiously on such representations. For example, we have the following theorem.

Theorem 7.3. *Let* $\varepsilon_d = \left(x + y\sqrt{d}\right)/2$. *If we are given a compact representation of* ε_d *and some positive integer* m, *the values of* x *and* y *modulo* m *can be determined in* $O(\log d)$ *operations on numbers of* $O\left(\log m + (\log d)^2\right)$ *bits.*

It is also possible to develop an algorithm to perform this same function which requires $O(\log m \log d)$ operations on numbers of only $O(\max(\log m, \log d))$ bits. The only real problem in doing this is the possibility that $(d_j, m) > 1$ for certain values of j. In order to circumvent this, we put $\lambda_j = xd_j + y\mu_j$ and select $x, y \pmod{m}$ such that

$$N(\mathfrak{h}_j) = |N(\lambda_j)|/d_j = \left|x^2 d_j + xy\left(\mu_j + \overline{\mu}_j\right) + y^2 N(\mu_j)/d_j\right|$$

is relatively prime to m. Since by (5.1) we know that $(d_j, \mu_j + \overline{\mu}_j, N(\mu_j)/d_j) = 1$, this can be done in no more than $O(\log m)$ operations. In the case that $(d_j, m) = 1$, we use $x = 1$, $y = 0$ and get $N(\mathfrak{h}_j) = d_j$. This process allows us to find a sequence of values of $\nu_i \in \mathcal{O}$ by (7.7) such that (7.6) can be used to find $\varepsilon_d \pmod{m}$. (See Jacobson and Williams [39].)

We conclude this section by mentioning that the idea behind expressing ε_d in compact representation was used recently by van der Poorten, te Riele and Williams [90] to verify the Ankeny-Artin-Chowla conjecture for all primes p ($p \equiv 1 \pmod 4$) up to 10^{11}. This conjecture states that if $D = p$, then $p \nmid u$.

8 The Subexponential Method

Undoubtedly, the most exciting recent development for accelerating the computation of R_d for large d has been the discovery of the subexponential method. Up to now all of the methods for doing this that we have discussed have been of exponential complexity, but as we shall see here, there is a subexponential algorithm for computing R_d if we are willing to accept the truth of certain generalized Riemann hypotheses. It is important to emphasize here that while the values of R_d that we find by this technique are very likely to be correct, they are only provably so if the GRH holds. The GRH is not just used to measure the complexity of computing R_d, but must also be assumed to establish rigorously the correctness of the value for R_d that the method produces.

The basic idea behind the subexponential method first occurred as Remark 2.15 in a technical report [52] (later a book chapter) by A. K. and H. W. Lenstra in 1987. Their idea was then elaborated by McCurley [60] and Hafner and McCurley [25]. It was originally intended as a technique to determine the class number of a quadratic field with negative discriminant; however, it can also be applied to real quadratic fields. As the details of this technique have been very well described elsewhere, we will provide only a brief sketch here.

We let $\mathcal{P} = \{p_1, p_2, \ldots, p_k\}$ be a set of rational primes such that $(d/p_i) = 1$ $(i = 1, 2, \ldots, k)$. We know that the ideal (p_i) must split into the product of two prime ideals $\mathfrak{p}_i$ and $\bar{\mathfrak{p}}_i$ in $\mathcal{O}$. We put $FB = \{\mathfrak{p}_1, \mathfrak{p}_2, \ldots, \mathfrak{p}_k\}$ and we suppose that the equivalence classes of the ideals in FB generate the class group Cl of $\mathbb{Q}\left(\sqrt{d}\right)$. For $\vec{v} = (v_1, v_2, v_3, \ldots, v_k) \in \mathbb{Z}^k$, we define $\mathfrak{p}_i^{-v_i}$ to be $\bar{\mathfrak{p}}_i^{v_i}$ $(v_i > 0)$ and

$$FB^{\vec{v}} = \prod_{i=1}^k \mathfrak{p}_i^{v_i}.$$

Let $\Lambda = \{\vec{v} \in \mathbb{Z}^k = FB^{\vec{v}} \sim (1)\}$. This is clearly a sublattice of $\mathbb{Z}^k$. For an introduction to the study of lattices, the reader is referred to any text on the geometry of numbers, such as Cassels [11, ch. 1]. Now let θ be the homomorphism

$$\theta: \mathbb{Z}^k \to \text{Cl}$$

defined by

$$\vec{v} \mapsto F[B^{\vec{v}}] \ .$$

θ is surjective and Λ is its kernel; thus,

$$\mathbb{Z}^k/\Lambda \cong \text{Cl}.$$

The recognition of this important fact began in the work of Pohst and Zassenhaus in 1979 [68] and appears in a more fully developed form in [69]. The important information that follows is that

$$|\det(\Lambda)| = |\,\text{Cl}\,| = h(d).$$

Let $\{\vec{v}_1, \vec{v}_2, \ldots, \vec{v}_n\}$ be a generating system for Λ and put

$$A_{k \times n} = \left(\vec{v}_1^T, \vec{v}_2^T, \ldots, \vec{v}_n^T\right).$$

Then the matrix A has rank k and there exists an integer matrix T such that multiplying A on the right by T will put A into Hermite normal form (see, for example, Newman [64]). That is,

$$AT = [0|H] \ , \tag{8.1}$$

where H is an upper triangular $k \times k$ matrix. The product of the elements on the diagonal of H is $h(d)$.

One of the problems in implementing this idea is guaranteeing that the classes of the ideals in FB will generate Cl. At the time that [25] was written it was known that under the GRH there exists an effectively computable constant c such that if $\mathcal{P}$ contains all the primes $\leq c(\log d)^2$, then FB will contain ideals whose classes will generate Cl. Later, in 1990, Bach [5] provided a value for c. In fact, he gives a table (Table 3, based on the size of d) of values of c_1 and c_2 such that if $\mathcal{P}$ contains all the primes $< (c_1 \log d + c_2)^2$, then FB can be used to generate Cl. This, unfortunately, is also only proved under the assumption of the GRH, but calculations done by Jacobson [32] suggest that Bach's constants are likely a good deal larger than they need be. This is important because keeping k small is essential for the effective execution of the algorithm. Thus, any substantial lowering of Bach's constants would be of enormous benefit to those utilizing the subexponential method.

One consequence of the development of the subexponential method is that under the GRH it is possible to show the existence of a short proof (verifiable in polynomial time) for the value of R_d (see Buchmann and Williams [10]). This, however, is a far cry from having a polynomial algorithm for computing R_d, even under the GRH. However, Buchmann [7] pointed out how the subexponential technique could be applied to the problem of computing R_d. His ideas were later modified and implemented by Cohen, Diaz y Diaz and Olivier [13], [14]. The important ingredient in all of this is that we must now compute a generator γ for any $FB^{\vec{v}} \sim (1)$ such that we have $FB^{\vec{v}} = (\gamma)$; we call this a *relation*. Suppose we have a set of n such relations

$$FB^{\vec{v_i}} = (\gamma_i), \tag{8.2}$$

with generators $\gamma_1, \gamma_2, \ldots, \gamma_n$. We put

$$\vec{r} = \frac{1}{2} \left(\log |\gamma_1/\bar{\gamma}_1| , \log |\gamma_2/\bar{\gamma}_2| , \ldots, \log |\gamma_n/\bar{\gamma}_n| \right)$$

and compute

$$\vec{r}T = (r_1, r_2, \ldots, r_n) \ ,$$

where T is the matrix in (8.1). The values of $r_1, r_2, \ldots, r_n$ are all real numbers and, in particular, r_i ($i = 1, 2, \ldots, n - k$) are integer multiples of R_d. Indeed, if we compute the real $gcd(r_1, r_2, \ldots, r_{n-k})$ (see Cohen [12, p. 289]), this is R_d.

Of course, it is possible that the set of relations that we produce may only generate a sublattice K of Λ. However, in this case we know that

$$|\det(K)/ \det(\Lambda)| \in \mathbb{Z}.$$

Also, we can (again under the GRH) get a sufficiently good approximation of $h(d)R_d$ by truncating the Euler product in (4.2) to produce a value h^* such that

$$h^*/2 < h(d)R_d < h^*. \tag{8.3}$$

Thus, if our computed value of $h(d)R_d$ lies in the interval between $h^*/2$ and h^*, we know that $K = \Lambda$ and our values for $h(d)$ and R_d must be correct (under the GRH).

The complexity of computing R_d by this technique was determined by Abel [1] in her doctoral thesis. Her result is that under the GRH the probabilistic running time to compute R_d is

$$\exp\left\{(1.44 + o(1))\,(\log d \log \log d)^{1/2}\right\} .$$

Very recently, Vollmer [92] noted that this result can be improved somewhat: Under the GRH there exists a probabilistic algorithm that with error probability ϵ, given in advance and independent of d, will compute R_d in expected time $\exp\left\{(\sqrt{2} + o(1))\,(\log d \log \log d)^{1/2}\right\}$.

Cohen et al. implemented their method and were able to compute values for R_d and $h(d)$ for values of d as large as 10^{45}. For example, at the author's request, they computed the regulator for the 38 digit (see [35])

$$D = 13208708795807603033522026252612243246 .$$

The value of RI in this case is .66971.

9 Computational Results

Recently, Jacobson [33] has made a number of improvements to the subexponential algorithm which greatly improve its efficiency. As these are fully described in [33], we will concentrate here on his main improvement — the use of sieving to compute relations.

The main bottleneck in running the subexponential method is the determination of enough relations (8.2) such that we finally get (8.3). Because of the Bach bounds, when d gets large we may have to compute many relations. Jacobson's approach was to apply the techniques of polynomial sieving, which have proved to be so effective in the problem of integer factorization (see, for example, Montgomery [62]). We first compute a normal basis for a primitive ideal $\mathfrak{a}$, where

$$\mathfrak{a} = \prod_{i=1}^{k_0} p_i^{v_i} = a\mathbb{Z} + \beta\mathbb{Z} ,$$

k_0 is small and v_i $(i = 1, 2, \ldots, k_0)$ is selected at random from $\{0, 1, -1\}$. This is done in such a way that $a = N(\mathfrak{a}) \approx \sqrt{d}/M$, where M is a given sieving radius. If $\gamma \in \mathfrak{a}$, then $\gamma = ax + \beta y$ $(x, y \in \mathbb{Z})$, and since $\mathfrak{a}$ contains γ, there must exist an ideal $\mathfrak{b}$ such that

$$(\gamma) = \mathfrak{ab}$$

Thus, if $\mathfrak{b}$ factors over FB, we have a relation. As before we get

$$N(\mathfrak{b}) = ax^2 + (\beta + \overline{\beta}) xy + (N(\beta)/a) y^2 = f(x, y) .$$

If we put $y = 1$, $f(x, 1)$ is a quadratic polynomial in x with integer coefficients. We attempt to factor $\mathfrak{b}$ over FB by sieving $f(x, 1)$ over the range $-M$ to M using the primes p_i $(i = 1, 2, \ldots, k)$. Of course, we make use of the observation that if $p | f(x, 1)$, then $p | f(x + mp, 1)$. This simple adjustment to the algorithm has resulted in a very considerable increase in its speed of computing relations. For example, the method of Cohen et al. could take as long as 4.46 hours to generate enough relations for a 39 digit d, but Jacobson's method could do this in only 8.07 seconds [33, Table 5.26]. Furthermore, as we can use many different ideals as our initial $\mathfrak{a}$, Jacobson's technique can be readily parallelized to run on several computers at once.

When the values of d become very large $(> 10^{70})$, we need well over 5000 prime ideals in FB. This means that the amount of time needed to determine R_d becomes very large. The reason for this, besides the computation of a large number of relations, is the difficulty of the linear algebra which prevents us from computing the class number and regulator simultaneously. The linear algebra required to compute only the class number, even with a large factor base, is much more manageable than that required to simultaneously compute R_d. Thus, it is best to use a much smaller factor base to determine a value S which is likely to be R_d, but at the very worst is an integral multiple of R_d, and then use the large factor base (as determined by the Bach bounds) to find a value for $h(d)$. If $Sh(d)$ falls in the range (8.3), then $S = R_d$ and we are done.

We should also mention that if $(d/p) = 1$ for as many of the small primes p as possible, then not only is $L(1, \chi_d)$ likely to be large and, as a consequence, RI is probably big, but also more prime ideals with small norm will appear in FB. As it is more likely that a small prime will divide a given number than a large one, this means that we would expect to find the required number of relations needed to compute R_d and $h(d)$ much more quickly than in the case where $(d/p) = -1$ for many of the small primes. To illustrate this we consider the 72 digit [38]

$$D_1 = 133007243922787512412600341028518035429251391005992761 39\backslash$$
$$9935498154029253.$$

We get $R_d = 6625291330661652053429358727545606.557249$, $h(d) = 4$, $L(1, \chi_d) = .145331$. These results required over 18 days of computing on a 296MHz SUN Ultra SPARC-II computer. However for the 80 digit

$$D_2 = 127794031002605867150254928246579160440674038637246970 39\backslash$$
$$7197773038860596550 53681$$

we get $R_d = 18287108921995753667199230265771142676945.486447$, $h(d) = 1$ and $RI = .551826$. These results required only 3.45 days [37] of computing on the same computer. The difference between D_1 and D_2 is that we have $(d/p) = -1$ for all primes $p \leq 337$ for D_1, where as for D_2 we have $(d/p) = 1$ for all primes $p \leq 239$.

Extrapolating from these and other results to larger values of d for which we expect to have large values of RI (the "easy" values of d) of 90 and 100 digits, we estimated that it would take about 112 days to compute R_d for a 90 digit d and well over five years to compute R_d for a 100 digit d on the same computer that we used for the smaller values of d. In fact, the technique was implemented by Jacobson in parallel on a cluster of 16 350Mhz Pentium II computers, and we found that for the 90 digit

$$d = 215224698103728400410483771240601671668634200915018506046263\backslash$$
$$91897771659159012655830863 1804$$

we get

$$R_d = 13141178379338133605434507674050601151666861 44.033218,$$

$h(d) = 1$ and $RI = .59718$. This computation was completed in 8 days. For the 101 digit value of

$$d = 130221941021903504103190853297932051273194641328847761633\backslash$$
$$6158366571379092583560263087397184669099836,$$

we computed

$$R_d = 3178025462317475553929176491549486361727631 63478260.945231,$$

$h(d) = 1$ and $RI = .57483$, a result that required 87 days of computing [36].

10   Conclusion

We have seen that the hard part of solving the Pell equation (1.1) is the problem of determining the regulator of $Q(\sqrt{D})$. We have also seen that we are on the verge of being able to do this (conditionally, at least) for values of d of up to 100 digits. This remarkable achievement, however, is accompanied by a regrettable loss of rigor. Thus, the problem of solving the Pell equation is still far from being solved in any computationally satisfactory way. Nevertheless, it should be mentioned that Hallgren [26] has recently shown that there exists a polynomial (in $\log d$) time *quantum algorithm* for finding the regulator of a real quadratic field of discriminant d. This is a very exciting result which essentially solves the main problem of this paper, but until quantum computers become available, we are left with many questions such as:

(1) Is there a deterministic polynomial (in $\log d$) time algorithm for finding R_d? If there is, then there is (under the GRH) a polynomial time algorithm for factoring d, something that is not widely believed.

(2) Is there an unconditional, deterministic algorithm for evaluating R_d that runs in expected subexponential time? There very likely is such an algorithm, but no one has any idea currently as to how to tackle this problem.

(3) Is there an efficient method for determining R_d, given an integral multiple S of R_d? The answer to this is yes if S/R_d is small. For suppose $S/R_d < B$. If we compute the reduced principal ideal $\mathfrak{a}$ such that $\delta(\mathfrak{a}) = S/q$ and find that $\mathfrak{a} = (1)$, then S/q must be an integral multiple of R_d; otherwise, it is not. As Shanks well knew, the complexity of finding a reduced principal ideal $\mathfrak{a}$, given a value for $\delta(\mathfrak{a})$ requires only $O\left(\log \delta(\mathfrak{a})\right)$ operations. Thus, this technique works well enough if B is small, but we have no idea of how to deal with this problem when B is large. Jacobson has observed that there is an unconditional method for finding R_d from S that will execute in $O\left(S^{1/3+\varepsilon}\right)$ operations.

(4) Is there an unconditional, deterministic algorithm for computing R_d that executes with a smaller time complexity function than $O\left(R_d^{1/2}d^\varepsilon\right)$? The information that we have presented previously suggests that this is most likely, but we are still, it seems, a long way from answering this question.

(5) Can Jacobson's method be made to run significantly faster? Almost certainly. There is much room for improvement, but the problems

involved such as reducing the Bach bounds,[7] performing the linear algebra, and computing a value for S, are not easy. See Jacobson [33] and Maurer [59] for further details.

(6) Can the number field sieve methods (see Lenstra and Lenstra [53]) be applied to the problem of computing R_d with the success that they have enjoyed with respect to the integer factoring problem or the discrete logarithm problem? So far no one knows how to do this, and it is believed by some experts that it may not be possible.

Thus, the current state of the art in solving the Pell equation is far from satisfactory. In spite of the enormous progress that has been made on this problem in the last few decades, we are still without answers to many fundamental questions. However, we are, it seems, beginning to understand what the questions should be.

References

[1] C. S. Abel, *Ein Algorithmus zur Berechnung der Klassenzahl und des Regulators reellquadratischer Ordnungen*, Ph.D. thesis, Universität des Saarlandes, Saarbrücken, Germany, 1994.

[2] W. Adams, *On the relationship between the convergents of the nearest integer and regular continued fractions*, Math. Comp. **33** (1979), 1321–1331.

[3] American Council of Learned Societies, *Biographical dictionary of mathematics*, vol. 4, Scribner's Sons, New York, 1991.

[4] A. Amthor, *Das Problema bovinum des Archimedes*, Zeitschrift für Math. u. Physik (Hist. Litt. Abtheilung) **25** (1880), 153–171.

[5] E. Bach, *Explicit bounds for primality testing and related problems*, Math. Comp. **55** (1990), 335–380.

[6] ———, *Improved approximations for Euler products*, Number Theory: CMS Proc., vol. 15, Amer. Math. Soc., Providence, RI, 1995, pp. 13–28.

[7] J. Buchmann, *A subexponential algorithm for the determination of class groups and regulators of algebraic number fields*, Séminaire de Théorie des Nombres (Paris), 1988–1989, pp. 27–41.

---

[7]Of course, even if the Bach bounds were considerably reduced, we still have to find a sufficient number of relations in order to guarantee (8.3). Thus, there is an interesting trade-off here that needs investigation.

[8] J. Buchmann, C. Thiel, and H. C. Williams, *Short representation of quadratic integers*, Computational Algebra and Number Theory, Mathematics and its Applications, vol. 325, Kluwer, Dordrecht, 1995, pp. 159–185.

[9] J. Buchmann and H. C. Williams, *On the infrastructure of the principal ideal class of an algebraic number field of unit rank one*, Math. Comp. **50** (1988), 569–579.

[10] ——, *On the existence of a short proof for the value of the class number and regulator of a real quadratic field*, Number Theory and Applications, NATO ASI Series C, vol. 265, Kluwer, Dordrecht, 1989, pp. 327–345.

[11] J. W. S. Cassels, *An Introduction to the Geometry of Numbers*, Springer, Berlin, 1971.

[12] H. Cohen, *A Course in Computational Algebraic Number Theory*, Graduate Texts in Mathematics, vol. 138, Springer, 1993.

[13] H. Cohen, F. Diaz y Diaz, and M. Olivier, *Calculs de nombres de classes et de régulateurs de corps quadratiques en temps sous-exponentiel*, Séminaire de Théorie des Nombres (Paris), 1993, pp. 35–46.

[14] ——, *Subexponential algorithms for class and unit group computations*, J. Symb. Comp. **24** (1997), 443–441.

[15] H. Cohen and H. W. Lenstra, Jr., *Heuristics on class groups*, Number Theory, Lecture Notes in Mathematics 1052, Springer, Berlin, 1984, pp. 26–36.

[16] ——, *Heuristics on class groups of number fields*, Number Theory, Lecture Notes in Mathematics 1068, Springer, Berlin, 1984, pp. 33–62.

[17] E. J. Dijksterhuis, *Archimedes*, Princeton University Press, Princeton, 1987.

[18] G. Eisenstein, *Allgemeine Untersuchengen über die Formen dritten Grades mit drei variabeln, welche der Kreistheilung ihre Enstehung verdanken*, J. Reine Angew. Math. **28** (1844), 289–374.

[19] G. Eneström, *Über der Ursprung der Benennung "Pellsche Gleichung"*, Bibliotheca Math. **3** (1902), 204–207.

[20] L. Euler, *Correspondance, Mathématique et Physique*, St. Petersbourg, 1843, P. H. Fuss (ed.), T1, pp. 35–39.

[21] P. Fermat, *Oeuvres de Fermat*, vol. II, Gauthier-Villars, Paris, 1891–1912.

[22] D. H. Fowler, *The Mathematics of Plato's Academy: A New Reconstruction*, Clarendon Press, Oxford, 1999.

[23] P. M. Fraser, *Ptolemaic Alexandria*, vol. 1, The Clarendon Press, Oxford, 1972.

[24] A. Granville and K. Soundararajan, *The distribution of values of* $L(1, \chi)$, to appear.

[25] J. L. Hafner and K. S. McCurley, *A rigorous subexponential algorithm for computation of class groups*, J. Amer. Math. Soc. **2** (1989), 837–850.

[26] S. Hallgren, *Polynomial-time quantum algorithms for Pell's equation and the Principal Ideal Problem*, unpublished manuscript, 2001.

[27] H. Hankel, *Zur Geschichte der Mathematik in Altertum und Mittelalter*, 2nd ed., Hildesheim, 1965.

[28] T. L. Heath, *Diophantus of Alexandria*, 2nd ed., Dover, New York, 1964.

[29] J. E. Hofmann, *Studien zur Zahlentheorie Fermats*, Abh. Preuss. Akad. Wiss. (1994), no. 7.

[30] L.-K. Hua, *Introduction to number theory*, Springer-Verlag, Berlin, 1982, pp. 326–329.

[31] M. J. Jacobson, Jr., *Computational techniques in quadratic fields*, Master's thesis, University of Manitoba, Winnipeg, Manitoba, 1995.

[32] _____ , *Experimental results on class groups of real quadratic fields (extended abstract)*, Algorithmic Number Theory - ANTS-III (Portland, Oregon), Lecture Notes in Computer Science, vol. 1423, Springer-Verlag, Berlin, 1998, pp. 463–474.

[33] _____ , *Subexponential Class Group Computation in Quadratic Orders*, Ph.D. thesis, Technische Universität Darmstadt, Darmstadt, Germany, 1999.

[34] _____, *Computing discrete logarithms in quadratic orders*, J. Cryptology **13** (2000), 473–492.

[35] M. J. Jacobson, Jr., R. F. Lukes, and H. C. Williams, *An investigation of bounds for the regulator of quadratic fields*, Experimental Mathematics **4** (1995), 211–225.

[36] M. J. Jacobson, Jr., R. Scheidler, and H. C. Williams, *The efficiency and security of a real quadratic field based-key exchange protocol*, Public-Key Cryptography and Computational Number Theory, Walter de Gruyter, Berlin, to appear.

[37] M. J. Jacobson, Jr. and H. C. Williams, *The size of the fundamental solutions of consecutive Pell equations*, Experimental Math. **9** (2000), 631–640.

[38] _____, *New Quadratic Polynomials with High Densities of Prime Values*, to appear.

[39] _____, *Modular Arithmetic on Elements of Small Norm in Quadratic Fields*, unpublished MS.

[40] P. T. Joshi, *The size of $L(1,\chi)$ for real nonprincipal residue characters χ with prime modulus*, J. Number Theory **2** (1970), 58–73.

[41] W. Knorr, *Archimedes and the measurement of the circle: a new interpretation*, Arch. Hist. Exact Sci. **15** (1975), 115–140.

[42] H. Konen, *Geschichte der Gleichung $t^2 - Du^2 = 1$*, Leipzig, 1901.

[43] R. Kortum and G. McNeil, *A Table of Periodic Continued Fractions*, Lockheed Aircraft Corp., Sunnyvale, CA, 1961.

[44] B. Krumbiegel, *Das problema bovinum des Archimedes*, Zeitschrift für Math. u. Physik (Hist. Litt. Abtheilung) **25** (1880), 121–136.

[45] J. C. Lagarias, *Succinct Certificates for the Solvability of Binary quadratic Diophantine Equations (Extended Abstract)*, Proc. 20th IEEE Symp. on Foundations of Computer Science, 1979, pp. 47–54.

[46] _____, *Succinct Certificates for the Solvability of Binary quadratic Diophantine Equations*, Technical Memorandum 81-11216-54, Bell Labs, Sept. 28, 1981.

[47] D. H. Lehmer, *A list of errors in tables of the Pell equation*, Bull. Amer. Math. Soc. **32** (1926), 545–550.

[48] ——, *On the indeterminate equation* $t^2 - p^2 Du^2 = 1$, Annals of Math. **27** (1926), 471–476.

[49] ——, *On the multiple solutions of the Pell equation*, Annals of Math. **30** (1928), 66–72.

[50] ——, *Guide to Tables in the Theory of Numbers*, National Research Council, Washington, 1941.

[51] P. G. Lejeune-Dirichlet, *Vorlesungen über Zahlentheorie*, 4th ed., Braunschweig, 1893.

[52] A. K. Lenstra and H. W. Lenstra, Jr., *Algorithms in Number Theory*, Technical Report 87-008, University of Chicago, 1987.

[53] A. K. Lenstra and H. W. Lenstra, Jr. (eds.), *The Development of the Number Field Sieve*, Lecture Notes in Mathematics, vol. 1554, Springer, Berlin, 1993.

[54] H. W. Lenstra, Jr., *On the calculation of regulators and class numbers of quadratic fields*, London Math. Soc. Lecture Note Series **56** (1982), 123–150.

[55] J. E. Littlewood, *On the class number of the corpus* $P(\sqrt{-k})$, Proc. London Math. Soc. **27** (1928), 358–372.

[56] S. Louboutin, *Explicit upper bounds for* $|L(1,\chi)|$ *for primitive even Dirichlet characters*, Acta Arith. **101** (2002), 1–18.

[57] M. S. Mahoney, *The Mathematical Career of Pierre de Fermat*, 2nd ed., Princeton, New Jersey, 1994.

[58] A. Martin, *Solution*, The Analyst **4** (1877), 154–155.

[59] M. Maurer, *Regulator approximation and fundamental unit computation for real quadratic orders*, Ph.D. thesis, Technische Universität Darmstadt, Darmstadt, Germany, 2000.

[60] K. S. McCurley, *Cryptographic key distribution and computation in class groups*, Proc. NATO ASI on Number Theory and Applications, NATO ASI series C, vol. 265, Kluwer, Dordrecht, 1989, pp. 459–479.

[61] R. A. Mollin and H. C. Williams, *Computation of the class number of a real quadratic field*, Utilitas Mathematica **41** (1992), 259–308.

[62] P. L. Montgomery, *A survey of modern integer factoring algorithms*, CWI Quarterly **7** (1994), 337–366.

[63] H. L. Nelson, *A solution to Archimedes' Cattle Problem*, J. Recreational Math. **13** (1981), 162–176.

[64] M. Newman, *Integral Matrices*, Pure and Applied Mathematics, vol. 45, Academic Press, New York, 1972.

[65] W. Patz, *Tafel der Regelmässigen Kettenbrüche für die Quadratwurzeln aus den Natürlichen Zahlen von 1 – 10000*, Becker and Erler, Leipzig, 1941.

[66] _____, *Tafel der Regelmässigen, Kettenbrüche und ihres vollständigen Quotienten für die Quadratwuzeln aus den Natürlichen Zahlen von 1 – 10000*, Akademie-Verlag, Berlin, 1955.

[67] O. Perron, *Die Lehre von den Kettenbrüchen*, 2nd ed., Chelsea, New York, undated.

[68] M. Pohst and H. Zassenhaus, *On unit computation in real quadratic fields*, Symbolic and Algebraic Computation, Lecture Notes in Computer Science, vol. 72, Springer, Berlin, 1979.

[69] _____, *Über die Berechnung von Klassenzahlen und Klassengruppen algebraischer Zahlkörper*, J. reine angew. Math. **361** (1985), 50–72.

[70] J. H. Rahn, *An Introduction to Alegbra*, T. Brancker (Tr.), London, 1668.

[71] R. Schoof, *Quadratic fields and factorization*, Computational Methods in Number Theory (H. W. Lenstra, Jr. and R. Tijdeman, eds.), no. 155, Part II, Math. Centre Tracts, Amsterdam, 1983, pp. 235–286.

[72] P. Schreiber, *A note on the Cattle Problem of Archimedes*, Historia Math. **20** (1993), 304–306.

[73] J. F. Scott, *The Mathematical Work of John Wallis*, London, 1938.

[74] Ch. J. Scriba, *John Pell's English edition of J. H. Rahn's Teutsche Algebra, For Dirk Struick*, R. S. Cohen et al., (eds.), pp. 261–274, Reidel, Dordrecht, 1974.

[75] C.-O. Selenius, *Kettenbruchtheoretische Erklärung der zyklischen Methode zur Lösung der Bhaskara-Pell-Gleichung*, Acta acad. Aboensis, math. phys. **23** (1963), no. 10.

[76] _____, *Rationale of the chakravala process of Jayadeva and Bhaskara II*, Historia Math. **2** (1975), 167–184.

[77] D. Shanks, *Review RMT30*, Math. Comp. **16** (1962), 377–379, also **23** (1969), 217, 219.

[78] ———, *Review RMT10*, Math. Comp. **23** (1969), 213–214.

[79] ———, *Class number, a theory of factorization and genera, 1969 Number Theory Institute*, Proc. Symp. Pure Math. 20, AMS, Providence, R.I., 1971, pp. 415–440.

[80] ———, *The infrastructure of real quadratic fields and its applications*, Proc. 1972 Number Theory Conf. (Boulder, Colorado), 1972, pp. 217–224.

[81] ———, *Systematic examination of Littlewood's bounds on $L(1, \chi)$*, Analytic Number Theory, Proc. Symp. Pure Math. (Providence, R.I.), vol. 24, AMS, 1973, pp. 267–283.

[82] ———, *Review RMT11*, Math. Comp. **28** (1974), 333–334.

[83] ———, *On Gauss and composition I, II*, Number Theory and Applications, NATO ASI series C, vol. 265, Kluwer Academic Press, Dordrecht, 1989, pp. 163–179.

[84] K. Shankar Shukla, *Acarya Jayadeva, the mathematician*, Ganita **5** (1954), 1–20.

[85] A. Srinivasan, *Computations of class numbers of real quadratic fields*, Math. Comp. **67** (1998), 1285–1308.

[86] P. Tannery, *L'Arithmétique des Grecs dans Pappus*, Mémoires Scientifiques, T.I., Toulouse, 1912–37, pp. 80–105.

[87] ———, *Sur la mesure de circle d'Archimède*, Mémoires Scientifiques, T.I., Toulouse, 1912–37, pp. 226–253.

[88] E. Teske and H. C. Williams, *A problem concerning a character sum*, Experimental Math. **8** (1999), 63–72.

[89] A. J. van der Poorten, *A Note on NUCOMP*, unpublished MS.

[90] A. J. van der Poorten, H. te Riele, and H. C. Williams, *Computer verification of the Ankeny-Artin-Chowla conjecture for all primes less than 100000000000*, Math. Comp. **70** (2001), 1311–1328.

[91] I. Vardi, *Archimedes' Cattle Problem*, Amer. Math. Monthly **105** (1998), 305–319.

[92] U. Vollmer, *Asymptotically fast discrete logarithms in quadratic number fields*, Algorithmic Number Theory - ANTS-IV (Leiden, The Netherlands), Lecture Notes in Computer Science, vol. 1838, Springer, 2000, pp. 581–594.

[93] W. Waterhouse, *On the Cattle Problem of Archimedes*, Historia Math. **22** (1995), 186–187.

[94] A. Weil, *Fermat et l'équation de Pell*, Collected Papers, vol. III, Springer, Berlin, 1979, pp. 413–419.

[95] ———, *Number Theory. An Approach Through History*, Birkhäuser, Boston, 1984.

[96] G. Wertheim, *Die Algebra des Johann Heinrich Rahn (1659) und die englische übersetzung derselben*, Bibliotheca Math. **3** (1902), no. 3, 113–126.

[97] E. E. Whitford, *The Pell Equation*, College of the City of New York, New York, 1912.

[98] H. C. Williams, *On mid period criteria for the nearest integer continued fraction expansion of $\sqrt{D}$*, Utilitas Math. **27** (1985), 169–185.

[99] H. C. Williams and J. Broere, *A computational technique for evaluating $L(1,\chi)$ and the class number of a real quadratic field*, Math. Comp. **30** (1976), 887–893.

[100] H. C. Williams and P. A. Buhr, *Calculation of the regulator of $\mathbb{Q}\left(\sqrt{D}\right)$ by use of the nearest integer continued fraction algorithm*, Math. Comp. **33** (1979), 369–381.

[101] H. C. Williams, R. A. German, and C. R. Zarnke, *Solution of the Cattle Problem of Archimedes*, Math. Comp. **19** (1965), 671–674.

Printed and bound by CPI Group (UK) Ltd, Croydon, CR0 4YY

26/10/2024

01779669-0003